The Digital Transformation of Product Formulation

In competitive manufacturing industries, organizations embrace product development as a continuous investment strategy since both market share and profit margin stand to benefit. Formulating new or improved products has traditionally involved lengthy and expensive experimentation in laboratory or pilot plant settings. However, recent advancements in areas from data acquisition to analytics are synergizing to transform workflows and increase the pace of research and innovation. *The Digital Transformation of Product Formulation* offers practical guidance on how to implement data-driven, accelerated product development through concepts, challenges, and applications. In this book, you will read a variety of industrial, academic, and consulting perspectives on how to go about transforming your materials product design from a twentieth-century art to a twenty-first-century science.

- Presents a futuristic vision for digitally enabled product development, the role of data and predictive modeling, and how to avoid project pitfalls to maximize probability of success
- Discusses data-driven materials design issues and solutions applicable to a variety of industries, including chemicals, polymers, pharmaceuticals, oil and gas, and food and beverages
- Addresses common characteristics of experimental datasets, challenges in using this data for predictive modeling, and effective strategies for enhancing a dataset with advanced formulation information and ingredient characterization
- Covers a wide variety of approaches to developing predictive models on formulation data, including multivariate analysis and machine learning methods
- Discusses formulation optimization and inverse design as natural extensions to predictive modeling for materials discovery and manufacturing design space definition
- Features case studies and special topics, including AI-guided retrosynthesis, real-time statistical process monitoring, developing multivariate specifications regions for raw material quality properties, and enabling a digital-savvy and analytics-literate workforce

This book provides students and professionals from engineering and science disciplines with practical know-how in data-driven product development in the context of chemical products across the entire modeling lifecycle.

Alix Schmidt is a senior data scientist in Dow's Core R&D Information Research team in Midland, Michigan. Alix earned a BS in chemical engineering at the University of Illinois Urbana–Champaign in 2009 and then joined Dow Corning initially as a process research engineer. Since then, Alix has held a variety of roles at Dow Corning and Dow and completed an MS in data science at Northwestern University. Alix has experience with polymer process research, high-throughput research, machine learning for manufacturing troubleshooting, and data-driven product development. Her interest and experience in materials informatics allow her to lead technical data science strategy at Dow, and she has presented and chaired at the AIChE spring meeting on this topic.

Kristin Wallace earned a BS in chemical engineering (2006) and an MS in applied science (optimization focus) (2008) at McMaster University. She has worked on a variety of analytics projects since joining ProSensus Inc. in 2018 as a project engineer in Burlington, Ontario. Her particular interest in product formulation using projection to latent structures (PLS) has led her to be involved with related consulting projects, contributing to the development of FormuSense (commercial software), authoring blogs and magazine articles, as well as presenting and chairing at several AIChE spring meetings. Prior to working at ProSensus, she spent five years designing and troubleshooting non-ferrous electric arc furnaces.

The Digital Transformation of Product Formulation
Concepts, Challenges, and Applications for Accelerated Innovation

Edited by
Alix Schmidt and Kristin Wallace

CRC Press
Taylor & Francis Group
Boca Raton London New York

CRC Press is an imprint of the
Taylor & Francis Group, an **informa** business

Designed cover image: © 2025 Shutterstock, PopTika and Yurchanka Siarhei

First edition published 2025
by CRC Press
2385 NW Executive Center Drive, Suite 320, Boca Raton FL 33431

and by CRC Press
4 Park Square, Milton Park, Abingdon, Oxon, OX14 4RN

CRC Press is an imprint of Taylor & Francis Group, LLC

Library of Congress Cataloging-in-Publication Data
Names: Schmidt, Alix, editor. | Wallace, Kristin, editor.
Title: The digital transformation of product formulation : concepts, challenges and applications for accelerated innovation / edited by Alix Schmidt and Kristin Wallace.
Description: First edition. | Boca Raton : CRC Press, 2025. | Includes bibliographical references.
Identifiers: LCCN 2024015904 (print) | LCCN 2024015905 (ebook) | ISBN 9781032474069 (hardback) | ISBN 9781032474076 (paperback) | ISBN 9781003385974 (ebook)
Subjects: LCSH: New products. | Product design—Data processing. | Artificial intelligence—Engineering applications.
Classification: LCC TS170 .D53 2025 (print) | LCC TS170 (ebook) | DDC 670.285—dc23/eng/20240524
LC record available at https://lccn.loc.gov/2024015904
LC ebook record available at https://lccn.loc.gov/2024015905

ISBN: 978-1-032-47406-9 (hbk)
ISBN: 978-1-032-47407-6 (pbk)
ISBN: 978-1-003-38597-4 (ebk)

DOI: 10.1201/9781003385974

Typeset in Times
by Apex CoVantage, LLC

Contents

SECTION 3 Predictive Modeling

Alix Schmidt and Kristin Wallace

SECTION 4 Optimization and Inverse Design

Alix Schmidt and Kristin Wallace

SECTION 5 Case Studies and Special Topics

Alix Schmidt and Kristin Wallace

Contributors

Erin Antono
Citrine Informatics
Redwood City, California, USA

Gabriele Bano
GlaxoSmithKline
Collegeville, Pennsylvania, USA

Massimiliano Barolo
CAPE-Lab – Computer-Aided Process Engineering Laboratory
Department of Industrial Engineering
University of Padova
Padova, Italy

Anne-Catherine Bedard
Core Research and Development
The Dow Chemical Company
Midland, Michigan, USA

Joan Borràs-Ferrís
Department of Applied Statistics and Operational Research and Quality
Universitat Politècnica de València
València, Spain

Birgit Braun
Core Research and Development
The Dow Chemical Company
Lake Jackson, Texas, USA

Luis Briceno-Mena
Core Research and Development
The Dow Chemical Company
Lake Jackson, Texas, USA

Marlene Cardin
ProSensus Inc.
Burlington, Ontario, Canada

Daniel Christiansen
Polyurethanes Product R&D
The Dow Chemical Company
Lake Jackson, Texas, USA

Jerome Claracq
Core Research and Development
The Dow Chemical Company
Hoek, Netherlands

Carl Duchesne
Department of Chemical Engineering
Université Laval
Quebec, Quebec, Canada

Pierantonio Facco
CAPE-Lab – Computer-Aided Process Engineering Laboratory
Department of Industrial Engineering
University of Padova
Padova, Italy

Christopher Farrow
Enthought, Inc.
Austin, Texas, USA

Alberto Ferrer
Department of Applied Statistics and Operational Research and Quality
Universitat Politècnica de Valencia
València, Spain

Michael C. Heiber
Enthought, Inc.
Austin, Texas, USA

Maxwell Hutchinson
Citrine Informatics
Redwood City, California, USA

Arthi Jayaraman
Department of Chemical and Biomolecular Engineering
and
Department of Materials Science and Engineering
and
Data Science Institute
University of Delaware
Newark, Delaware, USA

Mengjie Liu
Department of Chemical Engineering
Massachusetts Institute of Technology
Cambridge, Massachusetts, USA

Shizhao Lu
Department of Chemical and Biomolecular Engineering
University of Delaware
Newark, Delaware, USA

Kaiwen Ma
Core Research and Development
The Dow Chemical Company
Lake Jackson, Texas, USA

Sukrit Mukhopadhyay
Core Research and Development
The Dow Chemical Company
Midland, Michigan, USA

Alexander Nguyen
ProSensus Inc.
Burlington, Ontario, Canada

Anastasia Nikolakopoulou
GlaxoSmithKline
Cambridge, Massachusetts, USA

Daniel Palací-López
IFF Benicarlo S.L
Benicarló, Spain

Sean Paradiso
Citrine Informatics
Redwood City, California, USA

Sreekanth Rajagopalan
Core Research and Development
The Dow Chemical Company
Lake Jackson, Texas, USA

Benjamin Reiner
Core Research and Development
The Dow Chemical Company
Collegeville, Pennsylvania, USA

Alix Schmidt
Core Research and Development
The Dow Chemical Company
Midland, Michigan, USA

Joel C.J. Strickland
Intellegens
Cambridge, Cambridgeshire, United Kingdom

Kristin Wallace
ProSensus Inc.
Burlington, Ontario, Canada

Thomas M. Whitehead
Intellegens
Cambridge, Cambridgeshire, United Kingdom

Phillip F.D. Woolston
Domino Printing
Bar Hill, Cambridge, United Kingdom

Ou Yang
GlaxoSmithKline
Boston, Massachusetts, USA

Foreword

This book looks at a series of topics on leveraging data to accelerate the development of new formulations and products in the material, chemical, polymer, and food industries.

A major focus of my research at McMaster University (1972–2008) and a basis for founding ProSensus was developing advanced technologies for extracting information from the ever-increasing amount of "big data" arising in the process industries and using that information to improve the understanding, monitoring, and optimization of those processes. But, on working with companies on these problems, it also became obvious that the need for using "big data" was becoming critical in R&D departments, specifically for the rapid development of new and improved products.

It has been exciting for me to see the dramatic changes and utilization of new technologies behind product development over the past few decades. These changes have been brought on by the digitalization of R&D departments through connectivity tools to collect data from an ever-expanding set of new analytical characterization equipment; powerful databases to store, collate, and extract selected data; and advanced data analytic tools for modeling and optimizing potential product performance based on these data.

Product development has traditionally relied upon some highly educated individuals using their prior experience and scientific knowledge. However, that experience was often restricted to just the materials, formulations, and chemistry related to those development projects that they were individually involved with. The collective knowledge of the developers and their data were fragmented, rarely accessible through models and centralized databases, and hence not easily transferrable to other product developers. The experiments were usually based on simple experimental designs (such as factorials or mixture designs) on just the formulation ratios of a small set of selected materials and/or reactants. Alternative raw materials, the properties of all those raw materials, and the processing conditions used to formulate the product were rarely considered at the same time. Model-based optimization of the entire development process was almost never considered.

It is therefore not surprising that, as of a decade ago, a survey of large corporations found that while 93% of executives emphasized that their companies relied upon product innovation for long-term success, only 18% of those executives believed that their company derived a competitive advantage from its current innovation strategy (Koetzier & Alon 2013). Furthermore, they felt that the situation was getting worse and were less satisfied with their innovation performance then than they were in 2009. The major reasons behind that negative sentiment were that the traditional approaches to product development were based on much less information and data, on a lack of digitalization (and as a result, poor accessibility to prior development work), and on a lack of analytical tools that could handle the increasingly available data to expand the development process.

Fast forward a few years, and a more recent survey of C-suite executives showed a 200% increase in "levels of disruption" from 2017 to 2022, with nearly 75% of those surveyed having already integrated AI into their business for various applications, including product development (Daugherty et al. 2023).

Product quality is truly a multivariate property. A product is never characterized by just a single property but by a multitude of properties (usually between 2 and 20). Its quality is determined by the proper balance of all these properties. These properties, in turn, are dependent on many factors: (1) the choice of raw materials used to make a product; (2) the properties of each of the raw materials (a key to exploring promising new raw materials never before used by the developers); (3) the formulation ratios used to combine or react the raw materials; and (4) the processing conditions used to manufacture the product.

All these factors need to be considered simultaneously, not just one at a time, because they are usually highly interactive. Different material combinations call for different formulation ratios and different processing conditions in order to achieve an acceptable product. However, to consider all these data simultaneously involves large and more complex data sets, and hence the need for the advanced analytical tools and software outlined in this book.

As an example, 20 years ago I was involved with the development of functional polymer blends. The data involved seven final product quality variables, three classes of raw materials (rubbers, polypropylenes, and oils), and operating conditions for the blending process. The available database for modeling consisted of ~110 commercial and development polymer blends, and a database on the properties and costs of a large number of rubbers available for consideration most of which had never been used previously. The result from the modeling and optimization was the immediate development of several functional polymers with significantly improved quality all at lower costs (costs are almost always part of the optimization). Furthermore, each of these new functional polymers included at least one of the rubbers that the company had never used before.

Given the explosion in analytical instrumentation, some more recent projects have included several thousand measurements on each material and involved data from manufacturing as well as R&D. With such large complex data and trade-offs between materials, processing conditions, and costs, the need for advanced analytical and modeling tools suggested in the chapters of this book are not just beneficial but essential to improving the innovation and product development cycles.

As a final observation, I want to emphasize that for all these new advanced tools to make a major difference in a company's ability to develop better products faster, there must be an acceptance of the technology by the company's product developers. Too often I have seen resistance to using these tools because the developers feel that it is taking away their control of the development process. However, my experience with those developers who have adapted to these technologies has been quite the opposite. They are able to interact with the modeling and optimization results by using their extensive scientific knowledge to understand why certain materials and formulations suggested by the models and optimization make sense and then use those new insights to work with and guide these technologies to more rapidly develop better products. A product development chemist working at an international chemical company summarized her experiences with using these technologies for the development of new polymeric coatings by saying "I have been working in product development for 30 years now and this is the most exciting stuff I have ever been involved with."

I along with all the authors of this book hope that we can inspire other developers in a similar way.

John F MacGregor
Distinguished University Professor (Emeritus), McMaster University
Founder and Chairman, ProSensus Inc.

REFERENCES

Daugherty, Paul, Karthik Narain, and Jim Wilson. 2023. "Breakthrough innovation: Is your organization equipped for breakthrough innovation?" *Accenture*. https://www.accenture.com/content/dam/accenture/final/capabilities/technology/technology-innovation/imagery/Accenture-Breakthrough-Innovation.pdf

Koetzier, Wouter, and Adi Alon. 2013. "Why 'low risk' innovation is costly: Overcoming the perils of renovation and invention." *Accenture*. https://www.aefinfo.fr/assets/medias/documents/2/1/21601.pdf

Section 1

Getting Started

Alix Schmidt and Kristin Wallace

To have found yourself reading this page, it is likely that you have an interest in the future of chemical and material product development in the digital age. Perhaps you are a seasoned formulator who has noticed an influx of artificial intelligence (AI) buzzwords in your workplace and would like to understand what it means for you. Perhaps you are a materials science practitioner who has always been data-savvy, and you want to learn what it would take to model your products. Maybe you are a student or a newly graduated industrial chemist with skills and interest in programming, and you are on the lookout for ways to use these skills to accelerate your research. Or it could be that you are a university instructor asked to equip your students with an understanding of how data science is used in industrial product design, but you find that explanations of day-to-day challenges – let alone data itself – are hard to come by. Lastly, you may be a product development manager trying to understand how modern machine learning techniques, and AI, buzzwords aside, can help your team fend off the commoditization of your products. Regardless, we believe this text can answer your questions.

In the following chapters, you will read a variety of industrial, academic, and consulting perspectives on how to go about transforming your materials product design from a 20th-century art to a 21st-century science. The text is organized into sections that follow the lifecycle of a materials data science project: getting started and selecting an opportunity, preparing and enriching your data, predictive modeling of product performance, formulation optimization and inverse design, and finally a

DOI: 10.1201/9781003385974-1

closer look at some special topics and case studies. Within each section, there are chapters at a variety of technical depths, from beginner introduction to advanced practice. The reader is invited to pick and choose chapters as needed to support your digital transformation journey; section introductions will provide a roadmap to the structure of each section.

In Section 1, we provide the starting point for the journey. Chapter 1 discusses the possibilities and motivation for a fully digital materials science future. Chapter 2 narrows in to explain the scope of the book, set some baseline nomenclature, and contrast the challenges of applying data-driven methods in materials formulation to more well-known AI applications. Chapter 3 will close the section with a discussion of how to vet project opportunities for actionable decisions and high-quality data that will lead to success.

1 The Digital Transformation of R&D Labs

Michael C. Heiber and Christopher Farrow

1.1 THE BUSINESS CASE

Digital transformation has been an important trend and high priority for many businesses across all sectors of the economy. The central objectives for most are similar: to leverage advances in digital computing technologies to dramatically accelerate existing business processes and also identify fundamentally new ways of interacting with and delivering value to customers. It is not about any one technology in particular but about combining many disparate technology tools to develop categorically better products and services for existing and new emerging markets.

One common aspect of digital transformation is changing how customers find and engage with your offerings and putting systems in place that reduce "friction" in the buying process. This theme emerged first in business-to-consumer (B2C) markets with the rise of online shopping and features designed to make it easy for shoppers to find and buy the items they are looking for and also introduce additional items they did not even know they wanted. Excellent product search, product information pages, product Q&A, product recommendations, consumer reviews, fast and low-cost shipping, and one-click purchasing all make spending money easy. These customer-facing features are backed up by a diverse set of digital technologies (web GUIs, databases, secure data transfer, machine learning/AI, cloud computing, robotics, and many more).

At the same time, traditional business-to-business (B2B) purchasing could perhaps be characterized as a polar opposite, where it is often actively hard to make a purchase. Product suppliers purposefully withhold information about their products: the technical specifications, pricing, shipping information, etc., behind information request forms to limit the information that is available to competitors and also to collect contact information for sales and marketing. But over time, technical buyer expectations are shifting, and there is a strong driving force to provide more information to buyers faster to accelerate purchasing decision-making. Thus, there is an ongoing shift to make B2B purchasing look a lot more like B2C. As Figure 1.1 illustrates, there is a shift to provide a more direct, responsive, and deeper relationship between customers and supplier R&D teams. Chemical and materials companies leading in this arena are producing data-rich customer-facing websites that are portals to much deeper information and technical services that are backed by modern

DOI: 10.1201/9781003385974-2

FIGURE 1.1 Digital transformation can strengthen and deepen supplier-buyer relations at the technical level, creating opportunities for new digitally enhanced specialty products and services.

computing technologies. This has been especially prevalent for coatings formulation applications, where prominent examples include Dow's Paint Vision (https://www.dow.com/paintvision/), BASF's Lab Assistant (https://www.lab-assistant.basf.com/), and Evonik's Coatino (https://www.coatino.com/). Nevertheless, there is still a long way to go for this type of system to be pervasive in the materials and chemical industry.

In industries driven by scientific innovation, the digital processes and tools that are enabling this new paradigm of customer interactions start at the sales and marketing end of the business but extend all the way down to manufacturing and eventually R&D.

In specialty materials and chemical markets, close technical interaction between researchers and technical buyers is critical to business success. A supplier's product must fit very specific technical requirements in addition to manufacturing supply and cost constraints. There are complex, multifunctional property requirements, as well as a wide variety of processing constraints. In this scenario, the ability to iterate faster with potential customers, increase technical service quality, and increase the probability of meeting challenging product requirements can give suppliers a distinctive competitive advantage that leads to more product sales. And being able to do so while also decreasing the R&D cost of each product development cycle can increase the profit margins on each sale. This is especially important in mature business segments where multiple competitor products have emerged, and there is growing cost pressure.

In new, emerging product markets, the detailed factors are different, but the overall needs are similar. There is a need to be able to iterate on your product technology quickly with your partners and future customers. Faster, higher-quality development iterations provide the ability to consistently meet challenging technical milestones, which allow one to secure the critical partnerships that position oneself as a preferred supplier in the new market. Also, the ability to quickly optimize and standardize the product fabrication process at the lab and pilot scales reduces the risk of purchasing the wrong capital equipment when it comes time to make big investments in manufacturing capacity.

In order to be able to offer digitally enhanced services and superior products to customers, there is a strong driving force to accelerate innovation. However, this need to accelerate innovation comes from other places as well. Companies face a constant battle to stave off the commoditization of what were once high-margin specialty products. Key product innovations and/or supplementary engineering services can help sustain higher margins. At the same time, companies are looking forward to large emerging product markets where they can have the first-mover advantage in tackling urgent societal issues like climate change, sustainability, and human health.

As they pursue these business strategies in both old and new markets, there are new forms of digitally enhanced competition. Companies are no longer only competing with other large established materials and chemical suppliers. They are facing competition with large tech companies pursuing vertical integration strategies and bringing materials and chemical R&D in-house to keep up with the innovation demands of their rapidly evolving end products. Companies like Amazon, Apple, Google, Meta, Microsoft, Samsung, etc., all now have materials and chemical R&D teams and already have a wealth of digital talent and a strong digital culture in place, which can allow them to develop materials and chemical products in a very different way than more traditional suppliers. Traditional suppliers are also facing new competition from born-digital startup companies that are building their IP portfolio and company identity around the use of cutting-edge digital technologies. These smaller, more agile companies are often establishing valuable intellectual property in hot emerging markets faster than their much larger and slower-moving competitors. As Figure 1.2 illustrates, there is clearly an intensifying R&D innovation race with major business implications.

Big tech companies and born-digital startups may not be able or interested in manufacturing these chemical products, but they intend to control the valuable IP that they can then license to other players in the supply chain. Innovation-driven chemical and materials suppliers do not want to become glorified contract manufacturers. Contract manufacturing typically runs on much slimmer profit margins and requires far less R&D investment.

FIGURE 1.2 Traditional specialty materials and chemicals suppliers are now competing in an innovation race with big tech companies and born-digital startups, as well as each other.

In many areas, it really is an "innovate or die" situation.

Death in this situation typically means termination of emerging new business initiatives or divestment of existing business units to other companies more suited to serving commodity chemicals markets. Unfortunately, in both cases, it often means researchers are on the job market looking for new work. The point here is not to scare people but to emphasize the importance of understanding how the pieces fit together and why there is an elevated sense of urgency.

Altogether, there is a wave of R&D digital transformation happening all over the world and across all materials and chemicals product markets. Digital transformation is poised to help companies:

- Discover new materials and chemicals in strategically important areas faster and at lower cost
- Bring new, innovative products to market faster and at lower cost
- Iterate with customers during technical bidding process faster and at lower cost

- Increase agility and resilience in the face of dynamic changes in supply chain, regulatory landscape, and customer demands
- Create new digital products and technical services that add value to core non-digital products

And, yet, this transformation process has numerous challenges that make it distinct from digital transformation processes happening in other areas of the business and make it significantly more difficult and prone to failure.

> A wide variety of digital technologies and systems can be integrated into these complex R&D processes to help achieve the accelerated innovation and business growth goals, but choosing the right problems to tackle, the right tools for the job, and creating a robust implementation plan is not easy.

In the following sections, we'll describe some of these unique challenges on the journey from a human-centric lab to a digital-centric lab and finally conclude with recommended strategies and tactics for success that have been borne out of over two decades of industrial scientific computing and digital transformation experience at Enthought.

1.2 TRANSFORMING FROM A HUMAN-CENTRIC LAB TO A DIGITAL-CENTRIC LAB

The traditional R&D labs that have been the engine of innovation for companies around the world for generations are largely designed around human constraints. In these labs, almost all operations in a research workflow are executed by people: experiment planning, sample preparation, sample transfer, sample characterization, data transfer, data analysis, knowledge synthesis, and decision-making. These operations are executed on human-sized modular hardware (fume hoods, glassware, analytical instruments, personal computers, lab notebooks) with human interfaces (handles, buttons, knobs, gauges, displays). This modular design offers flexibility, so that different labs and processes could have different capabilities, and modules could be arranged in different configurations to support different workflows. And all of the modules shared a common interface, a single human researcher. However, while humans are intelligent, creative, and good at pattern recognition, they are also slow, need long periods of rest, and are susceptible to cognitive biases. In contrast, computers and robots are fast, need very little rest, and perform repetitive tasks much more consistently but are not nearly as creative or intellectually flexible. If we were to throw out our preconceptions of how an R&D lab should look and operate and then re-imagine what one might be like if it were built around the constraints of robots and computers, what would this digital-centric lab be like?

In a digital-centric lab, we still want to retain that modularity that allows reconfiguration, adaptability, and ad hoc operation. After all, this is not a manufacturing process that is intended to run exactly the same way over and over each day for years. There are of course operations that need to be highly repeatable, but we need flexibility to modify and rearrange the elements of a process as we learn more about a customer's requirements and the behavior of our future product over the design space. In such a system, modules with well-designed interfaces are critical. But these are no longer primarily human interfaces, we need digital interfaces that are called application programming interfaces (APIs) and robotic interfaces. An API allows different components in the system to talk to each other in a programmatic manner. It must receive commands from users or systems for what operation to perform, and how to perform it, and must return results data. There must also be standardized robotic interfaces for transferring different kinds of physical objects between modules. Altogether, this would allow research workflows to be highly reproducible with much less human bias and error and scriptable to enable automation that would dramatically increase the productivity of the human researchers who oversee and continually improve the system.

To enable the reconfigurability and flexibility that is so critical in an R&D environment, the APIs must not only support communication between modules but also be understandable by domain experts. With well-designed and well-documented APIs that follow domain-driven design principles, domain experts without degrees in computer science and deep expertise in software engineering can reconfigure, improve, and maintain integrated digital-centric workflows. They can also communicate with individual modules as needed for troubleshooting and ad hoc analysis.

As Figure 1.3 illustrates, these underlying modules are necessarily quite diverse in function, form, and origination. There are still components from traditional chemistry and materials science for performing experimental procedures, theoretical tools for physical simulations, computing hardware and software from computational engineering and computer science, algorithms from data science, and robotics tools.

> All of these pieces must be combined in a way that engages human experts who still have critical domain expertise to provide and must ultimately be responsible for and govern the system.

So, what are these digital-centric modules that are needed to assemble a working digital-centric lab? One can broadly put them into several categories that can help build a mental model that will aid in understanding the system design.

1.2.1 Physical Modules

- **Sample handling, dispensing, and storage** – Robotic equipment for moving materials (liquids, powders, pastes, and solid objects) around the lab, dispensing controlled quantities into containers, and then storing samples for later use

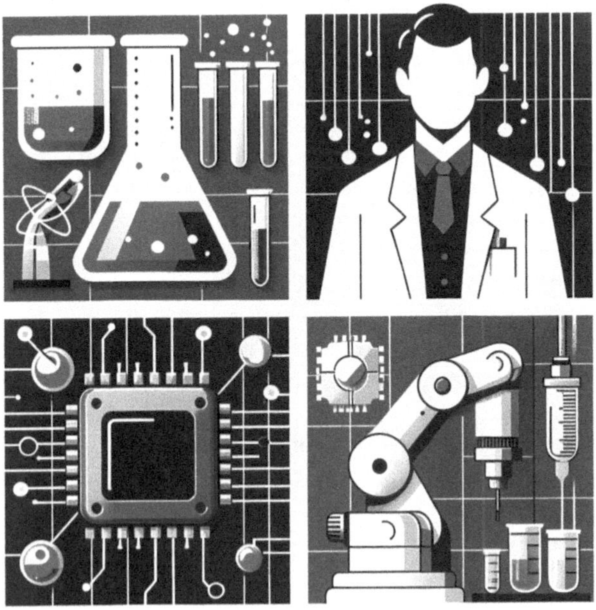

FIGURE 1.3 Many different types of technological components must fit together and work harmoniously in the digital-centric lab to enable researchers and business leaders to innovate much more rapidly.

- **Sample transformation** – Equipment that performs physical or chemical transformations on samples: mixing, reactions, separation, purifications, crystallization, temperature treatments, forming, etc.
- **Sample characterization** – Analytical equipment that measures physical, chemical, biological, thermal, and electrical properties

1.2.2 Digital Modules

- **Data storage** – Systems for organizing and storing data that comes from other modules so that it can be used later include data lake, data warehouse, data mesh, relational databases, and non-relational databases
- **Compute** – Systems for on-demand execution of basic to computationally intensive operations, including workstations, HPC clusters, and cloud computing instances

- **Theoretical simulation** – Computational tools that can be used to predict materials and chemical properties, including density functional theory, molecular dynamics, kinetic Monte Carlo, phase-field, and finite element
- **Machine learning (ML) tools** – Software that allows the development, deployment, tracking, and management of predictive ML models that are trained in order to assist with different decision-making tasks
- **Data analysis tools and applications** – Software for automated or semi-automated extraction of meaningful metrics out of raw sample characterization data coming from one or more analytical techniques
- **Project orchestration applications** – Software that organizes projects and product development campaigns and leverages historical project data to make more informed project management decisions
- **Decision-support tools** – Software that leverages trained ML models and AI algorithms to assist with R&D task planning and decision-making toward quickly and efficiently reaching defined objectives

To understand how these different types of modules would come together and interact to support a digital-centric specialty chemical workflow, let's walk through an idealized example of how a digital-centric lab project might play out.

1.2.3 Project Initiation

A request for proposal comes into the company from a potential customer through a web-based digital RFP system that guides the technical buyer to an approximate solution. The RFP system asks key technical questions up front to shorten the requirement gathering phase. The system then alerts relevant technical sales and business managers to the opportunity, and all of the information goes into an RFP database.

After an initial business-level approval, the project data and initial technical specifications are passed to the project orchestration app using its project creation API. Research managers are alerted to the opportunity and decide on how to staff the project. The project graphical user interface (GUI) in the app allows managers to assign project leaders and set both budgets and timelines. It also shows other similar projects executed in the past with the option to open a project summary to inspect who was involved and how things went previously to help with staffing and budgeting decisions. This same app is then used by project leads to enter in additional technical product requirements and constraints after talking details with customer technical buyers and the internal product R&D and production teams. All of this project information is saved into a project database that has been tailored for that particular product line.

1.2.4 Project Task Planning

Next, the project orchestration app has an experiment planning user interface, in which decision-support tools are used to recommend to the project leader which R&D operations to perform based on the defined objectives, budgets, timelines, and current state of data. It recommends when to perform DFT or MD simulations, when these data sources will provide the greatest amount of information per dollar, and it also recommends when to start performing physical experiments in the lab. For these experiments, it recommends which molecules to synthesize, how to perform the synthesis and purification, which ingredients to include in the formulation, and how to mix them. It can also take into account instrumentation details to suggest the optimal batch size and how sample conditions should vary over each batch.

During this process, it also recommends which intermediate characterization methods should be executed to validate that the right molecules have been synthesized with the right properties and that the formulations with the right properties have been created. In many cases, fast proxy measurements are executed every time a relevant sample is created, but in cases where a test is particularly slow and/or expensive, it is only performed as needed to validate the properties of the sample. In cases where intermediate characterization indicates poor results, the system can recommend when to terminate the investigation of a particular sample and when to continue because the value of fully characterizing the negative result is high enough to guide future experiments.

In all cases, the various recommendations are derived from AI algorithms, trained ML models, and all relevant R&D data that has been collected in the past. Recommendations are computed by performing numerical optimization of a very complex acquisition function over many dimensions, which is computationally intensive and requires the use of separate on-demand compute resources. To perform this operation, a compute instance is spun up to run the decision-support tool. The latest production ML model is downloaded from the model registry, and all relevant R&D data is retrieved from the simulation and sample characterization databases. The recommendations are calculated and the associated data are then communicated back to the project orchestration app, which displays the recommendations in a user-digestible form.

Initially, project leads will likely want to inspect and manually approve recommended tasks before they are performed. They may also want to occasionally override recommendations to create their own task plan, but as teams gain confidence in the algorithms and models, large sections of the project planning and execution are able to run completely autonomously, 24/7, with automated monitoring systems in place for safety and process excursion detection. Semi-automated workflows can also be set up in cases where expert assessment and judgment are still needed, and the system can notify experts when their input is needed. Well-designed GUIs within the project orchestration app can be developed to streamline the expert review process, so that human bottlenecks are minimized and human decisions are consistently informed by the relevant data.

1.2.5 Project Task Execution

Once a task plan has been determined, the project orchestration app communicates with the APIs of the relevant modules. Starting with theoretical simulations, the orchestration app runs the designated simulation software with the specified input parameters on compute modules. The simulation results are automatically analyzed using a data analysis tool running on the same compute module, and all raw and processed results are saved to a simulation database. As these results come in, the project orchestration app is alerted and dispatches ML model re-training and updated recommendation calculations to compute modules. The project task plan is then updated, given this new information to determine whether additional simulations are needed or whether the results are promising enough that more expensive experiments should be performed.

For experiments, the orchestration app tells sample handling modules to retrieve chemicals from storage and dispense specified quantities into assigned containers in the synthesis module. The orchestration app then specifies the reaction conditions, and the synthesis module executes the task and saves the metadata, process parameters, and online sensor readings to an experimental reaction database. A data analysis tool automatically reviews this process data and calculates the important final metrics, which are also added to the experimental database record. A sample handling module is then dispatched to collect the product sample.

From there, several sample aliquots are transported and dispensed into different analytical chemistry modules for chemical structure characterization. For each technique, an automated data analysis tool specifies measurement conditions, collects the raw spectral data, and performs peak analysis to quantify the structure and purity of the sample. All metadata, raw spectral data, and extracted metrics are saved to the experimental database and associated with the sample.

The experimental process continues from here to include various modules involved in formulation creation and characterization, as well as multiple levels of application sample creation and testing. This may even include part-level simulations near the end of the workflow. At defined intervals in the workflow, the ML models, task recommendations, and task plan are updated. Whether these updates are done after every data generation step in the workflow or at coarser intervals largely depends on the computational time needed to perform the updates relative to the time and cost of the next step in the previously determined task plan.

1.2.6 Project Monitoring

As this whole process unfolds, new regions of parameter space are explored, and promising new product solutions are revealed. Project leads monitor progress toward the project objectives in the project monitoring section of the project orchestration app. Here, one can see whether the properties of the "best" product are improving toward reaching the technical specifications provided by the customer over time as data is gathered and the research budget is consumed. If significant progress is not being made after some time, project leads can terminate an opportunity in order to free up resources to go after other sales opportunities. If good progress is being made and/or all objectives are met, project leads may decide to create a product candidate to send to the customer.

At this point, the project orchestration app tells the system to create several batches of the same candidate product and perform a more comprehensive characterization in order to more rigorously validate product performance and check process variability as well as gather additional unspecified properties needed by the customer to properly use the product candidate in their downstream process. If everything looks good, the project lead can have the product candidate sent to the customer for their own tests. If customers are able to accommodate it, multiple different candidate products can be sent to increase the likelihood that one or more products meet the requirements in that iteration.

Regardless, once customers have had a chance to evaluate the product candidate and learn more, themselves, about their own process and product that they are developing, they may add or modify the property requirements or process constraints. If the customer is willing to share their final success metrics and how each candidate performed, another higher-level ML model and recommendation system can be developed to even more efficiently satisfy the customer's needs. Ultimately, this system puts forth a candidate product that is selected by the customer and the path to full-scale manufacturing is eased by the fact that internal manufacturing constraints and processability/manufacturability/cost metrics were included among the project objectives set during project initiation.

1.2.7 Continuous Improvement

With this digital-centric lab up and running smoothly, winning more business, and ultimately saving R&D time and money, the role of product engineers and lab technicians shifts. Instead of being critical bottlenecks in operation of the product development process, their human disadvantages are now moved out of the way, and their human advantage of creative problem solving can be put to optimal use.

> The digitally transformed lab does not just achieve a new digital-centric operating state that carries everyone into the sunset; rather, it continuously improves through further innovation.

As Figure 1.4 emphasizes, being digital means continuously iterating and improving the digital-centric lab. This is not something you only focus on at the end but should be the engine for how you execute your digital transformation strategy.

Digitally savvy domain experts use the project orchestration API to perform ad hoc retrospective analysis on previous projects to identify process bottlenecks and sources of variability and uncertainty. This allows answering critical process improvement questions: is there a valuable characterization method that is the bottleneck to faster iterations? If so, could the development or purchase of a new proxy measurement provide some useful information at a much faster rate and reduce the need for the slower but more rigorous alternative? Are there new theoretical methods that could provide similar information at a lower cost? Is there a process or characterization method that introduces a high degree of variability or uncertainty? Does a particular piece of equipment break often or drift out of calibration? If so, do equipment or instrument vendors offer newer or better models that reduce these issues? Could an optimized preventative maintenance and calibration schedule improve overall performance?

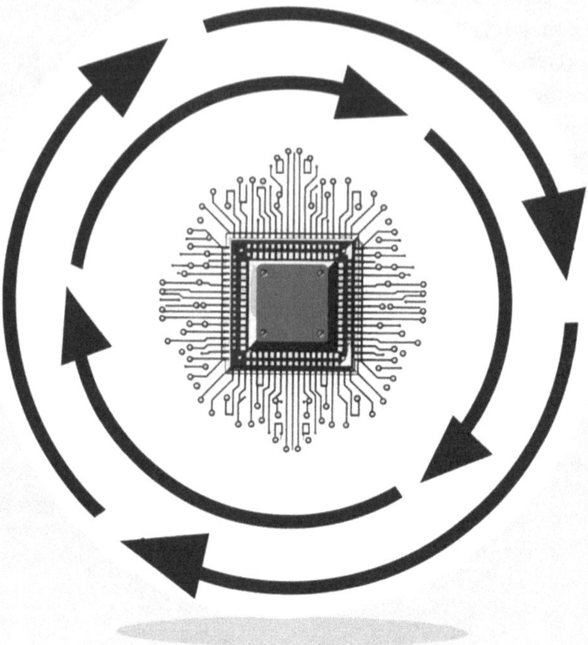

FIGURE 1.4 A drive for continued improvement is central to being digital. The individual modules and system as a whole must evolve as scientific learning evolves and as business needs evolve. A digital-centric lab is not something that you just buy, implement, and use.

In most cases, the predictive ML models that underpin the recommendation system will be trained using engineered input features. These features are created because domain experts have developed a mechanistic or semi-empirical understanding of composition-process-structure-property relationships, so they know which characterization data is important for predicting application performance. Embedding this knowledge into ML models through feature engineering allows them to generate better predictions with less data. Many of these engineered features are derived from the final metrics that are produced by the various data analysis tools. As this digital-centric R&D system collects data, experts periodically dig into the data at a deeper level to check their understanding and assumptions in order to improve the engineered features.

To make this easier, domain experts have established a product fingerprint and use a fingerprint explorer GUI application to explore these relationships. A product fingerprint is a vector containing only the most important calculated or measured structure and property metrics of a chemical or materials product. It contains the necessary information to distinguish a great product from a mediocre product from a bad product. Experts use the fingerprint explorer app to search over all samples made previously and quickly visualize the fingerprint data of specific samples and/or groups of samples. When looking at groups of samples, users can also see correlation analysis results between important input parameters and fingerprint metrics. If something in the data looks interesting or unexpected, users can drill down and see standard visualizations of the raw data together with a summary of the extracted final metrics, which are provided by each of the data analysis tools. This enables any domain expert to efficiently find samples worthy of more in-depth case analysis. Then, digitally savvy domain experts use the APIs directly in scripts to access the raw data and perform in-depth, ad hoc data visualization and analysis.

As new metrics are derived from existing measurement techniques or created from new ones, the sample fingerprint is updated. An important benefit of automated data analysis is that if the analysis algorithms are updated to produce better results or even completely new metrics, the updated algorithms can be quickly and systematically applied to all historical samples.

> The digital-centric lab will produce a much wider variety of different kinds of data and at a much faster rate than ever before. Well-designed data exploration, visualization, and analysis tools are critical for enabling researchers to supervise and improve the system.

As this digital-centric R&D system collects data and new engineered features are created, data science team members are using a machine learning operations (MLOps) system to monitor the performance of the ML models over time. They are also able to create and evaluate updated models with new engineered features as well as new model architectures. When improved models are created, data science team members can manage the testing and eventual deployment of these models into production. They also create retrospective optimization scenario benchmarks to help improve the acquisition functions and test how recent product optimization campaigns would have performed if different acquisition functions had been used. This is also done with updated value and cost estimates, things that are always difficult to quantify but can be improved over time with experience, market understanding, and good accounting.

While this description of a digital-centric lab may sound like a sci-fi fantasy, the vast majority of the technologies needed to reach this state exist. Many of these ideas have already been shown to work in proof-of-concept systems being developed by academic and government research groups around the world. Leading digital R&D companies have also started building digital-centric labs in some areas. Many of the previously mentioned digital-centric modules do not exist as off-the-shelf products but can be designed and built by companies with a strong digital vision and a financial business justification. Some of these modules may yet be too expensive for widespread adoption, and there are indeed some techniques that are very difficult and/or expensive to fully automate with robotics and may require manual input in the near-term, but there are no major fundamental science and engineering discoveries needed to do most of what has been described here.

In our experience, the greatest barriers to achieving the digital-centric lab of the future are actually non-technical.

1.3 THE IMPORTANCE OF PEOPLE AND NON-TECHNICAL CHALLENGES

Designing, implementing, and operating the digital-centric lab requires a new way of thinking about how R&D is done and how it is managed. It requires overcoming numerous challenging non-technical barriers that are often overlooked when researchers and engineers get excited to start implementing data-driven methods into their product development projects. Acknowledging that people and non-technical challenges are expected and understanding that significant effort will be required to address these challenges, is an important step in any digital transformation journey. This section will discuss:

- Creating and sharing data
- Fail fast mentality
- New digital skills
- R&D team composition
- Organizational change
- New business models
- Change management

Digital culture change starts with moving away from the mythology of the lone scientist making big discoveries in the lab toward a more collective and interdisciplinary way of problem solving. It requires a shift from individual researchers keeping secrets on how they do their work and "owning" the data for their assigned projects to an expectation of transparency and making both methods and results data as reusable across the organization as possible.

Researchers must realize that their role is not to produce experimental samples, results spreadsheets, or simulation files. As Figure 1.5 highlights, a lab chemist must realize they are ultimately producing data. This data, representing knowledge, methods, tools, and intellectual property, is the key product of R&D that provides a competitive advantage to the business.

To be continuously useful within an organization, data cannot be hidden away in spreadsheets, reports, and other files where it cannot be discovered or used.

Results data and associated metadata, knowledge, and methods need to be stored in appropriately structured databases. High-quality R&D data is a valuable corporate asset and must be treated accordingly. The custom software modules (tools, apps, and systems) created and owned by various teams across the company are also valuable

FIGURE 1.5 Comprehensive digital transformation changes the work of scientists from producing chemicals and materials to using and generating data that informs R&D and business decision-making.

corporate assets. Taking the extra time to make these things useful to others can be promoted within an organization by rewarding and promoting individuals based on these contributions and team business accomplishments that leverage these assets.

Leaders must also instill a culture of innovation that accepts failure as part of the process. Successes should be elevated to generate buy-in across the organization, stimulate ideas, and build momentum, but failures are critical to learning. Slow and expensive failures must be avoided, but fast and inexpensive failures must be part of the system. It must be ok for projects to fail without indicating personal failures.

> If projects are never failing, you are not taking big enough risks or you are not being honest and therefore missing out on a valuable learning opportunity.

Digital transformation requires taking on very new and sometimes risky projects to implement cutting-edge technologies into R&D processes. There are many potential reasons why a digital project might fail and a lot of them cannot be accurately predicted before the project begins, so it must be ok to fail fast and carry forward lessons learned to the next round of projects.

The digital-centric lab also requires people to develop a lot of new digital skills across all R&D staff (directors, team managers, researchers, and technicians). Training initiatives can be a good start, but this often also requires acknowledgment of the importance of digital skills as part of traditional roles.

R&D leaders must learn what digital technologies can do for their teams, how to choose the right projects to go after, how to build teams with the right mix of skills, and how to manage projects in a more Agile[1] fashion.

A significant number of researchers must develop programming, data analysis, data visualization, data science, and software engineering skills. The depth of these skills will vary significantly depending on specific teams and roles, but most researchers can greatly benefit from learning how to do programmatic data analysis and visualization.

However, individuals and teams that have both deep domain expertise and deep digital expertise are extremely valuable. They are uniquely able to navigate the complexities and tradeoffs when designing and developing scientific software tools. These tools must capture deep domain knowledge and complex research processes performed by senior experts in the lab, while also being robust, reliable, and extensible on the software side.

In most cases, these projects are most successful when driven by researchers who have augmented their domain expertise with digital skills instead of by classically trained software engineers who have learned some chemistry.

Researchers can start learning these digital skills in the classroom, but leaders must provide opportunities for individuals to put them into practice to solve real problems or the skills are never solidified and are often quickly forgotten.

Promoting a digital culture, digital skills, and managing R&D digital transformation often requires significant organizational change. Organizational restructuring is often needed to establish new job roles and teams, new reporting structures and methodologies, new funding structures, and new cross-team interaction patterns. Digital activities that were once firmly under the purview of corporate IT must have a much tighter connection with individual business units and labs. An explosion of new data, software, and hardware must now be appropriately managed to support local flexibility while maintaining security and strategic alignment across the organization. These corporate assets must be managed by establishing governance systems. A true culture of digital innovation supports creative, value-generating ideas from all corners of the organization, and systems must be established for soliciting

1 Agile is a popular software development methodology and set of principles that also excels with other types of engineering projects where there is a high degree of complexity, a large number of unknown factors at the beginning, and rapidly changing requirements. It will be discussed in greater detail in the next section.

and elevating good ideas. There must then also be funding and staffing mechanisms to support the pursuit of innovative ideas.

Finally, business leaders need to learn how the new capabilities afforded by digital technologies can be leveraged to grow their business in new ways. If all you are doing is producing the same products a bit faster, you are likely leaving large opportunities on the table.

Increased speed, agility, and data should allow deeper and more mutually beneficial ways to interact with customers through increased product customization and enhanced technical service.

It should enable companies to offer complete solutions instead of just a product, and it should lower the entry barrier for new prospective customers to sample your offerings and see if they are a good fit for their needs.

For example, let's imagine a new vision for the customer journey in a specialty chemical company that leverages all that digital transformation has to offer. Without any barriers, a technical buyer can navigate to a web app and view product categories and existing product grades with the ability to access industry-standard product specifications. For more information and assistance, the buyer can log into the web app and now access much deeper information about existing product grades as well as have the ability to specify unique product requirements and receive information about existing products that are close and see an estimate for if and when a new product grade can be made to meet their unique needs. Customers can request product samples with existing product grades shipped within days and new product candidates shipped within weeks.

The same web app can provide detailed process recommendations, manufacturing process simulation results, and part performance simulations. These can also include part-level optimization of multiple materials and multiple sequential processes thereby allowing co-optimization. Co-optimization can be made easier by recommending multiple material products and processes that are known to work well together, which can lead to bundling of materials offerings within your own product lines or with strategic external partners. There are also interfaces to share application requirements and testing data back and forth with the product development team to allow rapid product development iterations. The added value of this information and technical service system can be monetized using a digital subscription license model or by marking up the product itself.

In some cases, new digital products and services may be developed to solve customer problems in adjacent areas, resulting in new markets with no physical product at all.

In the rapidly evolving digital realm, business leaders need to keep an open mind to new ways of doing business that might look very different from how things have been done for many decades.

FIGURE 1.6 Without strong leadership support, through training, digital promotion, organizational restructuring, change management, and strong vision, doubt, fear, and discouragement will hinder transformation efforts.

Throughout all of these different types of change, there will always be doubters and detractors to digital transformation up and down the organization; people who think it will not work or who do not want to be forced to change. As Figure 1.6 illustrates, in the face of a big looming digital initiative, they may fear personal failure or see the process as a risk of unnecessarily disrupting a system that already works in an already profitable business. This fact must be tackled head-on by including change management as a critical part of any digital initiative. In most cases, R&D digital transformation projects simply cannot succeed without buy-in from senior leaders and senior technical staff. Projects need influential champions to get proper support from around the organization, and deep technical expertise is needed to solve the most valuable R&D problems at hand. Digital leaders must find ways to bring these critically valuable talents into the fold so that digitally hesitant members see where and how they can add value in the rapidly evolving digital-centric innovation machine.

1.4 STRATEGY AND TACTICS FOR SUCCESS

Given all of the things that can and should change, it can be challenging for leaders to set a short-, mid-, and long-term strategy that makes sense for their business. As Figure 1.7 highlights, digital transformation of an innovation-driven business can seem like an impossibly complex maze. There are so many complicated decisions to make along the way. A good digital strategy that acknowledges and confronts these challenges is necessary. This starts at the business level, where a good digital strategy should try to leverage existing strengths and integrate new digital capabilities that together enhance existing markets and help create new revenue streams. This will look different for every business, but it is important that this strategy remains agile and evolves as one learns more about customer needs, the limits of digital technologies, and internal capabilities. When beginning on the digital transformation journey, the short-term digital strategy will likely be similar to existing business

FIGURE 1.7 The digitalization of R&D is very different from that of other business processes. A variety of choices about handling complex scientific data, fusing data science and chemistry, and how and where to get started.

strategy with a plan to leverage the speed and efficiency enabled by digital technologies to reach existing objectives faster.

Once a business strategy has been set, a tactical plan must be implemented, one that informs how digital transformation should be implemented from top to bottom within a company. Executives at early movers in the materials and chemical industry, lacking examples from peer organizations, have typically tried adopting successful tactics from other business sectors, often at the advice of big business and IT consulting firms. The advice is to start with implementing digital infrastructure and enterprise software systems to digitize the existing business. These systems are often put into place to digitize and curate historical data, unify data entry for new data, automate business processes, and collect all data into centralized databases. With this structured data now organized, central data science and engineering teams are hired to clean and transform the data before finally developing predictive analytics that can be displayed to leaders in dashboards to assist with decision-making.

The problem is that this infrastructure-first approach frequently leads to business failure when transferred to R&D digital transformation in innovation-driven specialty materials and chemical companies. Implementation of monolithic enterprise systems across many business units and labs is too slow and the result is insufficiently flexible. In most cases, the compromises needed to force fit a general solution onto a diverse landscape of R&D activities lead to painfully awkward software and processes, a disillusioned workforce, and a data system that either completely misses important information or is impossible to efficiently navigate. There are numerous stories of companies spending tens of millions of dollars buying, building, and implementing digital tools over five years only to see no meaningful acceleration of the innovation process and no foreseeable return on investment. Even if there exists a more general enterprise solution to solve a number of the problems encountered in the lab, most organizations will have a very hard time anticipating their needs at a detailed enough level to make sound technical decisions early on in their R&D digital transformation journey. Such failures not only lose money but also create a chilling effect on willingness to fund future digital initiatives, even if the risk level is much lower, and can cause valuable digital talent to flee the organization. These challenges can set back digital transformation progress by several years and make it slow to catch up again.

R&D digital transformation requires a different approach that mitigates the risk of expensive, high-profile failures. Digital transformation of an apparel retailer, restaurant chain, or financial services provider should look a lot different than in a materials and chemical company. Even within the chemical industry, the digital transformation of a commodity chemicals business should look a lot different than in a specialty firm. In both cases, things are different because of how R&D drives the business. And even within a specialty chemical company, sales and marketing, logistics, finance, and accounting are very different from R&D. As depicted in Figure 1.8, job roles, processes, and data are all much more complex in R&D, which necessitates a different approach, different technology tools, different organizational models, and different digital skills.

The approach that consistently works best is to focus first and foremost on solving valuable problems. Choosing which digital initiatives to pursue early on sets the tone

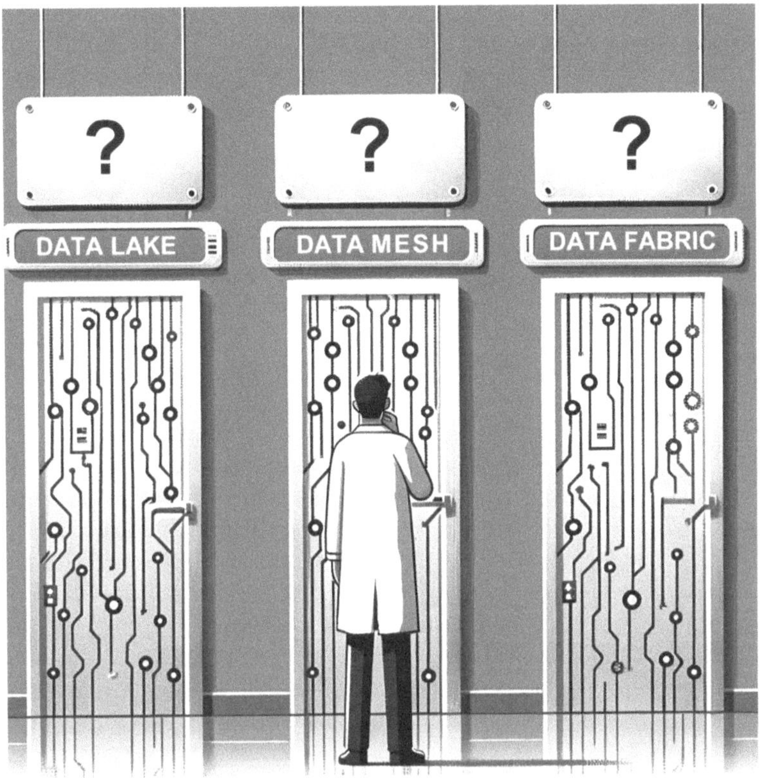

FIGURE 1.8 Seemingly equivalent choices for digitalization technologies or approaches can lead to different outcomes, and not all of them are desirable.

for the organization. Identify and select areas that are important to corporate business strategy and then work closely with the business leader, the lab manager, and researchers to identify specific bottlenecks in their workflows that can be alleviated with digital technologies.

> Projects should be chosen based on estimated impact, technical risk, and willingness of all people involved. The digital technology tools and infrastructure should be chosen after a clear problem has been identified, not before.

Projects may also need to be multifaceted and include the development and implementation of new software, hardware, digital skills training, modified workflows, and change management. All of these elements must be considered part of the solution. One does not simply buy a new piece of software or an automated instrument and become digitally transformed. It is a much larger and more complicated process.

When tackling complex R&D problems, it can be very difficult to know precisely what the solution should look like up front. R&D leaders need to become familiar and comfortable with the Agile development style, where there is less pre-planning and a more dynamic execution phase. Pre-planning should be used to clearly identify the problem, set focus on a solution approach, highlight the most value-generating capabilities, and identify any major risks and uncertainties. Agile solution development should rapidly seek to uncover potential unknown issues and clear up areas of uncertainty as soon as possible. It should also work to quickly validate value assumptions with real users. This allows the project to pivot the solution approach, scope, and technology choices quickly as new information comes to light. A key part of Agile development is regular releases of working tools and check-ins with the users and stakeholders involved to help escalate and resolve emergent issues as fast as possible. And if it becomes clear that a valuable solution can no longer be realized due to technical limitations, people, or budget, leaders and teams must be able and willing to put feelings aside and terminate a project, learn what they can from the experience, and move on to tackle the next problem. In most cases, there will be no shortage of valuable problems to go after, and lingering on a dead-end project serves no one in the long run. However, in successful projects, most check-in meetings serve to get critical feedback that helps re-prioritize the planned work, so that the most valuable tasks are always at the top of the list. This helps develop solutions that deliver the most concentrated value possible and the least amount of fluff. Agile development has become a mature practice in software development for dealing with complexity and can be adapted nicely to R&D digital transformation projects.

Through this iterative, value-focused Agile process, leaders, and R&D staff will start to see what is actually possible and the real impact of digital technologies.

The early projects should be able to show meaningful impact within six months and generate a return on investment within a year.

Early successes can be used to promote digital transformation across the organization and generate the buy-in to go after potentially longer and riskier projects. Early wins showing what has been improved, how much better things can be, and how different people participate in the process can alleviate the fears and reduce resistance from digital doubters. Soon you can have people requesting to be a part of the next digital initiative and wanting the same benefits and excitement for themselves, their teams, and their businesses. As the Agile development process plays out, the pieces become real and the team figures out how the pieces can be integrated. Soon a detailed vision for the full solution architecture and operational process of the complex digital solution becomes much more clear. A fully digital-centric lab, as described in Section 1.2, is not built in a month or even a year. It may take five+

years, but if done properly, the transformation process will deliver significant value along the way that justifies the continued and prolonged investment.

> At this digital technology inflection point, it's an exciting time to be an ambitious researcher or business leader looking to make a positive impact in the world.

2 Product Formulation Fundamentals

Kristin Wallace and Alix Schmidt

2.1 INTRODUCTION

The digital-centric lab envisioned in Chapter 1 is a holistic approach to accelerated innovation. At the heart of a digital-centric lab lies the focus of this book: product formulation bolstered by predictive modeling.

The authors of this book define product formulation as scientific activities aimed at designing new or improved functional compounds or mixtures to meet targeted quality properties.

The formulation of advanced materials and specialty chemicals can be extremely complex; therefore, many scientists and engineers have devoted entire careers to conducting such Research and Development (R&D) in highly focused application areas. Complexities aside, product formulation boils down to the strategic selection of three inputs: what ingredients to combine, in what ratio, and under what conditions to combine them. Outputs in a product formulation context quantify the function of the final product and are interchangeably referred to throughout this book as quality properties or performance properties. Product formulation inputs and outputs are illustrated in Figure 2.1.

Predictive modeling generally refers to the practice of using historical data or information to mathematically predict future events (IBM 2023).

Predictive modeling can be dated back to the early 1800s when Carl Friedrich Gauss successfully forecasted the position of the dwarf planet Ceres (Wikipedia 2023). Today, and largely thanks to the relatively recent explosion of data, predictive modeling has many useful applications spanning essentially every sector from retail to finance to manufacturing and everything in between. As the world rapidly evolves around the exciting concept of artificial intelligence (AI), the subsets of predictive modeling and

DOI: 10.1201/9781003385974-3

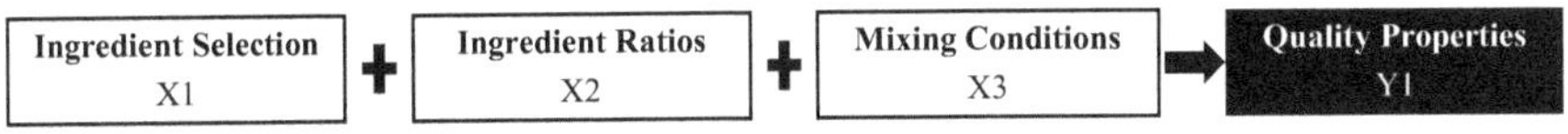

FIGURE 2.1 Product formulation inputs (X) and outputs (Y).

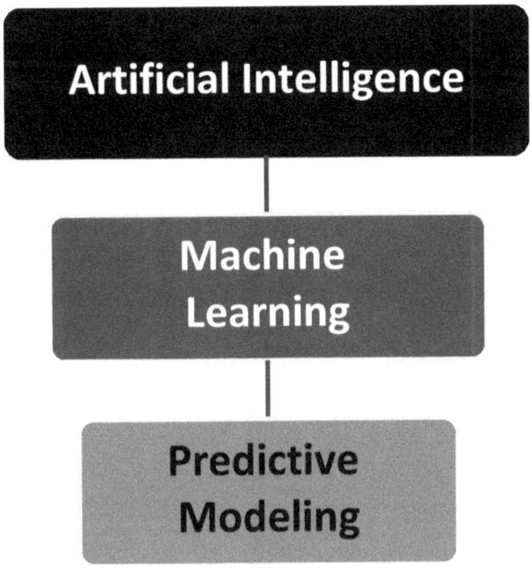

FIGURE 2.2 The hierarchical relationship between artificial intelligence, predictive modeling, and machine learning.

machine learning (ML) (see Figure 2.2) will become more prevalent than ever, driving efficiency and empowering automation and innovation in many sectors.

Predictive modeling plays an important role in the digital transformation of product formulation. Conventional product formulation typically entailed the mysterious art of combining traditional domain knowledge with intangible "know-how," accumulated through many years of practical experience, to develop single-factor Design of Experiments (DoEs). This work was conducted by a small number of highly specialized individuals, and the goal was to test hypotheses and explore new regions of a design space by scheduling these trials into a busy laboratory or pilot plant and then later interpreting the results. Today, the data-driven science of predictive modeling offers an accelerated product formulation workflow that also broadens resource capacities and reduces development costs. Formulators can complete their explorations *in silico* by testing ideas rapidly using the predictive model or even by specifying targets to an optimization routine coupled to the model. Either way, the end goal is for the data-driven tools to provide the experimentalist with a suggested first experiment that is consistently closer to the target so that lab synthesis and testing iterations are reduced.

2.2 SCOPE OF TEXT AND SUGGESTED FURTHER READING

To be clear, this book on formulation design does not include the related topic of molecular design. As the name suggests, the focus of molecular design is on defining individual molecular structures, whereas the focus of product formulation is on the combination of multiple functional ingredients. A common application of molecular design is drug discovery. "Computational approaches streamlining drug discovery" by Anastasiia Sadybekov and Katritch Vsevolod (Sadybekov and Katritch 2023) is suggested to readers who are interested in that topic.

This book strives to provide a beginner- to intermediate-level introduction to the main elements of a predictive formulation project. The objective is to provide the reader with a baseline understanding of the challenges to expect, proven approaches for success, the nomenclature to be used for further literature searching, and suggested starting points for programming your first inverse design model. The tasks involved in such a project, and their alignment to chapters within this text, are depicted in Figure 2.3.

In any one of the steps within Figure 2.3, there is significantly more technical depth that could be pursued. As this book provides just a survey of the relevant techniques, we also recommend the following references for further reading on covered topics (in addition to the references cited within each chapter), as well as suggested reading on subjects not discussed here.

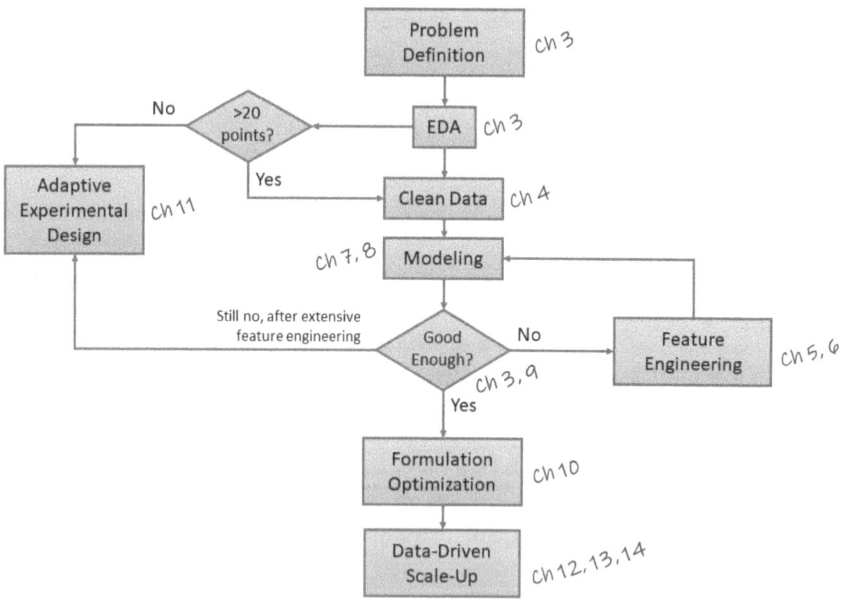

FIGURE 2.3 A formulation inverse design project flowchart, labeled with the chapters aligned to each modeling task: problem definition, EDA (exploratory data analysis), cleaning data, adaptive experimental design, predictive modeling, feature engineering, model validation, formulation optimization, and data-driven scale-up.

1) Empirical modeling principles such as bias-variance trade-off, train-validation-test splits, and derivations of statistical and machine learning algorithms:
   - "The Elements of Statistical Learning" by Trevor Hastie et al. (2013)
   - "A Few Useful Things to Know About Machine Learning" by Pedro Domingos (2012)
   - "Introduction to Data Mining" by Pang-Ning Tan et al. (2019)
   - "Machine Learning: A Probabilistic Perspective" by Kevin P. Murphy (2012)
2) Exploratory Data Analysis (EDA)
   - "Exploratory Data Analysis" by John Turkey (1977)
   - "Making Sense of Data: A Practical Guide to Exploratory Data Analysis and Data Mining" by Glenn Myatt (2006)
   - "Exploratory Data Analysis with R" by Roger Peng (2016)
   - "Pandas 1.x Cookbook: Practical Recipes for Scientific Computing, Time Series Analysis, and Exploratory Data Analysis using Python" by Matt Harrison and Theodore Petrou, 2nd ed. (2020)
3) In-depth discussion on the complementary nature of experimental design and traditional statistical modeling with data-driven machine learning
   - "Statistical Modeling: The Two Cultures" by Leo Breiman (2001)
   - "Data-driven Modeling and Learning in Science and Engineering" by Francisco J. Montáns et al. (2019)
4) In-depth discussion on design of experiments and how to collect new data
   - "Design and Analysis of Experiments" by Douglas C. Montgomery (2019)
   - "Design of Experiments: A Realistic Approach" by V.L. Anderson and R.A. McLean (1974)
5) Measurement Systems Analysis (MSA, a.k.a. Gage R&R)
   - "Measurement Systems Analysis Reference Manual" by Automotive Industry Action Group (AIAG) (2010)
   - "A Study in Measurement" by Neetu Choudhary (2017)

2.3 AUDIENCE AND BOOK CONTENTS

This book has been written as reference on the current state of data-driven product formulation for technical researchers with science, technology, engineering, and mathematics (STEM) backgrounds. Readers are assumed to have some basic experience in applying mathematical and statistical concepts to scientific research data but are not assumed to have an advanced education in ML or computer programming. This book aims to provide awareness of key topics and concepts, while emphasizing strategies to overcome common challenges in adopting a data-driven approach specifically to product formulation.

As an edited compilation, this book represents a diverse collection of experiences gathered by both academics and professionals working in related fields. Accordingly, while some chapters are relatively technical and detail-oriented, others are an informal summary of advice and practical tips. The editors hope that this contrast will allow readers to appreciate the spread of challenges that must be overcome and considerations that must be balanced in successfully transforming product formulation workflows within an organization. A grand total of 185 years of experience in data-driven materials formulation (collectively amassed by the group of authors) are packed into this book. We hope that the range of topics covered will allow all readers to find value regardless of your background, application area, and stage in your digital transformation.

The book is organized into 15 chapters which are grouped into five sections, as follows:

Section 1 Getting Started

Chapter 1 The Digital Transformation of R&D Labs
Chapter 2 Product Formulation Fundamentals
Chapter 3 Defining a Successful Predictive Formulation Project

Section 2 Preparing Your Data

Chapter 4 Challenges with Formulation Datasets
Chapter 5 Feature Engineering: Enhancing Your Data with Descriptors
Chapter 6 Machine Learning for Analysis of Structural Characterization

Section 3 Predictive Modeling

Chapter 7 Machine Learning Techniques for Prediction Properties of Formulations
Chapter 8 Modeling of Product Formulations Using a Latent Variable Approach
Chapter 9 Gaining Trust in Your Model

Section 4 Optimization and Inverse Design

Chapter 10 Introduction to Formulation Optimization
Chapter 11 Adaptive Experimental Design
Chapter 12 Inverse Design via PLS Model Inversion

Section 5 Case Studies and Special Topics

Chapter 13 Case Studies
Chapter 14 Special Topics
Chapter 15 Conclusion

Section introductions briefly provide helpful context for the upcoming chapter topics, while guiding readers through the lifecycle of a product formulation project. The reader is highly encouraged to read all section introductions, even if the proceeding chapters are not of interest.

The contents of this book reflect and address the unique challenges that surround applying machine learning methods to product formulation datasets. Unlike other successful and well-known applications of AI and machine learning, such as virtual personal assistants (Apple Siri, Amazon Alexa, etc.) and customer recommendation systems (within Netflix, Spotify, etc.), product formulation applications must overcome many technical hurdles such as collecting and structuring highly protected data, handling missing data, feature engineering and incorporating physics and mechanistic insights, and developing and validating predictive models that agree with governing scientific principles and domain knowledge. The identification of these formulation-specific modeling challenges and the recommendations to address them were borne from the aforementioned years of experience by the authors, and they pose the primary unique value of this text relative to popular machine learning educational resources.

2.4 NOMENCLATURE

In addition to the key terms described earlier in the Introduction, there are several recurring themes and concepts that appear throughout this book, often with slightly different nomenclature due to the diverse authors and the number of distinct scientific fields that interface in formulation inverse design. For clarity, definitions of some important themes and their interchangeable nomenclature are summarized as follows.

Model/Machine Learning (ML)/Data Science/Data-Driven/AI – The general concept of applying mathematics and statistics to historical data in order to extract value and make informed decisions.

Formulation/Mixture/Blend/Recipe – Data that captures ingredient selections, ingredient ratios, and mixing conditions for a single particular product.

Experiments/Samples/Observations/Rows – The set of unique identifiers that map formulation data to a single particular product. These identifiers differentiate products from one another in a historical dataset.

Variables/Features/Descriptors/Properties/Columns – Data that captures information about individual ingredients (such as density or molecular weight) or mixing conditions (such as temperature or mixing speed).

2.5 LEARNING GOALS

After reading this book, readers should be able to:

- Identify opportunities where business outcomes can be improved by implementing data-driven methods to product formulation
- Understand what foundational data behaviors are prerequisites for building a data-driven product development organization
- Understand the main steps in the modeling process, what challenges to expect, and where time is likely to be spent
- Have an awareness of what is possible, understand nomenclature used for literature searching, and know what advanced topics can be further studied to solve various problems
- Be able to contextualize AI/ML buzz from outside the materials industry to understand what is relevant and is not relevant to materials design problems, as well as understand the special challenges materials data pose
- Understand common barriers to translating technical success with data-driven methods to business success and how to address them
- Begin to explore your data, fit models, and make decisions based on data

Chapter 3 will tackle the first learning goal of identifying promising opportunities for data-driven product formulation.

REFERENCES

Anderson, V.L., R.A. McLean, 1974. *Design of Experiments: A Realistic Approach*. CRC Press. 1st ed. https://doi.org/10.1201/9781315141039

ASQC Automotive Division/AIAG, Automotive Industry Action Group. 2010. *Measurement Systems Analysis Reference Manual*. American Society for Quality Control. 4th ed.

Breiman, Leo. 2001. "Statistical Modeling: The Two Cultures (with Comments and a Rejoinder by the Author)." *Statistical Science* 16. https://doi.org/10.1214/ss/1009213726.

Choudhary, Neetu. 2017. "A Study in Measurement." *Quality Progress* 50 (9): 42–47.

Domingos, Pedro. 2012. "A Few Useful Things to Know About Machine Learning." *Communications of the ACM* 55 (10): 78–87.

Harrison, Matt, Theodore Petrou. 2020. *Pandas 1.x Cookbook: Practical Recipes for Scientific Computing, Time Series Analysis, and Exploratory Data Analysis Using Python*. Packt Publishing. 2nd ed.

Hastie, T., R. Tibshirani, J.H. Friedman. 2013. *The Elements of Statistical Learning: Data Mining, Inference, and Prediction*. Springer.

IBM. 2023. "What Is Predictive Analytics". *IBM*. Accessed on Nov 11, 2023. https://www.ibm.com/topics/predictive-analytics

Montáns, Francisco J., Francisco Chinesta, Rafael Gómez-Bombarelli, J. Nathan Kutz. 2019. "Data-Driven Modeling and Learning in Science and Engineering." *Comptes Rendus Mécanique* 347 (11): 845–855. ISSN 1631-0721. https://doi.org/10.1016/j.crme.2019.11.009. (https://www.sciencedirect.com/science/article/pii/S1631072119301809)

Montgomery, Douglas. 2019. *Design and Analysis of Experiments*. Wiley. 10th ed.

Murphy, Kevin P. 2012. *Machine Learning: A Probabilistic Perspective*. MIT Press.

Myatt, Glenn. 2006. *Making Sense of Data: A Practical Guide to Exploratory Data Analysis and Data Mining*. Wiley-Interscience.

Peng, Roger. 2016. *Exploratory Data Analysis with R*. Leanpub.

Sadybekov, Anastasiia V., Vsevolod Katritch. 2023. "Computational Approaches Streamlining Drug Discovery." *Nature* 616 (7958): 673–685.

Tan, Pang-Ning, Michael Steinbach, Vipan Kumar, Anuj Karpatne. 2019. *Introduction to Data Mining*. Pearson Education Limited.

Turkey, John. 1977. *Exploratory Data Analysis*. Pearson.

Wikipedia. 2023. "Carl Friedrich Gauss". *Wikipedia*. Accessed on Nov 11, 2023. https://en.wikipedia.org/wiki/Carl_Friedrich_Gauss

3 Defining a Successful Predictive Formulation Project

Alix Schmidt

As you might expect, the outcome of the first efforts at digital transformation set the stage for an organization's journey for years to come, and selecting the right project can pay off in new revenue, culture change, and accelerated technology adoption. Even if your project case is already determined, knowing what questions to ask to shape your problem at the start can increase your probability of success. We think of data science as the act of making information machine-readable, and then applying learning algorithms to identify useful patterns. This is, of course, the exciting technical component of the work, but the data science task most related to the success of your project is clearly defining what you are trying to accomplish and correctly framing it in terms of your data. While it might sound trivial, selecting a good use case for data-driven product design is easier said than done.

There is a sort of chicken-and-egg problem that comes with "deciding to do data science," because that sentiment is essentially starting with a solution and then searching for a problem – not typically a recipe for success. However, transforming the way your team works is unlikely to happen without a solution-driven push, especially when researchers are unaware of the power of data-driven product design. The correct way to handle this situation is neither to push forward and tackle your most important problem without considering its suitability to machine learning nor to pick a case with a plethora of available data but no clear objective. Your job is to identify some candidate projects in the overlap – ones that have a clear path of action from better decisions to value *and* have enough high-quality data for your solution to be effective in making those decisions. Among those options, choosing a project with supportive subject-matter experts (SMEs) will also present a much easier path to success. Getting this project selection right may be your most important choice in your journey of digital transformation (Figure 3.1).

Let us further consider the parts of the objective. You need to find projects that have:

1) a clear path of action from better, faster decisions to value,
2) enough high-quality data for your solution to be effective in making those decisions, *and*
3) SMEs who are neither skeptics of AI/ML nor have unreasonably high expectations of what is possible.

DOI: 10.1201/9781003385974-4

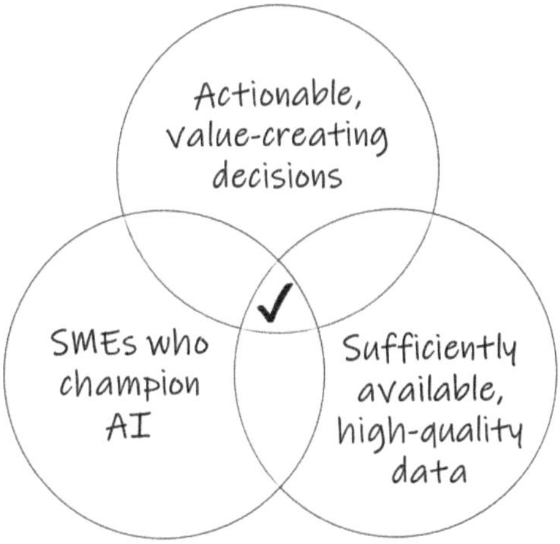

FIGURE 3.1 The most successful data-driven formulation projects will be at the intersection of actionable decisions, supportive subject-matter experts (SMEs), and high-quality data.

3.1 DEFINING YOUR ANALYTICS PATH TO VALUE

The first criterion is a clear path of action from better decisions to value creation. This itself has two parts – value and path of action. "Value" can mean many things: higher revenue, lower costs, faster time to market, etc. Fight the urge to assume that doing fewer experiments will directly lead to more "value," and work with your whole product team to clearly specify the opportunity. Is faster enough, or do the products need to be higher performing as well? Does getting to market faster give you access to new opportunities, and if so, do those opportunities look similar to or different from your current products? Can the rest of your commercialization process match the speed required?

Model predictions must be *actionable* to be valuable

Value-creating opportunities must be actionable, meaning you have a well-understood connection between controllable variables in your system and your value targets, for example, "if we can achieve performance 10% higher than current, we can secure a contract with XYZ Customer for new sales" or "if we can shorten development times from 7 experimental cycles to 2, we can work with the 6 customers we turned away last year." Consider Figure 3.2 on how different types of analytics are used, and what portion of the decision-making process they address. For a given business opportunity, what are the current bottlenecks in that decision today? Identifying how far down the path of automation you need to go before your data-driven tool debottlenecks this decision and thus creates a noticeable difference

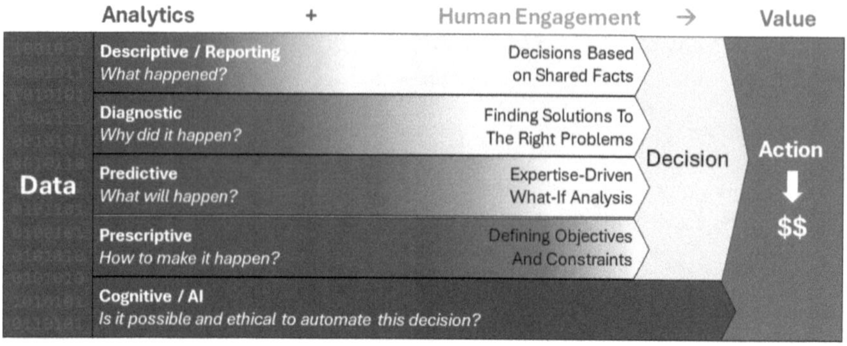

FIGURE 3.2 Comparison of the roles of various types of analytics – descriptive, diagnostic, predictive, prescriptive, and cognitive – in the decision-making process. Besides cognitive/artificial intelligence that automates decisions, each type requires some degree of human engagement by consumers of the analytics to realize the value. The nature of this interaction is listed and the proportion of the decision-making by the human vs. the analytics is indicated by the proportion of light vs. dark along the decision path.

to your stakeholders is a useful exercise. Generally, the further down in the figure you go, the more comprehensive and high quality your data must be, and the more advanced your analytics techniques may be, but the easier or faster it is for consumers of the data to make the decision. What is the right choice of complexity + value for your organization to engage in right now? Do you have employees who are skilled in the data and analytics techniques required to execute the work effectively? Are there projects that can create value at various points along stages of analytics in a multigenerational plan?

It is helpful when considering how your analytics will create value to list out the relevant variables in a qualitative sense: what new information do you need to complete your decision in a way that avoids the current bottleneck or challenge? What variables do you need to describe/predict/prescribe to make better decisions? What variables can you control? What variables do people currently use, and/or what is missing?

Explicitly talking through these questions with the team can be quite enlightening and is very useful to get all team members on the same page on the technology and what is known. You may even uncover where know-how is missing or where experts disagree with each other – valuable places for machine learning to help. To accomplish this exercise, you might choose to document a team brainstorm of the variables present in your target problem area; one version of this exercise could look like Table 3.1.

After you have exhausted the potential sources of variation in the system, the team can inspect each variable and discuss the quality of the associated data. This task surfaces potential challenges in data digitization, aggregation, reliability, accuracy, and relevance. An example is provided in Table 3.2.

This exercise is very useful to understand if your system is set up such that your model will solve the problem that is actually blocking the current decision process,

TABLE 3.1
Variable Brainstorming Sheet

Controllable Input Variables	Fixed Input Variables	Target Outcomes	Variables Expected to Correlate with Target Outcomes
Cure temperature	Catalyst activity	Tensile strength	Crosslink Density
Catalyst type, amount	Filler amount	Elongation at break	Filler Amount
Crosslink density	Filler type	. . .	Filler Type
Diluent viscosity	Water content		Degree of Poly.
Degree of Poly.	Test ASTM		. . .
Volume per batch	Mixing shear force		
. . .	Annual production kg		
	. . .		

TABLE 3.2
Variable Inspection Sheet

System Variable	Relevant Data	Is Data Available and Sufficient?	Expected Relation to Target Variable	Is the Variable Controllable or Fixed?
Cure temperature	Max press temp	No – lab technicians do not currently record analog reading	None	Controllable
Filler amount	Filler wt. %	Yes	Proportional to Tensile	Fixed
Filler Type	Filler name	No – need descriptors, e.g., surface area	??	Fixed
Water content	Incoming material certificate of analysis	No – unmeasured for some ingredients	Proportional to rheology	??
. . .				

or if it is an exciting scientific inquiry but will ultimately struggle with adoption because the solution is not on the critical path to better decisions. You can also begin to understand how to structure your model; which variables are X's and which are Y's? X's can be manipulated (e.g., formulation ratios), fixed (e.g., ingredient properties), or uncontrolled measured variables, but Y's such as product quality properties are a dependent result of the input variables and system noise – controlled only indirectly. Do you want to predict multiple Y's at once? Do you want to predict all Y's but only optimize on some? This process will begin to define how to apply machine learning algorithms toward actionable, value-creating decisions.

Part two of your project selection objective is to have enough high-quality data for your solution to be effective in making those decisions. This question is not often overlooked because it is the main technical exercise of data science. Given a set of valuable business opportunities, how do you determine which one is most technically feasible?

3.2 HOW MUCH DATA IS "ENOUGH"?

One of the most common questions data scientists receive from researchers interested in utilizing modeling is "do I have enough data?" However, this question is fully dependent on what they are trying to accomplish. Unsupervised analysis? Explanation, prediction or optimization? Extrapolation or precise exploitation? How much non-linearity? What is the dimensionality (i.e., how many variables)? Collinearity or confounding variables? Continuous, ordinal, or categorical data? It is important to remember that a high volume of data does not guarantee a high volume of information. What is the difference? Well, the learning algorithms work by capturing patterns between the information in the data, that is, signals of the underlying mechanism, but not all data is informative to those patterns. One of the reasons that Industrial data science is still challenging even in a world of automated machine learning methods is that the rate-limiting step is not finding and training the right model algorithm but instead designing the precise alignment of the right data to the right problem. The data must include:

- an overlap between where you have data and where you want to use the model
- features/variables that capture relationships to your target variables
- enough information for the algorithm to learn the patterns with sufficient accuracy and precision

A high volume of data does not guarantee a high volume of information.

Next, we will discuss each in more detail – how these requirements affect the amount of data you need and how to assess the quality of a candidate data set using Exploratory Data Analysis (EDA). Note that if you need a comprehensive background on the techniques involved in EDA there are many excellent references available (Kelleher et al. 2020; Tukey 1977; Myatt 2006; Peng 2016); this chapter will demonstrate some of these techniques in the context of answering the aforementioned questions for your specified predictive formulation problem but does not provide in-depth explanations of the techniques.

For learning algorithms to help you make decisions with data, your *data information space must sufficiently cover your decision space.*

With even a moderate number of variables, the dimensionality of your data can rapidly lead to completely empty spaces of the data – where there are no samples with a given combination of X and/or Y values. If this space is where you are primarily interested in using the model, then the rest of the data in other spaces may end up being irrelevant or will require significant extrapolation. Extrapolation can be risky for empirical models and often requires the incorporation of a domain-based fundamental model to be successful. This situation regularly occurs when combining historical data sets from different formulators, labs, or product lines; there will be clusters of product development around a certain region or archetypal material, but sparse or empty spaces between them. If your model goal is to predict outcomes in the empty space to prioritize new research areas, a predictive formulation model may not be the right fit; instead, you may need to shift your question to using a model to prioritize which new experiments in the empty space are most informative (this topic will be discussed in Chapter 11).

Alternatively, if your goal is to accelerate new product tweaks in the populated regions, then clustered data is still good for modeling, and in fact you may get better results splitting the data back out into those smaller, more focused sets. This is what is meant by overlapping your data with your decision space; you can have a lot of data, but if it is not data *about what you are trying to do*, it has minimal value.

Some initial exploratory data analysis of your data space can go a long way toward understanding your modeling goal; in this case, simple scatterplot visualizations are a good tool. A scatterplot matrix of your performance variables, such as that shown in Figure 3.3, will show what performance combinations have been explored. What region of the performance space will create value for your team? What goals seem reasonable? For example, in Figure 3.3, although a formulation with a high Y2 value and a low Y3 value might be very attractive, it is clear that there is a correlation between those two variables; since the historical data lies only along that correlation, a model probably will not be able to achieve that goal without new adaptive experimental design.

For the formulation space, that is, what combinations of ingredients have been tested previously, since there are likely many more than ten ingredients, you will need to reduce the dimensionality of the data in order to easily visualize it. Go-to techniques here are Principal Components Analysis (PCA) (Hotelling 1933) and Uniform Manifold Approximation and Projection (UMAP) (McInnes et al. 2018). For further background and examples, see da Silva et al. (2016) and Ivosev et al. (2008). Figure 3.4 shows a visualization of the PCA scores for a formulation data set. The score values themselves have no interpretable meaning, but new theoretical formulations can be projected onto the score space to understand their relative uniqueness. Figure 3.4 is typical of real industrial data sets, with most research concentrated around a few archetypal materials, and lines of inquiry pushing out in a handful of directions. Are there any unique product areas of interest?

The next requirement is that you have variables (also known variously as features, predictors, or descriptors) that have a relationship to your target variables. This sounds obvious, as uncovering these relationships is the entire objective of machine learning, but *how well* the signals of the underlying behavior in your system are captured by the features is critically important. Though it is more common in the

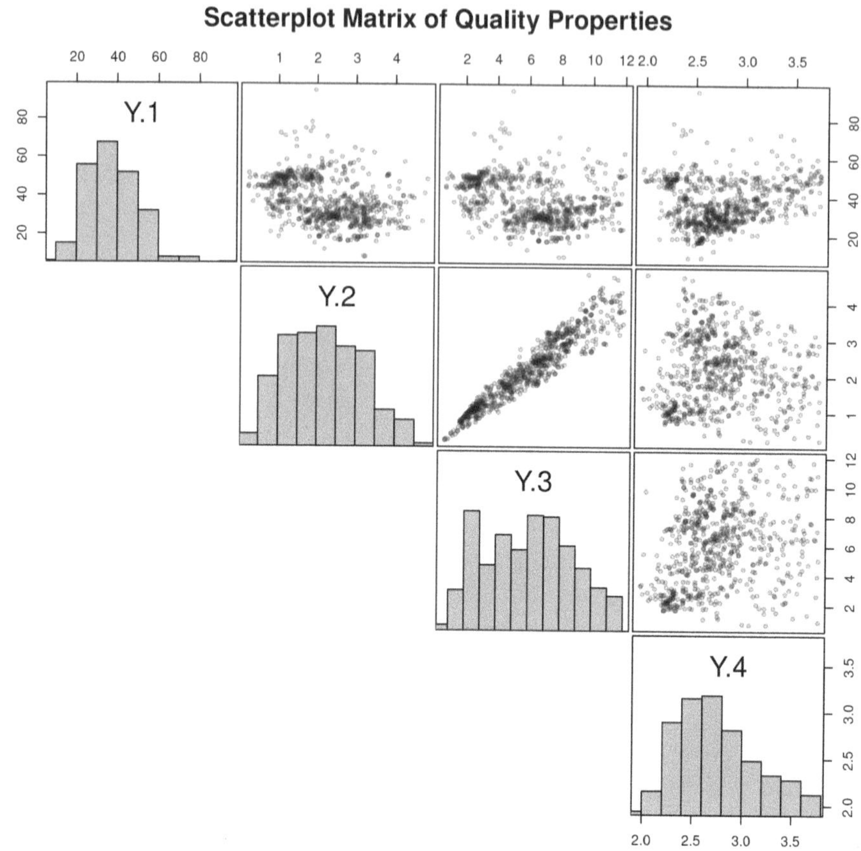

FIGURE 3.3 A scatterplot matrix and histograms of the various quality variables for a formulation data set.

process chemometrics space than the product design space, it is possible that important predictors are not being documented in the experimental data. In these cases, no matter which algorithm you use or how many new samples you collect, predictions will always be inaccurate because the predictive information is just not there. Examples of this include experimental "noise" such as lab temperature and humidity, unknown differences between ingredient lots ("why does last year's batch work better?"), or test method metadata (one lab measured at 25 degrees Celsius and another at 40, but neither documented it).

Another consideration is redundant variables, which can make a data set look larger and more useful than it truly is. If you have five variables that are all actually describing the same thing then you only have one variable with novel information and four redundant ones. If variables are partially but not fully correlated, then there may be some interaction relationship between them that is predictive of the outcome. To assess how correlated your features are, you can calculate the correlation matrix. With a limited number of variables, one can inspect the correlation matrix

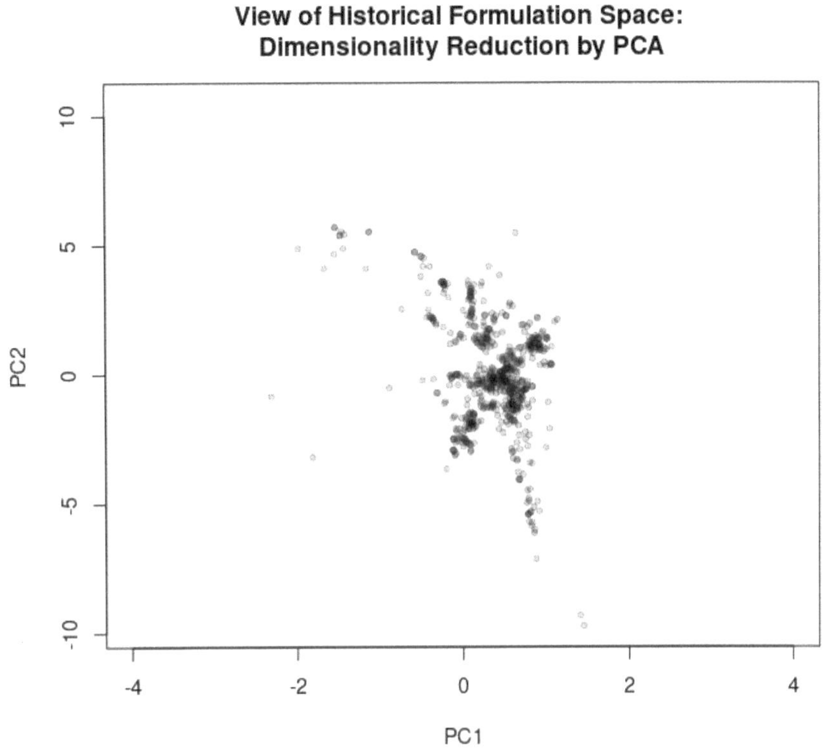

FIGURE 3.4 A score plot of the first two components of a PCA model of sample formulation data.

as a numerical table, but when the number of variables grows large, reading individual numerical values is inefficient. In these cases, the data is typically presented as a heatmap, with a color gradient used to depict the range of correlation values. Gu et al. provide excellent examples using very large genomics data (2016). If there is interest in going beyond linear correlation, there are plenty of methods to explore around non-linear correlation metrics such as mutual information-based methods with roots in information theory; Battiti (1994) and Peng et al. (2005) provide good starting points.

Model predictions are at the mercy of measurement noise

Lastly, this is also where measurement noise comes in. Understanding the size of your measurement variability within each feature versus the expected size of the effect of that feature versus how widely you are prepared to vary it is critical. Anyone with experience using statistics has an intuitive understanding of a model's limitations in dealing with noise, such as the difficulty of detecting variable effects that are smaller than the variable's measurement noise. It is a misconception that an algorithm branded as "machine learning" can somehow overcome this fundamental challenge. If you have access to measurement systems analysis results giving the variance of your data (or repeat measures within your data set), you can then compare this measurement noise to the variance of values within your data set. From an EDA perspective, classical summary statistics such as variance and range are helpful, along with univariate histograms to inspect the distribution of the data. These tools are also useful to help identify which features are normally distributed and which are not, which significantly affects their behavior in learning algorithms and may need to be accounted for in model development.

An initial pass on modeling the data should be conducted during the exploratory data analysis to check for "modeling feasibility," that is, the presence of some initial correlating variables. At this stage, you do not need to worry about making a train/test split of the data; measuring and avoiding overfitting can be addressed during formal model development, but if there is catastrophic underfitting then your next steps will be different. If the prediction feasibility appears low as in Figure 3.5 at left (e.g., <30% of variance explained), it may be much more valuable to spend time collecting additional data in the form of descriptor variables or improved measurements than it is to test more samples or explore better model algorithms. Examining the correlation matrix and training some preliminary ordinary least squares (OLS) regression, partial least squares (PLS) regression, and random forest models is a great place to start.

Demonstrating modeling feasibility brings us to the last data requirement: the data contains enough information to predict outcomes *with the accuracy and precision that you need* to achieve your goal. If the information space covers the decision space but is sparse, then your concern is if you have enough *samples* to capture the complexity of the underlying process. In low-dimensional, linear systems this could be fairly small – a few tens of points. But in more complex systems, you need more information. At the other extreme, in a highly non-linear system where your goal is

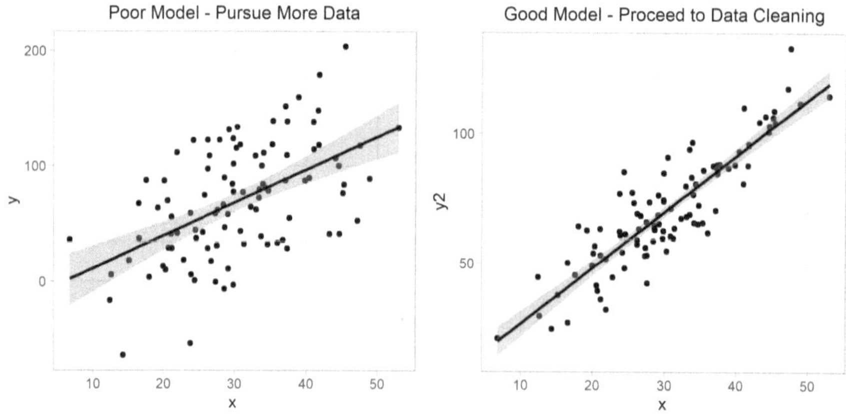

FIGURE 3.5 Two parity plots showing the actual vs. predicted values for two OLS regression models, a poorly performing model (left), and a model indicating good feasibility (right).

to globally optimize, hundreds to thousands of points may be necessary to achieve a model that is "good enough." Most cases are in between and depend on how well the signal is mapped to data.

Data can be "information sparse" in several different ways. The "curse of dimensionality" is perhaps the most intuitive (Altman and Krzywinski 2018). This principle affects all data sets, but it shows how even in the best-case scenario – a limited number of normally distributed numerical variables – as you add more variables, the density of the data in a given region rapidly decreases. Figure 3.6 provides a visual example, where at only four dimensions, the learning algorithm has only 8% of the data available to model behavior in a given region.

So, in general, it is useful to minimize the number of variables you are using to only those which are useful in the final model. This process of feature selection is one of the first steps of modeling, but to start you should include as much data as possible

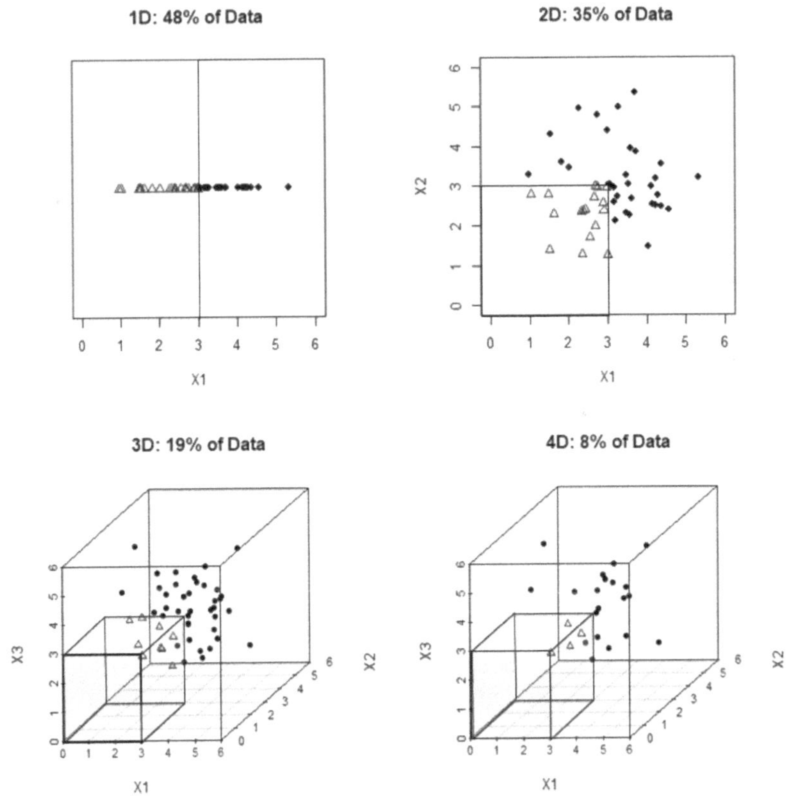

FIGURE 3.6 Four scatterplots showing the amount of data in the region below the mean of each dimension (triangles), where X1, X2, and X3 are simulated data normally distributed with mean = 3 and variance = 1 (you can consider the 4th dimension to be time, where only the first half of data is plotted). Top left: 48% of the points in X1 < 3. Top Right: 35% of points in X1 and X2 < 3. Bottom Left: 19% of points X1, X2, and X3 < 3. Bottom Right: 8% of points X1, X2, and X3 < 3 and time < 50%.

to avoid missing any unknown effects, especially if your engineering knowledge of the system is incomplete.

Data can be sparse in another way as well. At this point, we can note what type of variables you are likely to have for a data-driven formulation project, starting with the one set of data that you will always have: the formulation recipes themselves. As will be further described in Chapter 4, the formulation ratio data set at its most basic contains the experiment ID, the ingredients used (i.e., their names as text), and the amount of each ingredient used (normalized to weight fraction or percent). When used in modeling, each ingredient is a data column, each formulation is a row, and the respective weight fractions are the values. For industrial formulation problems with several hundred products having 10–15 ingredients each, this leads to a data table in which the vast majority of each column and each row are zeroes. There may even be ingredients that were only used one or two times, meaning there is insufficient variation of this feature to effectively model it. The variable summary statistics and univariate histograms from EDA will help to identify this situation, and the team may need to decide to collect more experimental data if including that ingredient is critical to success. Chapter 4 will continue with further details on work that may be needed to build and clean your final data set.

Because highly imbalanced data is very difficult to accurately model, we very often need to supplement the formulation data set with additional information such as properties of the ingredients (often called "descriptors"). These variables are usually more information-rich, with distributions closer to normal. Some examples are ingredient viscosity, surface tension, pH, or any other properties relevant to how the ingredient materials function in the application. The SMEs will be able to identify the data they are currently collecting but keep in mind that there may be useful data that was previously overlooked or too difficult to access. Brainstorm with the team – are there public data sources that have variables that could be related to the problem? Is there known data in an unstructured form that could be mined for modeling? Should we generate new data based on SME hypotheses? It is often quite time-consuming to collect this descriptor data, so the availability of high-quality descriptors could be a significant factor in project selection.

If needed, you can go even further and design new variables based on uncovered non-linear relationships (e.g., adding log or inverse terms, variable interactions) or mechanistic models. You could even incorporate advanced data such as images or time series, modeling those as needed to determine the informative features. In machine learning parlance, this exercise is called "feature engineering," meaning you are designing the most useful set of variables and variable transformations for your system. This process is also a key opportunity to engage your subject matter experts and incorporate their know-how into your data. Chapters 5 and 6 will go into detail on feature engineering and creating descriptive features from complex analytical data of your materials, respectively. In this way, you can begin with whatever data is already available, create initial models, and then work toward more complex, enriched data sets as needed until your model reaches the accuracy required for success.

There is one piece of data that has a particularly high leverage on the technical feasibility of your project, and that is your performance data, or Y data.

A model can never be more accurate than the measurement system of your quality variables.

It is very important to carefully evaluate the relationship between your Y variables and success of the project. Questions to ask include:

- How confident are you in the data? How much measurement noise is there?
- How closely related is this variable to the value-creating decision? If you are able to optimize on this variable (or variables), will you achieve your goal? Or are there other considerations that need to be made simultaneously but that aren't captured in a quality variable?
- To be successful, do you only need the Y prediction(s), or is there other information users also need to move forward with the proposed solution? Are there constraints on any variables that affect how they should be treated?
- How closely related to the in-application performance of the material is your Y measurement? In the best case, both you and your customers utilize the same industry standard test. At the worst case, performance is qualitative and could vary between customers – how to calibrate their expectations to your quantitative data?

A misstep in any of these items could result in a model that is extraordinarily accurate but useless to your product design team.

A final, and very common, problem is that no data on failed experiments is recorded. This usually occurs when a particular formulation cannot be physically formulated (e.g., incompatible ingredients) or the formulated product cannot be tested in the usual way (e.g., does not perform well enough for the test to function properly or is outside of calibration range). If these formulations are not included in the data, it may be difficult for the model to infer the boundaries of the design space, resulting in proposed new experiments that are not viable. Always encourage lab researchers to document experiments that do not formulate well or result in materials that cannot be physically tested.

In summary, not all historical or legacy data is as valuable as we think it is. This is why some feasibility studies to measure the actual usefulness of the data set are a helpful first step in selecting your project. Initial exploratory data analysis on candidate data sets using the techniques presented here can help elucidate how extensive the data is, how well distributed over the decision space it is, and possibly some insight into the complexity of the underlying model.

One subtle aspect of the question of enough data is: how accurate does the model need to be to provide the decision support level you identified in the evaluation of the business opportunity? In other words, if you could magically acquire additional data to improve the model, how would you know when it is "good enough" and you can stop? For this challenge, you need to clearly define how your model is used to create value, and then compare the value-creating state to the current state. Essentially, how much better do we need to predict the future in order to achieve success?

In the technical sense, models are often judged on how much of the variation in the data can be explained by the model, typically R^2. There is not a single value of R^2 that is considered to be "good enough"; it depends on how much variance you can explain in your current process. To illustrate, consider two examples using the simplest empirical model – a univariate regression. In an undesigned medical study, there may not be any of the variance of the response currently explained (e.g., current state $R^2 = 0$). Having enough data to regress a single variable with 95% confidence may be useful to the project even if the overall model only explains a small portion of the response variability, such as $R^2 = 0.5$. In this example, the modeling exercise is successful for medical researchers because they have the statistical power to conduct a hypothesis test and identify the direction of the signal (i.e., positive or negative). Because the baseline knowledge was so low, the accuracy required of the data-driven approach to create value is also fairly low. Contrast this with the use case of analytical instrument calibration, which is fundamentally a data-driven model making a useful prediction. When calibrating the instrument signal to your sample standards, an R^2 of 0.5 would be completely unacceptable. An R^2 near 99% is usually needed for calibration cases, since the baseline state is 100% of variance explained in the known standards. This thought experiment showcases the ambiguity around what is "good enough." Most often for predictive formulation, we are somewhere between those two extremes.

In projects where the current baseline is a human decision, each person has a mental model based on their experience that informs what they decide to do. For product development teams that are not using any data-driven tools, this mental model is your baseline to beat. However, keep in mind that each individual will have a different effective R^2. For example, new employees may be significantly less accurate in guessing how to formulate a new product to hit a desired specification, leading to many iterations in the lab, while a 20-year veteran formulator may be done right away. What success looks like is thus a critical conversation with stakeholders to set your technical feasibility. Does the model need to be more accurate than the most senior expert, because all new product opportunities go through them? Or would being more accurate than the most inexperienced half of your formulators be valuable by accelerating a large enough portion of the team's output? What happens when your expert formulator retires? Does "the art" go with them? If so, capturing some of the patterns that a person has (possibly unconsciously) captured in their head into a model may be very useful to newer employees.

Ultimately, you won't get a final answer on technical feasibility until you begin the project and dig into technical details, but you can get a ballpark estimate of the technical scale of the challenge broadly by having these conversations on how the size of the data compares to the size of the decision. Even in high dimensional formulation problems, we are often surprised by how little data is actually needed to uncover predictive relationships that are useful to development scientists.

All of these pieces together – the amount of informative data, the overlap between the data and the decision, the complexity of the underlying data-generating mechanisms versus the accuracy of the model required to create value – tell us about the feasibility and complexity of a particular candidate project. Some of them you will not fully understand until the team is well into modeling. Red flags in any of the areas described earlier should merit a conversation with stakeholders about the risks of the research. Alternatively, a high-quality data set with a straightforward modeling goal presents a great quick-win opportunity to show the value of data-driven approaches.

3.3 SETTING THE TEAM UP FOR SUCCESS

This brings us to the final component of selecting a good project: have SMEs who are neither skeptics of AI/ML nor have unreasonably high expectations of what is possible. The importance of people in digital transformation was discussed in Chapter 1, but this topic cannot be overstated. The fact is, your domain experts will be involved in the ugly nitty-gritty of AI/ML: the data cleaning, the mediocre models, and the unreasonable solutions. Modelers will be iterating back and forth with subject-matter teammates on questions about the data countless times, and all will be investing significant time. So, having a partner who is open-minded and excited about the use of AI/ML, understands that the process is a journey and may be generational, and has the time and patience to put in a lot of work on the data will make your project go much more smoothly and increase the chance of success. A skeptical or fearful SME can significantly slow down the project by not being forthcoming with information, or they can even reduce trust and adoption of the completed tool by bad-mouthing the process. If you have several value-creating, technically feasible project options, it is best to work with a supportive team. More discussion will be provided in Chapter 9 on how to work with your stakeholders throughout the project to build trust in the model.

When identifying team members, it is helpful to be clear about how your subject-matter experts and their domain knowledge relate to your project. Ultimately, most product design projects do not fully automate a decision, and thus you are likely to be serving formulators information such as candidate experiments for their review (i.e., "decision support"). This level of analytics relies on the users' assessment of how *useful* the outputs from your model are, and thus it is critical to come to a common understanding of what users *expect* the tool to be able to do. Do users want model solutions to replicate their current understanding but faster or teach them something completely new about their products? There is a thin line between designing a tool that provides solutions intriguingly different from those that users would normally come up with versus proposing solutions so far-fetched that users dismiss the tool or potentially even all data-driven approaches. Different domain experts may have different opinions on this balance.

You also need to set expectations on whether your users view model interpretability or model accuracy as the higher priority. Are they willing to sacrifice accuracy for an understanding of the patterns captured in the model, or does your application require the highest accuracy possible? For example, if the model is to allow your customers to test how your products function in their system before buying, you want the highest accuracy possible so as not to mislead them. In this case, you may even want the model user *not* to be able to interpret the model so that they continue to use your tool and products. In contrast, if the model is to be used within your team to recommend experiments with completely new performance profiles, your researchers are likely to want to know why those materials work differently. Allowing your

SMEs to interpret how the model is making predictions (and lining that up with their experience and hypotheses) can be very powerful in building their trust in the tool. Although model-agnostic variable interpretation techniques are growing in popularity, it can still be difficult for researchers to build an intuition about why a black-box or non-linear model produces the results it does. So, understanding what your model needs to deliver for the project to be successful can help guide your technical direction in the project.

Ultimately, you need to understand your chosen SME partners and their level of skepticism or excitement about AI. If they are skeptical, extra attention to educating them about the process of data science is critical, because during development the model *will* be wrong, possibly ridiculously so, but you need their feedback to fix it. Make sure your team are people who understand that machine learning is a process not a magic wand and use their learnings on the project to help the rest of the organization understand myth versus fact.

3.4 KICKING OFF THE PROJECT

As you can see, there are very many things to consider when starting a data-driven formulation project, many of which you won't know before you start. Because of this, it is highly recommended to adopt a phased or Agile approach to the project, as discussed in Chapter 1. An approach to iteratively delivering results and insights such as that of Figure 3.7 will allow you to incrementally de-risk the project, learn more about what it will take to create value, and allow your team to re-evaluate your chance of success as you go.

This approach allows you to build trust in the data-driven approach as you go, or back out of the project quickly if the accuracy you need just is not there. Chapter 9 will provide details on how to verify model performance to support these decisions as you move through the project. The following are some initial considerations when planning your project.

Carefully consider the amount of time you are willing to spend collecting data. It is expected that data collection and cleaning will take much longer than model development, potentially 80% of the development phase of the project. And, as will be demonstrated in Chapters 5 and 6, it is possible to conduct research that adds a

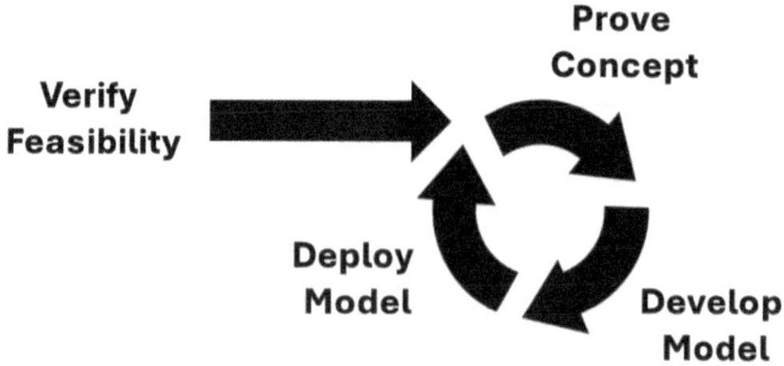

FIGURE 3.7 A schematic of an iterative approach to pursuing an AI project.

very significant amount of understanding to your system and your model, but this time-consuming work will also extend how long it takes before your project can demonstrate value. Remember that your model only needs to be accurate enough relative to your baseline to create value. You will reach a point of diminishing returns where incrementally more data has a minimal impact on the overall success of the project but could significantly extend the timeline and budget. Start small with a minimally viable set of data, and once you have demonstrated success with that, use the value created to justify additional data collection efforts. Examples of this approach include starting with just one product class or one performance target.

It is helpful during model development to create a mock-up user interface or simple deployment. This is because it helps you ensure that your model is aligned with the value-creating actions that the users will take. Recall that for "human-in-the-loop" AI, you need to provide researchers with the correct, useful information to improve their decision, so directional feedback on usability is very valuable. Sometimes, small changes in model software functionality represent huge changes in how the model or optimization problem is set up, so it is much better to know those earlier rather than later. This is very much aligned with the now industry-standard software development approach called Agile development. Along these lines, a variety of references are provided: Denning provides a strategic view of Agile (2022), Thompson provides advice on machine learning project management in general (2023), and Carter and Hurst (2019) and Collier (2022) provide explanations of using Agile for analytics projects.

Lastly, keep in mind that model deployment is essentially software engineering and may require completely different skills and resources than the model development portion of the project. Initial approval from users on the UI design and functionality from the model development iterations is very helpful to have in order to scope the software development project appropriately.

With these things in mind, you can lay out a plan with your team on what you are trying to learn at each phase of the project. One possible example:

- Feasibility stage: Check for signals in the data; evaluate size of opportunity.
- Proof-of-concept stage: Start with the readily available data, even if your eventual goal is broader. Is your model sufficient or do you need more data?
- Development stage: Focus on model success before worrying about implementation (e.g., data engineering, software development), but do get user feedback on results and usability if possible, such as with a mock-up application. Can you demonstrate proof-of-value? What can the model do and not do? How complex will deployment need to be?
- Checkpoint: Is this model valuable enough to invest in deployment?
- Deployment stage: Design how users will interact with the model, and what front-end and back-end functionality the tool must have to stay useful and relevant.

With the guidance provided in this chapter, you can be well on your way to a successful data-driven formulation project. You should have an idea of what data you have available, know that the data is of reasonably high quality, and have a clear definition of what you are going to do with your eventual model. To move forward and really dig into your problem, the next step is all about data preparation.

REFERENCES

Altman, N., and M. Krzywinski. 2018. "The Curse(s) of Dimensionality." *Nature Methods* 15: 399–400. https://doi.org/10.1038/s41592-018-0019-x.

Battiti, R. 1994. "Using Mutual Information for Selecting Features in Supervised Neural Net Learning." *IEEE Transactions on Neural Networks* 5 (4): 537– 550. https://doi.org/10.1109/72.298224.

Carter, Eric, and Matthew Hurst. 2019. *Agile Machine Learning*. New York: Apress.

Collier, Ken. 2022. *Agile Analytics: A Value-Driven Approach to Business Intelligence and Data Warehousing*. Boston: Addison-Wesley.

Da Silva, Renato R.O., Paulo E. Rauber, and Alexandru C. Telea. 2016. "Beyond the Third Dimension: Visualizing High-Dimensional Data with Projections." *Computing in Science & Engineering* 18 (5): 98– 107. https://doi.org/10.1109/MCSE.2016.90.

Denning, Stephen. 2022. *Age of Agile*. New York: AMACOM.

Gu, Zuguang, Roland Eils, and Matthias Schlesner. 2016. "Complex Heatmaps Reveal Patterns and Correlations in Multidimensional Genomic Data." *Bioinformatics* 32 (18): 2847–2849. https://doi.org/10.1093/bioinformatics/btw313.

Hotelling, Harold. 1933. "Analysis of a Complex of Statistical Variables into Principal Components." *Journal of Educational Psychology* 24 (6): 417. https://doi.org/10.1037/h0071325.

Ivosev, Gordana, Lyle Burton, and Ron Bonner. 2008. "Dimensionality Reduction and Visualization in Principal Component Analysis." *Analytical Chemistry* 80 (13): 4933–4944. https://doi.org/10.1021/ac800110w.

Kelleher, John D., Brian Mac Namee, and Aoife D'arcy. 2020. *Fundamentals of Machine Learning for Predictive Data Analytics: Algorithms, Worked Examples, and Case Studies*. 2nd ed. Cambridge: MIT Press.

McInnes, Leland, John Healy, and James Melville. 2018. "Umap: Uniform Manifold Approximation and Projection for Dimension Reduction." *arXiv.* https://doi.org/10.48550/arXiv.1802.03426.

Myatt, Glenn. 2006. *Making Sense of Data: A Practical Guide to Exploratory Data Analysis and Data Mining*. New York City: Wiley-Interscience.

Peng, Hanchuan, Fuhui Long, and C. Ding. 2005. "Feature Selection Based on Mutual Information Criteria of Max-Dependency, Max-Relevance, and Min-Redundancy." *IEEE Transactions on Pattern Analysis and Machine Intelligence* 27 (8): 1226–1 238. https://doi.org/10.1109/TPAMI.2005.159.

Peng, Roger. 2016. *Exploratory Data Analysis with R*. Research Triangle Park, NC: Lulu.com.

Thompson, Simon. 2023. *Managing Machine Learning Projects: From Design to Deployment.* Shelter Island, NY: Manning.

Tukey, John. 1977. *Exploratory Data Analysis*. London: Pearson.

Section 2

Preparing Your Data

Alix Schmidt and Kristin Wallace

As you can see, there is a lot to think about when engaging in data-driven formulation projects, not the least of which is data quality. Once you have selected a product system to model, the next step is to refine and enhance the data set. While the current iteration of the AI hype cycle was kicked off by "big data" and huge machine learning models trained with internet data, there are rarely, if ever, materials data sets large enough to take advantage of those advances. The fact is, because lab experimentation and subsequent data processing is still primarily a manual or semi-automated process, the cost of collecting experimental data is high enough that most projects suffer instead from "small data" challenges.

In the small data regime the quality of each data point matters, and as such, time invested in thoroughly cleaning and enriching your data pays off in modeling accuracy later on. In fact, don't be surprised if the majority of time for your project is spent on this single task. Section 2 focuses on this data preparation process. First, Chapter 4 reviews the structure of the data sets used in a formulation problem, and the author highlights some common data quality problems learned from many years of industry cases. Some of these problems can be addressed with new data but some are inherent challenges that should be considered when selecting model algorithms. It is highly recommended that all readers consult Chapter 4 when beginning a new modeling effort.

As discussed in Chapter 3, to conduct efficient data science, you want to collect as much data as needed to achieve the required accuracy but no more, lest the team fall

DOI: 10.1201/9781003385974-5

into a rabbit hole of pursuing one new variable after another. Chapters 5 and 6 present increasingly complex ways to add new data for modeling, with the intention that the modeler engages with the most accessible techniques first and moves forward only when additional accuracy is needed. Specifically, if the cleaned formulation data set alone is not sufficient for solving the business problem, then the data can be further enhanced by transforming variables or adding additional subject-matter-based descriptors to inform the system. This process – feature engineering – is the subject of Chapter 5. Note that feature engineering pertains only to defining new variables (i.e., columns in the data set), not to collecting additional samples (i.e., rows in the data set). Finally, Chapter 6 presents techniques for applying machine learning to advanced materials characterization data that can provide a detailed description of the system behavior in a formulation.

4 Challenges with Formulation Datasets

Kristin Wallace

4.1 INTRODUCTION

Complex formulations are common in a variety of chemical- and material science-based industries. Manufacturers can address rising raw material costs, unwanted inventory costs, supply chain shortages, and process complexities by reformulating existing products with fewer, cheaper, and more readily available ingredients. However, formulators must carefully select numerous ingredients in hopes of achieving multiple performance targets while balancing cost and ingredient availability and adhering to environmental or other relevant regulations.

Formulation knowledge is typically well-protected by manufacturers in competitive industries. In many cases, the formulation knowledge is the bread of the business. As Industry 4.0 and digital transformations continue to migrate knowledge from the minds of a small number of highly experienced individuals to databases, data lakes, and formulation software platforms, manufacturers are adopting new strategies for protecting their competitive edge. Signing trade secret agreements is no longer sufficient – data-leaks and cyber-attacks are significant threats; therefore, restrictive IT policies and advanced IP protection are being put into place by many organizations. Although these actions are justified, they also inhibit innovation.

While R&D professionals begin to recognize and explore the potential of consolidating data to build powerful models, they are likely to encounter opposition from colleagues who are responsible for prioritizing knowledge and data security. Furthermore, manufacturers interested in hiring a consultant to assist with data-driven product development must face endless meetings, levels of authorization, and NDA revisions in order to move forward. Much thought will be given to what steps must be taken to ensure that your data doesn't end up in the wrong hands. Do you need to anonymize your variable names? Obfuscate the numerical data?

The effort required to prepare a dataset for predictive modeling is often underestimated. However, if data collection challenges can be overcome, there is significant value in unlocking the secrets in your data. Custom and commercial formulation tools, algorithms, and platforms can significantly accelerate product development timelines.

There is certainly no shortage of initial challenges when one is tasked with making better use of data for product development applications. This chapter discusses many common data preparation challenges, including identifying all relevant types of formulation data, locating and accessing all available historical data points, and

DOI: 10.1201/9781003385974-6

then compiling disparate data into a large and well-structured format, identifying and resolving common data quality problems, and evaluating the suitability of the dataset for predictive modeling.

4.2 DATA AWARENESS

How is it that an Oreo© cookie tastes exactly the same every single time you bite into a fresh one? Why is a new architectural paint more resistant to kitchen spills than its predecessor? The key to making either a consistent product or developing a new product is one and the same – it comes down to controlling variation. As in any input-output system, product formulation inputs can be actively minimized or intentionally introduced in order to obtain a desired output.

What are the possible sources of input variation in a product formulation context? Typically, there are three types (or blocks) of formulation input data: ingredient properties, formulation ratios, and process conditions (Cardin et al. 2015). Outputs in a product development context are generally referred to as "quality properties" (Cardin et al. 2015) or "performance properties," although industry-specific terminology, such as Critical Quality Attributes (CQAs) in pharmaceuticals, also exist. Awareness of these four types of data serves as a useful starting point in identifying which types (and which variables) are relevant to your specific application. It is good practice to create a list of sought-after data, organized by data type, prior to starting data collection.

Definitions and examples of each block of formulation data follow.

4.2.1 Ingredient Properties

There are multiple factors that affect the decision of which ingredients to combine in a formulation. Ingredients may be selected based on expected function, supply chain availability, familiarity, or cost. Domain experts often rely on intuition to select candidate ingredients that are likely to result in a desired product. However, much of this intrinsic knowledge can actually be captured by quantitatively characterizing the ingredients.

> "***Ingredient properties***" refer to any characteristics about the ingredients that may contain variation. These can be physical properties or chemical properties and are highly dependent upon the industry and nature of the materials.

Some examples of ingredient properties include molecular weight, density, viscosity, melt temperature, and particle size. Other sources of variation in the ingredients include non-numerical information such as supplier and batch number. Strategies for including non-numerical data in a predictive model are discussed in this chapter and in Chapter 6. Finally, cost can also be considered an ingredient property; however, cost data should generally be excluded from modeling, and only used for simulation/optimization.

TABLE 4.1
Example Ingredient Properties

Ingredient	Supplier	Amylose Content (wt%)	Mn (× 10^{-3} g/mol)	Mw (× 10^{-3} g/mol)	Mz (× 10^{-3} g/mol)	PDI (Mw/Mn)(-)	Relative Unit Cost ($/unit)
Starch1	Company A	70	41	214.5	463.7	5.23	40
Starch2	Company A	23	62.3	531.4	821.2	8.53	45
Starch3	Company B	0	90.1	722.2	1040	8.02	40
Starch4(new)	Company B	0	246	316	377	1.3	46
Starch5(new)	Company C	70	62	123	190	2	39
Starch6(new)	Company C	28	148	219	296	1.5	42
Starch7(new)	Company C	35	38.9	203.9	448.5	5.24	38
Starch8(new)	Company C	12	56.6	613.5	853.1	10.84	47
Starch9(new)	Company D	0	103.9	761.1	1000	7.33	44

Source: Adapted from Muteki (2006)

An example of ingredient property data is included as Table 4.1. This data has been adapted from Muteki et al. (2006).

4.2.2 Formulation Ratios

In addition to selecting which ingredients to combine in a formulation, how much of each ingredient to use must also be determined. To illustrate the importance of ingredient amounts, let us consider the simplest case of combining two unique ingredients. If these ingredients are combined in a 50:50 ratio, the resulting product quality will be significantly different than if the same two ingredients are combined in a 1:99 ratio.

"***Formulation ratios***" refer to the relative quantities of all ingredients used within a single formulation/experiment/product. Formulation ratios can be defined in several different ways including percent, fraction, or another consistent quantity such as mass.

An example of formulation ratio data is included as Table 4.2. This data has been adapted from Muteki et al. (2006).

4.2.3 Process Conditions

Another source of variation in product formulations are the conditions under which ingredients are combined.

TABLE 4.2
Example Formulation Ratios

Formulation	Ingredient	Mass Fraction	Ingredient Class	Product Type
Exp 1	Starch1	0.5	Starch	4
Exp 1	Additive1	0.5	Additive	4
Exp 2	Starch1	0.41	Starch	4
Exp 2	PLA	0.18	Lactic Acid	4
Exp 2	Additive1	0.41	Additive	4
Exp 3	Starch1	0.41	Starch	4
Exp 3	PLA	0.18	Lactic Acid	4
Exp 3	Additive1	0.41	Additive	4
Exp 4	Starch1	0.33	Starch	1
Exp 4	PLA	0.33	Lactic Acid	1
Exp 4	Additive1	0.33	Additive	1

Source: Adapted from Muteki (2006)

"***Process Conditions***" refer to the conditions under which ingredients are combined.

Depending on the industry and application, the number and nature of relevant process conditions varies greatly. In a food example, baking time and oven temperature are common process conditions that would clearly impact the quality of the final baked good. Other examples of process conditions include mixing speed, ambient conditions, and flow rates. Process conditions can also include non-numerical information such as the chemist who completed the experiment, and the equipment or facility where the experiment was completed.

An example of process condition data is included as Table 4.3. This data has been adapted from Muteki et al. (2006).

4.2.4 Quality Properties

The output (Y) variables in a product development dataset are referred to as "**Quality properties**" and include any measurement or quantification related to the functional performance of the final product.

Again, the industry and application dictate the number and nature of quality properties. Some common examples in chemical or polymer formulations include viscosity or hardness, respectively. Quality properties can also be non-numerical, such as a Pass/Fall indicator.

TABLE 4.3
Example Process Conditions

Formulation	Molding Temperature (°C)	Location
Exp 15	150	Lab A
Exp 16	170	Lab A
Exp 17	150	Lab A
Exp 18	170	Lab A
Exp 19	150	Lab B
Exp 20	170	Lab B
Exp 21	150	Lab B
Exp 22	170	Lab B
Exp 23	150	Lab B
Exp 24	170	Lab B
Exp 25	150	Lab A
Exp 26	170	Lab A
Exp 27	150	Lab B
Exp 28	170	Lab B

Source: Adapted from Muteki (2006)

TABLE 4.4
Example Quality Property Data

Formulation	Tensile Strength (MPa)	Tensile Modulus (MPa)	Elongation at Break (%)	Density (g/cm³)
Exp 1	0.88	18.59	10.10	1.37
Exp 2	3.72	162.30	10.10	1.34
Exp 3	3.25	166.10	7.60	1.33
Exp 4	9.70	629.10	6.59	1.31
Exp 5	7.55	489.10	4.02	1.32
Exp 6	22.80	1522.00	7.59	1.30
Exp 7	23.80	1469.00	8.00	1.30
Exp 8	20.00	1430.00	4.21	1.29
Exp 9	3.03	134.50	11.00	1.35
Exp 10	8.73	568.90	7.30	1.32

Source: Adapted from Muteki (2006)

An example of quality property data is included as Table 4.4. This data has been adapted from Muteki (Muteki 2006).

With a good grasp of what types of data, and possibly which specific variables, are relevant to a formulation project, one is ready to begin data collection.

4.3 DATA COLLECTION

Collecting all relevant formulation data, as an intermediate step toward creating a consolidated and comprehensive database, is not a trivial task. The existence of data silos and data masking are likely to hinder data collection; however, the effort put forth in addressing these early challenges will directly impact the effectiveness of a predictive model built on this formulation data.

4.3.1 Data Silos

Although manufacturers generate and store enormous volumes of data, it typically resides in silos. There are many factors that can lead to the existence of data silos within an organization.

Conventionally, product development research and knowledge were trusted to a very small number of highly experienced individuals, housed in an R&D department. These domain experts spent long careers completing countless, well-planned experiments. Unfortunately, the infrastructure was not available to support documenting these experiments thoroughly and in a manner that would promote knowledge transfer in the future. Instead, experimental procedures and results were documented in a fragmented and ad-hoc manner – collected on anything from a paper notebook to a digital slide deck or spreadsheet on a local hard drive. Often key information such as measurement units was missing, and data from "failed" experiments were typically discarded, since the foresight to retain this valuable information was lacking.

In addition, although more advanced approaches are available today, many R&D professionals still rely heavily on traditional Design of Experiment (DoE) principles to guide new product discovery, by investigating the impact of variation in a single variable at a time. Often, R&D departments are tasked with answering a very specific question as quickly as possible, and their investigation generates a small dataset that is later tucked away or forgotten about after the study has concluded and the next urgent question is upon them. This repeated workflow results in a large number of small datasets being generated over time. There is no continuity to the format, structure, or location of these datasets, and often it is difficult to find a simple list of studies that have been completed to date, which is only the first clue for then locating the data itself.

Thankfully, most organizations now understand the potential value of their data, and research departments have accordingly prioritized creating central repositories or databases containing product development data. However, data-leaks and cyber-attacks are increasingly common. As a result, restrictive IT policies and advanced IP protection are being put into place by many organizations, while a message of "Zero Trust" is circulated, creating a corporate culture that significantly inhibits innovation.

Although the existence of data silos can hinder initial data collection, the reward of being able to apply data-driven methods to a consolidated dataset is well worth the upfront effort.

4.3.2 Data Masking

Once the disparate data has been located, data masking may present itself as the next challenge that must be overcome before consolidating the dataset. As an added level of protection, it is a relatively common practice for manufacturers to anonymize variable names and/or obfuscate the data values prior to storing or releasing a dataset either internally or externally.

Although it is possible to build and apply a predictive model on masked data, these practices are likely to hinder project timelines and cause confusion and frustration for the individual or group of individuals attempting to analyze the data. It is difficult to do sanity checks on masked data, since there is no context. Examples of anonymized and obfuscated data are provided in this section to further illustrate the difficulties they can cause.

In cases where data masking is deemed necessary, it is critical to document the implemented transformations and make this information available to individuals who are involved in building, interpreting, and applying predictive models.

Table 4.5 provides an example of quality property data before and after anonymizing the variable names. While a formulation name may provide some context as to the date or chronological order of experiments, an anonymized formulation name lacks such insight. This context may be important for selecting training and validation groups (see Chapter 9 for further discussion on training and validation groups), as well as data that should be grouped into sub-models or excluded entirely. Any such metadata is valuable, and where possible, it should be recorded and included in a product formulation dataset.

Similarly, when the quality property names are anonymized, the ability to confirm expected relationships, such as a negative correlation between Tensile Modulus and Elongation at Break, is lost.

Another form of data masking is to center and scale the data values. Mean-centering and scaling is a common data-preprocessing step in many data analysis algorithms, but if data is provided to the person completing the analysis with some transformations already completed, it may be difficult to screen the "raw" data for outliers.

For example, in Table 4.6, the number average molecular weight (Mn) is mean-centered and scaled. Without the knowledge as to what transformations have been implemented, it becomes difficult to detect if the original data included non-sensical values or other outliers, since the transformed values do not have a physical meaning.

After all silos have been searched, and any data masking has been addressed, the data collection phase can be considered "complete." However, it is not uncommon to retreat back to data collection if exploratory data analysis or preliminary predictive models reveal shortcomings in the dataset. Do not be discouraged if this the case!

4.4 DATA CONSOLIDATION

Merging data records can be a difficult and onerous task, and as a result, data integration is a critical component of many data management platforms and services.

TABLE 4.5

Example of Quality Property Data before and after Anonymizing the Variable Names

Original Data					Anonymized Nomenclature				
Formulation	Tensile Strength (MPa)	Tensile Modulus (MPa)	Elongation at Break (%)	Density (g/cm^3)	Formulation	Y1	Y2	Y3	Y4
Exp 1	0.88	18.59	10.10	1.37	i	0.88	18.59	10.10	1.37
Exp 2	3.72	162.30	10.10	1.34	f	3.72	162.30	10.10	1.34
Exp 3	3.25	166.10	7.60	1.33	c	3.25	166.10	7.60	1.33
Exp 4	9.70	629.10	6.59	1.31	d	9.70	629.10	6.59	1.31
Exp 5	7.55	489.10	4.02	1.32	e	7.55	489.10	4.02	1.32
Exp 6	22.80	1522.00	7.59	1.30	b	22.80	1522.00	7.59	1.30
Exp 7	23.80	1469.00	8.00	1.30	g	23.80	1469.00	8.00	1.30
Exp 8	20.00	1430.00	4.21	1.29	h	20.00	1430.00	4.21	1.29
Exp 9	3.03	134.50	11.00	1.35	a	3.03	134.50	11.00	1.35
Exp 10	8.73	568.90	7.30	1.32	j	8.73	568.90	7.30	1.32

Source: Adapted from Muteki (2006)

TABLE 4.6
Example of Ingredient Property Data before and after Centering and Scaling

Ingredient	Mn ($\times 10^{-3}$ g/mol)	(Mn – Mean) SD
Starch1	41	–0.85
Starch2	62.3	–0.51
Starch3	90.1	–0.07
Starch4(new)	246	2.42
Starch5(new)	62	–0.52
Starch6(new)	148	0.86
Starch7(new)	38.9	–0.88
Starch8(new)	56.6	–0.60
Starch9(new)	103.9	0.15
Mean	94.31	0.00
SD	62.72	1

Source: Adapted from Muteki (2006)

Data management is extremely important for industries that generate truly "big data." However, most R&D or product development datasets are relatively small in comparison to the volume of data available on many continuous manufacturing processes. For example, a polyethylene manufacturer may be interested in analyzing two years' worth of hourly data for several thousands of variables, while in contrast, a longstanding R&D department may only be approaching 1,000 data points on fewer than 100 variables collected over decades. However, as advances and adoption of high-throughput experimentation continue, R&D data will become "bigger."

While a certain amount of automation may be possible through commercial software or custom algorithms, there is also typically a certain amount of manual intervention required in order to consolidate disparate data. Filling the role of a "data janitor" is about as glamorous as it sounds, but for those who embrace this necessity and carefully and comprehensively consolidate product formulation data, the opportunities unlocked for accelerated innovation are well worth the investment.

This section discusses various challenges encountered during data consolidation, including examples, and provides recommendations for overcoming consolidation challenges.

One aspect of data consolidation involves merging multiple records for a single type of formulation data (ingredient properties, formulation ratios, process conditions, or quality properties). For example, ingredient property data may be dispersed across multiple spreadsheets or Certificate of Agreement (CofA) documents from various suppliers. Locating, and consolidating all data from one source into an organized structure, is an essential step toward preparing for predictive modeling.

Several challenges may arise when merging multiple records for one type of formulation data, including nomenclature and measurement unit inconsistencies. These and other issues are discussed in further detail in the section on consolidation challenges.

The most powerful predictive models for product development applications simultaneously treat *all* sources of variation or, in other words, all blocks of data (ingredient properties, formulation ratios, process conditions, and quality properties) (Wallace and Cardin 2021). Therefore, another important aspect of consolidating disparate data is to connect multiple types of formulation data to one another.

The most effective way to connect different data blocks is to ensure consistent nomenclature is used for all linking variables. Linking variables (a.k.a. IDs, primary keys, or lookup variables) are those that occur in multiple data blocks. While some variables will occur in only one of the data blocks, others will occur in two to three blocks, and no variables should occur in all four blocks. Ingredient names are typically linking variables. For example, "Starch1" is a linking variable since it occurs in both the ingredient properties (Table 4.1) and formulation ratios (Table 4.2). Formulation names are another example of linking variables, since they occur in the formulation ratios (Table 4.2), process conditions (Table 4.3), and quality properties (Table 4.4).

If possible, planning experiment documentation to ensure proper linking variables can be extremely helpful in accelerating future use of the data. Data integrity of these linking variables is also critical, so using standardized identification numbers and preventing typos will also significantly accelerate data preparation.

4.4.1 Data Structure

Since no variables will occur in all four data blocks, the use of linking variables and a careful data structure is needed. Figure 4.1 illustrates the typical data structure employed for product development modeling.

An advantage of the data structure illustrated in Figure 4.1 is the inclusion of X in addition to X^T. X^T is a subset of X. While X^T includes only the ingredients that have been used in past formulations, X additionally includes new ingredients available for consideration in future formulations. The inclusion of X in the model structure makes it possible to predict quality properties for formulations that use one or more new ingredients (Muteki et al. 2006).

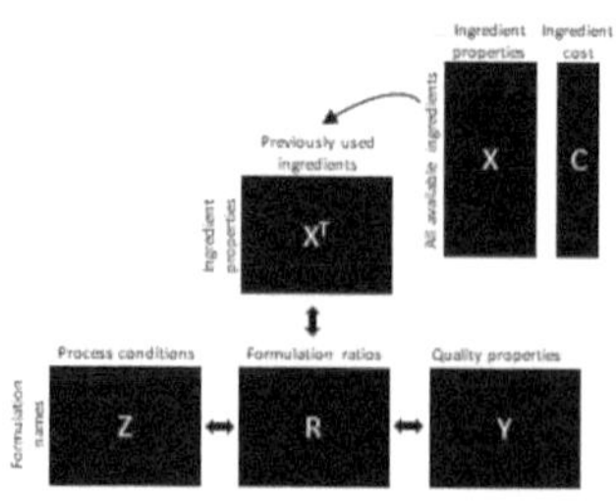

FIGURE 4.1 Connected data block structure. Adapted from Muteki et al. (2006).

4.4.2 Mixture Properties

A supplemental task in data consolidation is the calculation of mixture properties. Mixture properties (RX) are a key concept of the PLS model structure, as discussed in Chapter 8. Mixture properties are calculated by combining ingredient properties (X^T) and formulation ratios (R) using appropriate mixing rules. Although mixture properties may appear more related to feature engineering (see Chapter 5) than data consolidation, it is highly recommended that they be calculated at this stage since they often expose insufficiencies that may be otherwise overlooked. The concept of mixture properties is further introduced in this section to provide the necessary background for discussing consolidation and data quality challenges later in this chapter.

When mixture properties are available and sufficiently characterize the ingredients, formulation ratios can be excluded from a model. This variation of the model structure, as shown by Figure 4.2, is very powerful since it allows for new ingredients to be considered in future formulations, so long as the ingredient properties (X) are known.

Mixture properties can be calculated per ingredient class or globally (across all ingredient classes). This flexibility is helpful in cases where some ingredient classes have more missing data than others, when the ingredient property measurements differ between ingredient classes, or when the formulator only wishes to consider new ingredients in specific ingredient classes.

Often, the "ideal mixing rule" can be applied, but more complex mixing rules can be used when formulator knowledge warrants it. Under the ideal mixing rule, an overall mixture property is calculated as the weighted sum of material properties of individual ingredients included in the formulation.

This approach is clearly illustrated with a simple example. Suppose a formulation contains 30 wt% of Starch1 and 70 wt% of Starch2. If an ingredient property such as weight-average molecular weight of both ingredients is known, an overall mixture property can be calculated using ideal mixing rules, as shown in Equation 4.1:

$$\begin{aligned} MW_{mix} &= r_{Starch1} MW_{Starch1} + r_{Starch2} MW_{Starch2} \\ &= (0.3 \times 214.5) + (0.7 \times 531.4) \\ &= 436.33 \end{aligned} \quad \text{(Eq 4.1)}$$

Equation 4.1 can be generalized to Equation 4.2 or Equation 4.3.

$$x_{mix} = \sum r \times x \quad \text{(Eq 4.2) (Muteki et al. 2006)}$$

$$X_{mix} = R \bullet X \quad \text{(Eq 4.3) (Muteki et al. 2006)}$$

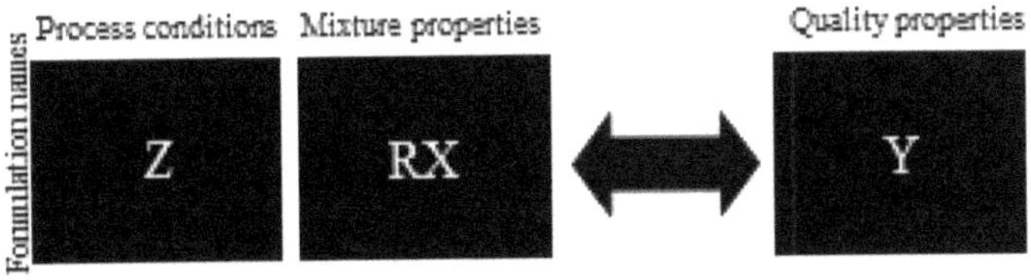

FIGURE 4.2 Revised data structure, incorporating mixture properties (RX). Adapted from Muteki et al. (2006).

4.4.3 Consolidation Challenges

A number of challenges may arise when consolidating multiple data records. Some of the most common issues are discussed in the following section (Wallace et al. 2022).

Inconsistent Nomenclature or Typos

Automating the consolidation of multiple data records is often inhibited by inconsistent nomenclature (naming conventions for each variable). A simple example is shown in Table 4.7.

In this example, two data records exist for an ingredient property (starch molecular weight). Since the two records use different abbreviations for molecular weight, it is easy to incorrectly create two unique variables when consolidating these records, even though they should actually be treated as a single variable.

Typically, some level of human intervention is needed to identify variables across multiple records or sources that should be consolidated into a single variable. Some options for consolidating inconsistently named variables include:

- Renaming all variables consistently prior to consolidation
- Creating consolidation mapping rules
- Using specialized tools available in commercial product development software (such as “combine rows/columns” in FormuSense)

For those just beginning data acquisition (rather than consolidating existing records), creating and distributing a standardized template, which includes nomenclature, is recommended to prevent this issue from arising in the future.

Duplicate Formulations and Variables

Another common consolidation challenge includes the presence of duplicate formulations and/or duplicate variables across multiple data records. *Duplicates* and *replicates* are distinctly different issues, which require different treatment when consolidating data. Duplicates are typically resultant from unintentional data-entry errors or from merging multiple data records that used all or some of the same data.

An example of duplicate formulations and duplicate variables in quality property data is provided as Table 4.8. Suppose that an experiment (“Exp 2) was included in two different data records, for use in two different historical analyses. One study included all four quality properties, while the other study only considered density. When these records are initially consolidated, the result may be that two rows (Exp 2) and two columns (Density) are named identically.

One may consider the following guidance for consolidating duplicates:

- Take an average of duplicate values, ignoring non-numerical or missing data (as shown in Table 4.8).
- Determine which duplicate is more correct or trustworthy, and ignore the duplicate value.
- Rename rows or columns to unique identifiers if the data entries are in fact replicates, not duplicates.

TABLE 4.7

Example of Inconsistent Nomenclature in Ingredient Property Data (a & b), the Incorrect Consolidation of These Variables (c) and the Correct Consolidation of These Variables (d)

a. Data Record 1		+	b. Data Record 2		→	c. Incorrect Consolidation			→	d. Correct Consolidation	
Ingredient	**Mw**		**Ingredient**	**mol wt**		**Ingredient**	**Mw**	**mol wt**		**Ingredient**	**MW**
Starch1	214.5		Starch4(new)	316		Starch1	214.5	missing		Starch1	214.5
Starch2	531.4		Starch5(new)	123		Starch2	531.4	missing		Starch2	531.4
Starch3	722.2		Starch6(new)	219		Starch3	722.2	missing		Starch3	722.2
			Starch7(new)	203.9		Starch4(new)	missing	316		Starch4(new)	316
			Starch8(new)	613.5		Starch5(new)	missing	123		Starch5(new)	123
			Starch9(new)	761.1		Starch6(new)	missing	219		Starch6(new)	219
						Starch7(new)	missing	203.9		Starch7(new)	203.9
						Starch8(new)	missing	613.5		Starch8(new)	613.5
						Starch9(new)	missing	761.1		Starch9(new)	761.1

Source: Adapted from Muteki (2006)

TABLE 4.8
(a) Example of Duplicate Formulations and Variables in Quality Property Data, and (b) Consolidated Quality Property Data after Treating Duplicate Formulations and Variables

a. Example of quality property data with duplicate formulations (Exp 2) and duplicate variables (Density) present

Formulation	Tensile Strength	Tensile Modulus	Elongation at Break	Density	Density
Exp 1	0.88	18.59	10.10	1.37	missing
Exp 2	3.72	162.30	10.10	1.33	missing
Exp 2	missing	missing	missing	missing	1.35
Exp 3	3.25	166.10	7.60	1.33	missing
Exp 4	9.70	629.10	6.59	1.31	missing
Exp 5	7.55	489.10	4.02	1.32	missing

b. Example of quality property data after consolidating the duplicate formulations (Exp 2) and the duplicate variables (Density)

Formulation	Tensile Strength	Tensile Modulus	Elongation at Break	Density
Exp 1	0.88	18.59	10.10	1.37
Exp 2	3.72	162.30	10.10	1.34
Exp 3	3.25	166.10	7.60	1.33
Exp 4	9.70	629.10	6.59	1.30
Exp 5	7.55	489.10	4.02	1.33

Source: Adapted from Muteki (2006)

Replicate Formulations and Variables

While duplicate data entries are typically unintentional, replicates tend to be intentional.

In the case of formulations, a particular experiment may have been run multiple times using the same configuration each time (ingredient selections and ratios, as well as process conditions). Formulation replicates can play an important role in building and validating a predictive model for product development applications; however, each formulation must be given a unique identifier.

Ingredient properties, process conditions, and quality properties may also include replicates. For example, an important quality property may be intentionally measured by two different experimental procedures or two different pieces of lab equipment.

An example of formulation and quality property replicates is included in Table 4.9.

Suppose that Exp 1 has been replicated to confirm the initial results and that density is documented for two different measurement methods. These replicates can be resolved by giving the two Exp 1 formulations unique, but related, identifiers such as "Exp 1.1" and "Exp 1.2." In addition, a new descriptor variable can be added for the

TABLE 4.9
(a) Example of Replicate Formulations and Variables in Quality Property Data, and (b) Consolidated Quality Property Data after Treating Replicate Formulations and Variables

a. Example of quality property data with replicate formulations (Exp 2) and duplicate variables (Density) present

Formulation	Tensile Strength	Tensile Modulus	Elongation at Break	Density	Density
Exp 1	0.88	18.59	10.10	1.37	1.38
Exp 1	0.88	18.59	10.10	1.36	1.39

b. Example of quality property data after replicates formulations (Exp 1) have been uniquely named, and replicate variables have been consolidated.

Formulation	Tensile Strength	Tensile Modulus	Elongation at Break	Density
Exp 1.1	0.88	18.59	10.10	1.37
Exp 1.2	0.90	18.43	10.15	1.36

Source: Adapted from Muteki (2006)

purpose of identifying replicates. Alternatively, one can consider removing replicate formulations if one of the entries can be identified as being the most correct or representative of all related replicates.

The density replicates in this example can either be treated by giving each measurement method a unique identifier, such as "Density Measurement Method 1" and "Density Measurement Method 2," or by consolidating them into a single variable by following the same guidance as provided for duplicates:

- Take an average of replicate values, ignoring non-numerical or missing data.
- Determine which replicate is more accurate/precise/trustworthy, and ignore the replicate value (as shown in Table 4.9).

Omissions (Missing Formulations and Variables)

While nomenclature inconsistencies, duplicates, and replicates present themselves fairly readily when consolidating multiple data records, identifying omissions can be more difficult. An omission is when an entire row or column, including its header, are missing from any one of the four blocks of data sources (ingredient properties, formulation ratios, process conditions, and quality properties).

The most common omissions are as follows:

- An ingredient that occurs in the formulation ratios but does not occur in the ingredient properties (omitted column).
- A formulation that occurs in formulation ratios but does not occur in process conditions and/or quality properties (omitted column).

Prior to building a predictive model, a product development dataset should be screened carefully with the aim of identifying omissions such as missing formulations and missing ingredients. If omissions are identified, placeholders for these formulations or variables should be created until the data is located and appended to the dataset.

In some cases, you may still be able to proceed with building a predictive model, even in the presence of omissions by taking some strategic modeling decisions. Some possibilities include the following:

- *Selecting a model structure that is unaffected by the omissions.*

For example, if one or more ingredients have been omitted from the ingredient properties, a model can be built on formulation ratios (R) and process conditions (Z) to quality properties (Y) (and where mixture properties (RX) are excluded). This will not always be a helpful solution, depending upon the project objectives as the resulting model in this case would not allow for the consideration of new ingredients.

- *Removing formulations that do not occur in all of the desired data sources.*

For example, if the size of the dataset is sufficiently large, you may be able to remove the affected formulations from all data sources and still have enough historical variation to build a useful model. Again, this will not always be a helpful solution, depending on the size and characteristics of the dataset.

Inconsistent Measurement Units

Since product development data has conventionally been collected in a fragmented manner over many years, undocumented and/or inconsistent measurement units across consolidated variables are common.

An example of ingredient property data with inconsistent measurement units is included as Table 4.10 and Figure 4.3. This data has been adapted from Muteki (2006).

In the example of Table 4.10, there is a significant jump in molding temperature between Experiments 18 and 19. A line plot, as shown in Figure 4.3, is a simple and helpful tool for locating measurement unit inconsistencies. However, noting that experiments 15–18 were completed in Lab A while Experiments 19–23 were completed in Lab B, it may be difficult to discern if this jump is attributed to the different lab/equipment, or if it is a measurement unit inconsistency. Tracking down documentation or the individual who completed the experiment can be difficult but is necessary. The alternative, of including a variable with inconsistent or unconfirmed measurement units in a predictive model, can be dangerous since data-based models may find correlations in flawed data that are not causal.

Where possible, creating and distributing a standardized data template that includes measurement units for each variable can prevent this issue from arising.

It should be clear that data consolidation is not a trivial task, nor can it be 100% automated. Further, after all data blocks have been successfully consolidated, one should begin the process of assessing data quality prior to building a predictive model.

TABLE 4.10
Example Process Conditions with Inconsistent Measurement Units

Formulation	Molding Temperature	Location
Exp 15	150	Lab A
Exp 16	170	Lab A
Exp 17	150	Lab A
Exp 18	170	Lab A
Exp 19	302	Lab B
Exp 20	338	Lab B
Exp 21	302	Lab B
Exp 22	338	Lab B
Exp 23	302	Lab B

Source: Data adapted from Muteki (2006)

Experiments 15–18 are recorded in degree Celsius while experiments 19–23 are recorded in degree Fahrenheit.

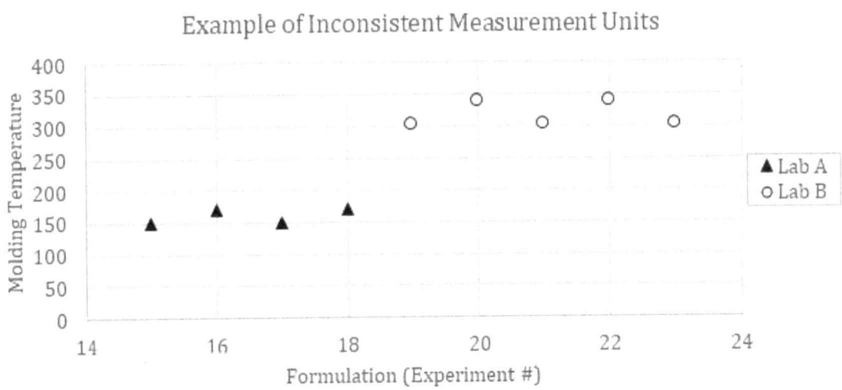

FIGURE 4.3 Example process conditions with inconsistent measurement units. Experiments 15–18 are recorded in degrees Celsius while Experiments 19–23 are recorded in degrees Fahrenheit. Data adapted from Muteki (2006).

4.5 DATA QUALITY

Owing to chemistry complexities, product development activities that follow a conventional approach (involving trial-and-error or Design of Experiments (DoE) with numerous physical experiments) tend to become iterative and resource-intensive. By contrast, building a predictive model on historical formulation data, interpreting the model, and performing simulations (applying the model to predict the outcome of new formulations) can drastically reduce the number of required physical experiments and, ultimately, accelerates the time required to develop a new or reformulated product. Assessing data quality for suitability to predictive modeling is not a trivial task, but the rewards are well worth the upfront effort.

In addition to challenges that arise specifically when consolidating multiple data records, there are additional data quality challenges that are commonly found within a single data block or within a consolidated dataset (Wallace et al. 2022). This section discusses several of these challenges.

4.5.1 Issues Common to All Data Sources

While some dataset challenges typically affect a particular data block (formulation ratios, ingredient properties, process conditions, or quality properties), others are commonly found among any one of the four blocks. Challenges that are common to all data blocks are discussed first before examining issues that are unique to a single block.

Small Datasets

Although the adoption of digital-centric labs (see Chapter 1) is quickly changing the landscape of product development datasets, the amount of data currently available to a formulator interested in building a predictive model is still relatively "small."

Although some data-based modeling methods (such as PLS) can still extract value from a handful of data points, others cannot. Moreover, in general, the more data points the better. Building a valuable predictive model from a small dataset (≤ 25 data points) is possible if sufficient variation exists; however, small datasets are susceptible to overfitting. Model validation and input from domain experts to gain trust in the model (see Chapter 9) is extremely important for models built on small data.

Missing Data

Missing data is distinctly different than the previously discussed omissions. Recall that an omission is an entirely missing row or column of data, including the header. By contrast, missing data refers to individual values that are missing for a particular variable.

An example of process condition data with some missing values is included as Table 4.11. This data has been adapted from Muteki (2006).

While some predictive modeling methods (such as PLS) can handle a substantial amount of missing data (~30–40%), others cannot, so locating missing data may become a roadblock. Imputation is a common strategy used to handle missing data; however, this approach should be applied cautiously, especially to small datasets.

Highly Correlated Data

While a large number of X and Y variables may exist for one product development dataset, there are typically much fewer degrees of freedom due to the underlying chemistry; in other words, the data is highly correlated therefore not every X variable can move independently of all other X variables, and not every Y can move independently of all other Y variables.

TABLE 4.11
Example Process Conditions, with Missing Data for Experiments 16, 21, 26, and 27

Formulation	Molding Temperature (°C)	Location
Exp 15	150	Lab A
Exp 16		Lab A
Exp 17	150	Lab A
Exp 18	170	Lab A
Exp 19	150	Lab B
Exp 20	170	Lab B
Exp 21	4	Lab B
Exp 22	170	Lab B
Exp 23	150	Lab B
Exp 24	170	Lab B
Exp 25	150	Lab A
Exp 26	n/a	Lab A
Exp 27		Lab B
Exp 28	170	Lab B

Source: Adapted from Muteki (2006)

Fortunately, some modeling approaches, such as PLS, are particularly well-suited to highly correlated data.

Insufficient Variation

Data-based models rely on variation. Product development relies on the intentional variation of one or more degrees of freedom. However, quality control and standard operating procedures conventionally oppose variation. As a result, many product development datasets include variables that lack variation. Some common examples of insufficient variation include:

- Using the same ingredient, in the same ratio, across an entire dataset (zero variation).
- Infrequent use of an ingredient, and/or only using an ingredient in the same ratio whenever used (insufficient variation). An example of this is included as Figure 4.4.
- Holding a process condition (such as molding temperature) constant across an entire dataset (zero variation).

It is good practice to exclude variables that have insufficient variation from a predictive model as they are likely to contribute to overfit. However, models that exclude ingredients with insufficient variation (as seen in Figure 4.4) require caution when

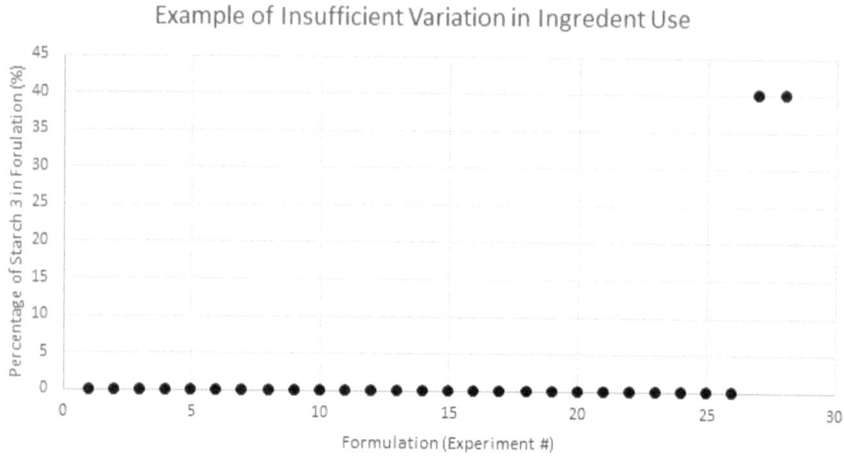

FIGURE 4.4 Example of insufficient variation in formulations – as shown for the ingredient use of Starch 3. Data adapted from Muteki (2006).

being applied for the purpose of simulation or optimization. This is a relatively complex scenario that may require iteration and additional steps to ensure that real-world constraints are met. For example, if every historical formulation contains 20% water, that variable would not be included in a predictive model, yet simulation and optimization should have this ingredient amount to handle the sum to 100% constraint. These special considerations and checks are handled in FormuSense.

It is difficult to provide general guidance on an appropriate threshold for acceptable or insufficient variation that is universally applicable to all datasets. Some software platforms, such as FormuSense, prevent variables with an extremely small standard deviation (≤ 1e–8) from being included in a predictive model. However, user discretion is needed to determine if additional variables should be excluded.

Misuse of Zero (0)

Product development datasets frequently misuse the value zero (0). Often, a zero is erroneously used when the intended meaning is that the data is missing. Using zero in place of a blank, or string such as "missing" or "n/a," is problematic since the value zero is often meaningful in the context of a specific variable.

An example of using zero in ingredient property data is provided as Table 4.12. In Table 4.12, there are three ingredients (Starch3, Starch4(new), Starch9(new)) for which the amylose content is reported as zero. Knowing that amylose is one of two molecules that constitute starch, it is unlikely that these values are true zeros. It is more likely that the value of amylose was unknown or that it was below the detectable limit for the equipment or procedure used to measure amylose.

In the situation where a measured value is below the detectable limit, that limit should be reported for context. A discussion on the treatment of inequalities in product development datasets is included later in this chapter, in the section titled "Issues Affecting Quality Properties."

TABLE 4.12
Example Ingredient Properties with Possible Misuse of Zero, along with Different Possible Interpretations of Zero

		Possible Misuse of Zero		Possible Interpretation of Zero
Ingredient	**Supplier**	**Amylose Content (wt%)**		**Amylose Content (wt%)**
Starch1	Company A	70		70
Starch2	Company A	23		23
Starch3	Company B	0	→	0 (true zero)
Starch4(new)	Company B	0	→	missing (unknown)
Starch5(new)	Company C	70		70
Starch6(new)	Company C	28		28
Starch7(new)	Company C	35		35
Starch8(new)	Company C	12		12
Starch9(new)	Company D	0	→	≤ 5 (below detectable limit)

Source: Adapted from Muteki (2006)

Significant Figures

It is common to find inconsistencies in the number of significant figures for any single variable in product development dataset. Table 4.13 provides an example of ingredient property data with a range of 1–7 significant figures.

TABLE 4.13
Example Ingredient Properties with Inconsistent Significant Figures for a Single Variable (PDI)

Ingredient	Supplier	PDI
Starch1	Company A	5.230004
Starch2	Company A	8.53
Starch3	Company B	8.02
Starch4(new)	Company B	1.3
Starch5(new)	Company C	2
Starch6(new)	Company C	1.5
Starch7(new)	Company C	5.24
Starch8(new)	Company C	10.84
Starch9(new)	Company D	7.33

Source: Adapted from Muteki (2006)

Ideally, the number of significant figures should be consistent for any particular variable and should reflect the precision of the equipment or procedure used to obtain the measurements. Inconsistencies in the number of significant figures within a single variable can cause a variety of challenges with building a predictive model. The later section titled "Issues Affecting Formulation Ratios" discussed one such challenge known as a "sum-to-1 violation."

Data Entry Errors

Data entry errors are unavoidable for any datasets that involve human input of data. Data entry errors refer to a variety of issues such as a value that is the wrong order of magnitude, a single incorrect digit within a multi-digit value, or an incorrect direction (+/–).

Prior to model-building, visual inspection of raw data line plots is an effective tool for identifying significant data entry errors.

Data entry errors that are not detected during the data preparation stage may still be detected later on by outlier detection tools relevant to the selected modeling method. For example, confidence limits and model validity metrics (such as Hotellings T^2 and Squared Prediction Error) are effective tools for locating additional data entry errors for PLS models embedded in the FormuSense software platform. For further information on PLS outlier detection tools, refer to Chapter 8.

Non-Numerical Data

In addition to numerical data, natural language descriptive or qualitative information can be a valuable component of a product development dataset. An example of non-numerical information is provided in Table 4.10, where "Lab A" or "Lab B" is included as a process condition variable.

Non-numerical variables are commonly found in most product development data blocks, and some examples are provided next.

- Ingredient Properties
    - domain expert knowledge – such as rating or categorizing the "ease of processing" each ingredient
    - first principles or theoretical information – such as capturing predominant chemical mechanisms or the ingredient molecular structure or shape
- Process Conditions
    - Discrete indicators – such as the laboratory, technician, and equipment
- Quality Properties
    - Qualitative (as opposed to quantitative) quality ratings – such as grading a product based on visual inspection completed by a human

Although non-numerical information can be more difficult to include in a predictive model, there are strategies for including this information passively or explicitly. An example of using non-numerical data passively is to color plots and visualizations using this information, as shown in Figure 4.4. Variables that are retained for passive

TABLE 4.14
Example of Converting a Non-Numerical Process Condition Variable into a Binary Categorical Variable

		Descriptor (N = 2)	→	Categorical (N − 1 = 1)
Formulation	Molding Temperature	Location		Lab A
Exp 15	150	Lab A		1
Exp 16	170	Lab A		1
Exp 17	150	Lab A		1
Exp 18	170	Lab A		1
Exp 19	302	Lab B		0
Exp 20	338	Lab B		0
Exp 21	302	Lab B		0
Exp 22	338	Lab B		0
Exp 23	302	Lab B		0

Source: Adapted from Muteki (2006)

uses such as plot coloring are referred to as *descriptor variables* in FormuSense but may have alternative nomenclature in other platforms.

Non-numerical variables can alternatively be converted into an appropriate number of binary variables, and these new variables can then be included explicitly in a predictive model. This process has different names in different fields, such as creating "categorical variables" in the FormuSense platform, "dummy variables" in statistics, or "one-hot encoding" in machine learning. For a non-numerical variable with N unique values, N-1 categorical variables are needed to capture all information. An example of converting a non-numerical process condition variable with two unique values (Lab A, Lab B) into 1 binary categorical variable is provided as Table 4.14.

4.5.2 Issues Affecting Ingredient Properties

Earlier discussions in Data Consolidation on data structure and mixture properties implied the importance of ingredient property data. However, this topic is worth revisiting for a more detailed discussion.

Recall that it is the inclusion of ingredient properties in the model data structure that provides the ability to predict the impact of new ingredients on quality properties. Therefore, the presence, accuracy, and precision of ingredient property data is highly important.

Non-numerical (qualitative) ingredient properties can be equally as important as numerical ingredient properties. Since non-numerical variables are effective ways to capture domain-expert knowledge, spending time documenting this information is a worthy investment that can improve:

- model fit and model predictive abilities
- model applicability (i.e., ability to consider new ingredients)
- knowledge transfer (by capturing intrinsic "know-how" and insight that has previously been undocumented)

However, locating ingredient property data is not a trivial task. And locating accurate and precise ingredient property data is even more difficult. The following are some challenges in collecting high-quality ingredient property data.

Available Properties Poorly Characterize the Ingredient

Many product development datasets do not initially include any ingredient properties, and the first-pass effort to populate ingredient properties often relies on locating material safety data sheets (MSDS) from each ingredient supplier. While MSDS data can be helpful in some contexts (such as its intended use – as a reference for materials handling and safety), this information is often not accurate and/or precise enough for materials discovery applications since it tends to be static. In order to take advantage of the adapted model structure depicted by Figure 4.2, which uses mixture properties in place of ingredient ratios, the ingredient properties used to calculate those mixture properties must sufficiently characterize the ingredients. If the documented ingredient properties are irrelevant to the measured quality properties, or if the ingredient properties are inaccurate or imprecise, the model fit and model predictive abilities will suffer. For more discussion on this topic, the reader is directed to Chapter 3.

Readily available MSDS data can still be a useful starting point for pilot projects that aim to assess the feasibility and suitability of predictive modeling for your product development application. However, the most useful ingredient properties are those that can be measured on a semi-regular basis, through the same equipment or procedure for all ingredients within a single ingredient class. In the absence of a standardized testing procedure, ingredient properties are susceptible to suffering from nomenclature and measurement unit inconsistencies, such as those described in the earlier "Data Consolidation" section.

4.5.3 Issues Affecting Formulation Ratios

Formulation ratios commonly present two unique challenges when preparing a dataset for predictive modeling. This section discusses sum-to-1 violations and nested ingredients.

Sum-to-1 Violations

While it is fairly obvious that the sum of ingredients for any single historical formulation must sum to 100 (when specified as percentages) or to 1.0 (when specified in fractions), formulation ratio data often does not obey this physical constraint. Missing ingredients, typos, inconsistent measurement units, or significant figures can lead to these "sum-to-1" violations in a dataset.

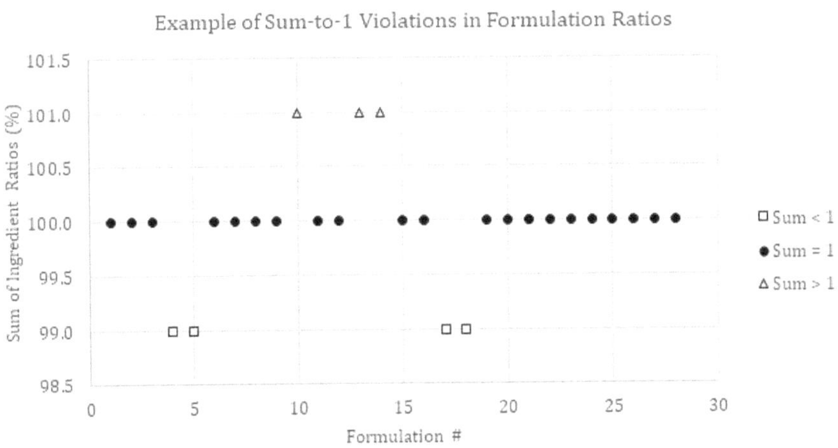

FIGURE 4.5 Example of a sum-to-1 violation in formulation ratios. Adapted from Muteki (2006).

Fortunately, as shown in Figure 4.5, identifying formulations with sum-to-1 violations is relatively simple when formulation ratios have been specified in percentages or fractions.

When formulation ratios have been specified by an amount (such as mass or PHR), instead of as percents or fractions, detecting sum-to-1 violations is much more difficult. In these cases, a manual review of all formulation ratio data may be required. Outlier detection tools (such as HT2 and SPE for PLS modeling) can sometimes be useful for detecting significant formulation ratio errors; however, these tools are not available until after a model has been built.

Compounds with Nested Ingredients

"*Compounds*" are materials that are listed as a single ingredient in formulation ratios, but which are themselves comprised of multiple ingredients. The constituents of a compound can be referred to as "*nested*" ingredients.

For example, and as shown in Table 4.15a, Additive3 is a compound in Exp 9. In Table 4.15b), we see that Additive3 is a combination of Additive1 and Additive2; therefore, formulation ratios for Exp 9 can be "exploded" to include Additive1 and Additive2 directly, as shown in Table 4.15c.

The presence of compounds with nested ingredients is common in formulation datasets. Compounds can simplify inventory requirements and physical formulation tasks by reducing the number of unique ingredients that must be available and handled.

However, the treatment of compounds with nested ingredients should be considered carefully before proceeding with building a predictive model. Compounds can be treated as unique ingredients like any other ingredient, or they can be "exploded" as shown in Table 4.15c, with the ingredient amounts propagated proportionally through to the original formulation. The decision of how to treat compounds is based

TABLE 4.15
(a) Example Formulation Ratios Including a Compound, (b) Composition of the Compound, and the (c) "Exploded" Formulation Ratios

a. Formulation Ratios Including a Compound			→	b. Composition of Compound			→	c. "Exploded" Formulation Ratios		
Formulation	Ingredient	Mass Fraction		Compound	Nested Ingredients	Mass Fraction		Formulation	Ingredient	Mass Fraction
Exp 9	Starch1	0.39		Additive3	Additive1	0.89		Exp 9	Starch1	0.39
Exp 9	PLA	0.17		Additive3	Additve2	0.11		Exp 9	PLA	0.17
Exp 9	Additive3	0.44				1.00		Exp 9	Additive1	0.39
		1.00						Exp 9	Additive2	0.05
										1.00

Source: Data adapted from Muteki (2006)

TABLE 4.16
Example of Formulation Ratios Including Additive1 as a Unique Ingredient

Formulation	Ingredient	Mass Fraction
Exp 8	Starch1	0.23
Exp 8	PLA	0.54
Exp 8	Additive1	0.23
		1.00

Source: Data Adapted from Muteki (2006)

on the availability of ingredient property data that sufficiently characterize the nested ingredients and compounds, and the presence of those same nested ingredients as direct ingredients in other formulations. Moreover, the desire to select the nested ingredients as unique ingredients in future formulations dictates the treatment of nested ingredients when preparing for predictive modeling.

In the example of Table 4.15, if Additive1 and Additive2 only occur in the dataset together as nested ingredients of Additive3, *and* the ingredient property data for Additive3 sufficiently characterizes the compound, *and* there is no desire to use Additve1 or Additive2 independently in future formulations, there is no need to "explode" the formulation ratios. However, consider another scenario where Additive1 occurs as a nested ingredient of Additive3 in Exp 9 *and* Additive1 also occurs in another formulation as an independent ingredient (see Exp 8 in Table 4.16), *and* there is a desire to use Additive1 independently of Additive2 in future formulations. In this case, it may be beneficial to "explode" the formulation ratios for Exp 9, assuming that the ingredient properties for Additive1 and Additive2 sufficiently characterize these materials. This discussion highlights the complexities of data preparation and modeling configurations; a significant level of human involvement is needed to navigate these decisions, rather than relying on automated machine learning.

4.5.4 Issues Affecting Mixture Properties

As discussed earlier, mixture properties are powerful input variables in a predictive model that provide the ability to predict the impact of new ingredients on quality properties. Unfortunately, mixture properties typically have the largest amount of missing data in a product development dataset. This is precisely why it is important to calculate mixture properties during data consolidation, rather than waiting for a formal feature engineering phase of the project.

Amplified Missing Data

Recalling Equation 4.1, for any formulation that uses an ingredient that contains a missing ingredient property, the overall mixture property for that ingredient class cannot be calculated. So, although the amount of missing data for an ingredient property may be small (Table 4.17), if the ingredients with missing properties have been used often (Figure 4.6), the calculated mixture property may be much higher (Figure 4.7).

To further understand the cause of this amplified missing data, further consider the example where the molecular weight of Starch1. Revisiting Equation 4.1, we see that the mixture property for a formulation that uses Starch1 in addition to Starch2 cannot be calculated.

$$\begin{aligned} MW_{mix} &= r_{Starch1} MW_{Starch1} + r_{Starch2} MW_{Starch2} \\ &= (0.3 \times \text{missing}) + (0.7 \times 531.4) \\ &= (\text{missing}) + (371.98) \\ &= \text{missing} \end{aligned} \qquad \text{(Eq 4.1 – revisited)}$$

It is important to note that zero (0) should NOT be used in place of a missing ingredient property for the purposes of calculating mixture properties. Refer to the section on "Issues Common to all Data Sources" for further discussion on the misuse of zero.

If Starch1 is seldomly used in past formulations, the mixture molecular weight property will only be seldomly missing. However, if Starch1 is used in the majority of the historical formulations, the starch mixture molecular weight will contain a large amount of missing data, which may limit the ability to include this property as

TABLE 4.17
Example of a Single Missing Ingredient Property for an Ingredient That Is Shown to Be Used Frequently (Figure 4.6) and That Amplifies into a Significant Amount of Missing Data in the Related Mixture Property (Figure 4.7)

Ingredient	Mw ($\times 10^{-3}$ g/mol)
Starch1	missing
Starch2	531.4
Starch3	722.2
Starch4(new)	316
Starch5(new)	123
Starch6(new)	219
Starch7(new)	203.9
Starch8(new)	613.5
Starch9(new)	761.1
% missing	**1/9 = 11%**

Source: Adapted from Muteki (2006)

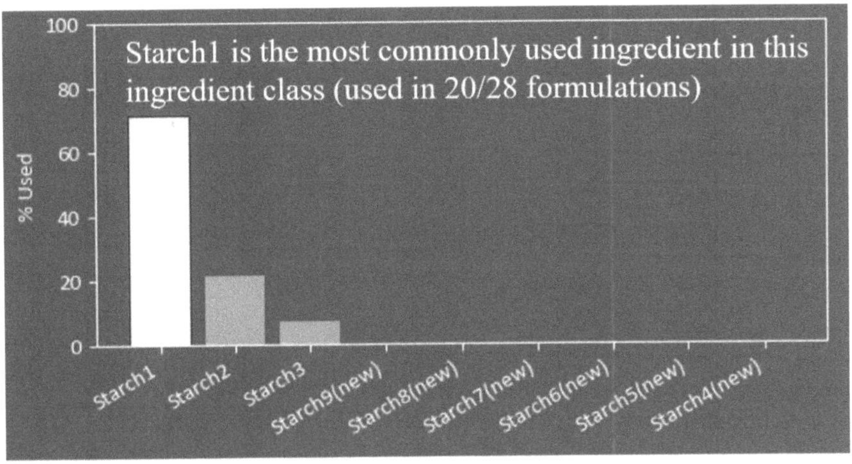

FIGURE 4.6 Example of a single missing ingredient property (Table 4.17) for an ingredient that is used frequently. Adapted from Muteki (2006).

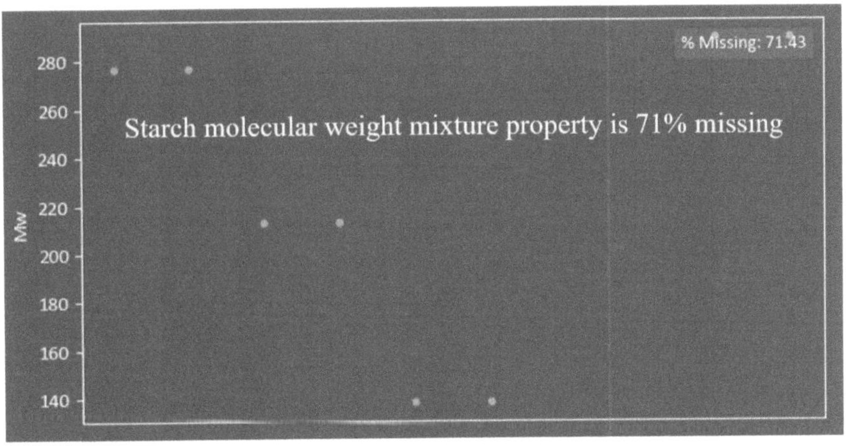

FIGURE 4.7 Example of how a single missing ingredient property (shown in Table 4.17) for an ingredient that is used frequently (shown in Figure 4.6) amplifies into a significant amount of missing data in the related mixture property. Adapted from Muteki (2006).

an input to the predictive model. If the model fit is poor without this mixture property and/or there is a desire to consider new starches in future/digital simulations, then it is critical to locate the missing ingredient properties (Starch1 molecular weight in this example) prior to moving forward with building the predictive model. As a general guideline extracted from experience, a mixture property variable that is more than 50% missing should not be included in a predictive model.

4.5.5 Issues Affecting Process Conditions

In addition to continuous numerical data, process conditions commonly include discrete indicators.

Discrete Indicators

Refer to the section on "Issues Common to All Data Sources" (non-numerical data) for examples and recommended treatment of discrete (non-numerical) data in process conditions.

4.5.6 Issues Affecting Quality Properties

Quality properties are the most important component of a product development dataset. This information characterizes the performance of the final product, and it is structured as the output (Y) of a predictive model. As such, data integrity and data preprocessing are extremely important for this block of data. Accurate and precise quality properties are essential for developing a reliable predictive model since a model can never be more accurate than your Y variable measurement. For example, and as further discussed in Chapter 9, a quality property that has a measurement error of +/– 20% cannot be predicted within +/– 5%. When evaluating the performance of a predictive model, it is important to set expectations based on the integrity of the quality property data.

In addition to continuous numerical values, there are a variety of other data formats or conditions that commonly occur in quality property data such as the presence of inequalities, ranges, trajectory and time-limited test results, qualitative/observation-based properties, and other non-numerical data (such as images). These issues are discussed in the following sections.

Inequalities (Below Detectable Limit)

Inequalities are common in quality property data since the equipment or methods used to measure performance properties often have a lower limit for detection. An example of quality property data inequalities is presented in Table 4.18. This data has been adapted from Muteki (2006).

Inequalities are difficult to include in predictive models like PLS, and some software platforms like FormuSense will automatically treat inequalities as missing data. Therefore, it is recommended that all inequalities should be replaced with an appropriate continuous value before model-building. Domain experts should be consulted on an appropriate replacement value for inequalities. In the example of Table 4.18, one might consider replacing "< 5" for Experiments 3 and 9 with a continuous value somewhere between 0 and 5.

Ranges

Ranges can occur in quality property data when the measurement equipment or test procedure has low precision. Depending upon the spread of the quality data, the presence of ranges can limit the usefulness of a predictive model. However, many times ranges do provide sufficient information to build a useful model.

TABLE 4.18
Example Quality Properties with Inequalities

Formulation	Tensile Strength (MPa)
Exp 3	< 5
Exp 4	9.70
Exp 5	7.55
Exp 6	22.80
Exp 7	23.80
Exp 8	20.00
Exp 9	< 5
Exp 10	8.73

Source: Adapted from Muteki (2006)

TABLE 4.19
Example Quality Properties (a) with Ranges and (b) Possible Treatment of Ranges

(a) Quality Property Data with Ranges			(b) Possible Treatment of Ranges		
Formulation	Y1 Tensile Modulus (MPa)	→	Y2 Lower Tensile Modulus (MPa)	Y3 Average Tensile Modulus (MPa)	Y4 Upper Tensile Modulus (MPa)
Exp 1	0–50		0	25	50
Exp 2	150–200		150	175	200
Exp 3	150–200		150	175	200
Exp 4	600–650		600	625	650
Exp 5	450–500		450	475	500
Exp 6	1500–2,000		1500	1550	2000
Exp 7	1000–1,500		1000	1250	1500

Source: Adapted from Muteki (2006)

An example of quality property data with ranges is included as Table 4.19a. This data has been adapted from Muteki (2006). As mentioned in earlier in this section, precise quality properties are essential to building a reliable predictive model.

As with inequalities, it is difficult to include ranges directly in a predictive model. As shown in Table 4.19b, a relatively simple treatment is to use one or more of the lower limit of the range, average of the range, and upper limit of the range.

Trajectory Data and Right-Censored Tests

Another challenge with quality properties is the treatment of data from trajectory and right-censored tests. Trajectories can be treated as several continuous output variables by selecting meaningful and discrete "check-points" from within the trajectories. However, this can lead to additional complications when some products do not achieve all "check-points," as can be the case for poor-performing products that are measured under a right-censored (time-limited) test.

In some cases, extrapolation or the insertion of another meaningful value, as determined by subject-matter experts, can be used to populate "missing" data in trajectory data that has been converted to discrete check-points.

Qualitative (Not Quantitative) Data

While most product development datasets will contain quantitative (numerical) measurements of key quality performance properties, qualitative (non-numerical) data is also relatively common. For example, formulations or products may be categorized into different product grades, based on visual appearance characteristics that are assessed by human inspection.

As with non-numerical input properties, this data can be converted into an appropriate number of binary variables for explicit inclusion in a predictive model. Otherwise, this non-numerical information can be retained for passive uses such as plot coloring. For further discussion on the treatment of non-numerical data, refer to the section titled "Issues Common to All Data Sources" ("Non-numerical data").

Image Data

Machine vision inspection systems are increasingly common in laboratory and manufacturing environments. Machine vision can be used to reliably quantify both visual and thermal quality properties through the use of color, 3D, polarization, or thermal hardware. In product development applications, machine vision data may be used to quantify characteristics such as:

- the size, shape, color, and texture of a product
- the presence, size, shape, color, and texture of defects
- the count and distribution of key ingredients within a product
- the molecular structure and assembly of structures within the product

Image processing algorithms can be applied to images of finished products to give quantifiable results for defined image features, and these results can in turn be used as quality properties in a predictive model. More on this topic is discussed in Chapter 6.

4.5.7 Data Quality Assessment Procedure

As evident, numerous data quality issues may exist within a consolidated product formulation dataset. The following procedure is recommended to identify data quality issues prior to building a predictive model.

1. **Calculate key metrics that summarize the entire dataset**. Summary metrics give an overall idea of the size and character of the dataset. Recommended summary metrics for a formulation dataset include:
   - number of formulations
   - number of ingredients and ingredient classes
     i. total number of ingredients and ingredient classes in the dataset
     ii. number of ingredients that have and have *not* been used previously per each ingredient class
     iii. number of ingredients and ingredient classes per formulation (min, median, average)
   - number of input variables (process conditions, formulation ratios, mixture properties) and output (quality properties)
     i. number of numerical variables
     ii. number of non-numerical variables
   - number of output variables
     i. number of numerical variables
     ii. number of non-numerical variables
2. **Calculate and visualize statistics on individual variables**. Recommended statistics for individual variables include:
   - specifically for non-numerical variables:
     i. generate binary categorical variables
   - For each input and output variable, review:
     i. line plot of raw data
     ii. distribution histogram
     iii. minimum, median, maximum value
   - percent specifically for formulation ratios, review:
     i. how frequently each ingredient is used (bar plot and/or line plot)
     ii. how frequently each ingredient class is used (bar plot and/or line plot)
     iii. sum-to-1 violations (line plot)
3. **Calculate and visualize mixture properties.** For each mixture property, review:
   - line plot of generated mixture property
   - percent missing

This procedure will help identify data quality issues, such as insufficient variance or a large amount of missing data. One can expect to spend more than 50% of project timeline on data preparation. Typically, the later phases of model-building, simulation, and optimization represent less than 50% of project efforts. Fortunately, platforms such as FormuSense, which have been developed specifically for product formulation applications, can expedite data preparation. In fact, the data assessment procedure described earlier is relatively automated in FormuSense.

4.6 CONCLUDING REMARKS

As described in Chapter 4, there are numerous challenges that may be encountered when locating and consolidating product formulation data. In addition, identifying and resolving data quality problems is often a somewhat manual and time-consuming task. One should neither undervalue the importance of these steps in a product formulation project nor underestimate the resource requirements to complete these tasks with care. As with any data-driven project, garbage in equals garbage out; therefore, it cannot be overemphasized that data preparation tasks are extremely important. Although preparing a dataset can be a tedious task, this is the foundation that will support the predictive model. The level of effort put into data preparation will directly inhibit or enhance the usefulness of the predictive model and the success of the digital transformation project.

Chapter 5 continues the discussion of overcoming dataset challenges, with a focus on ingredient properties since they are typically the most difficult properties to locate. Recalling that the inclusion of ingredient properties enables the calculation of mixture properties, and, ultimately, the ability to consider new untested ingredients in future formulations, ingredient properties deserve added effort to ensure that all key ingredients are sufficiently characterized.

For further understanding of general dataset challenges, refer to Section 12.2 for a case study using the starch blending data that was used throughout Chapter 4.

REFERENCES

Cardin, M., M.-J. Bruwer, J.F. MacGregor. 2015. "Using Multivariate Modeling and Optimization for Rapidly Developing New and Improved Elastomer Products." Presented October 13–15, 2015 at the F*all 188th Technical Meeting of Rubber Division*, ACS at Cleveland, OH.

Muteki, K. 2006. "Mixture Product Design Using Latent Variable Methods." PhD diss., McMaster University.

Muteki, K., J.F. MacGregor, T. Ueda. 2006. "Rapid Development of New Polymer Blends: The Optimal Selection of Materials and Blend Ratios." *Industrial & Engineering Chemistry Research* 45 (13): 4653–4660. https://doi.org/10.1021/ie050953b

Wallace, K., M. Cardin. 2021. "Accelerating Product Development Testing by Digitalizing Formulator Knowledge." *Rubber World Magazine* 263 (4): 28–36.

Wallace, K., M. Cardin, A. Nguyen. 2022. "Dataset Considerations for Rapid Product Development Applications." Presented April 10–14, 2022 at the *AIChE Spring Meeting* at San Antonio, TX.

5 Feature Engineering
Enhancing Your Data with Descriptors

Daniel Christiansen, Jerome Claracq and Sukrit Mukhopadhyay

5.1 INTRODUCTION

The rise in quantity and diversity of digital chemical data and improved accessibility to machine learning (ML) tools has made modeling material properties from composition – the amounts of each component in a material formulation or mixture, referred to as "ingredient ratios" in Chapter 4 – an achievable task (Rodrigues Jr. et al. 2021; Himanen et al. 2019; Vasudevan et al. 2019, 2021). However, little literature has demonstrated models trained solely on composition to predict material properties. This is due to several inherent weaknesses in the technique. Composition-data-driven ML primarily fails to incorporate physics or process conditions when describing the material composition (Vasudevan et al. 2019). This drawback may be lessened for large datasets due to relationships between composition and modeled properties embedded in the data. However, there are few material property datasets large enough for this task. An alternative to composition-data-driven modeling is replacing or supplementing formulation ratio data with material physical and chemical descriptors. Such descriptor engineering methods overwhelm the materials ML space and have been successfully demonstrated for capturing material physics, structure, and other parameters during the data enrichment and modeling processes. Such descriptors can come from various sources and benefit model performance and interpretability. This chapter illustrates some useful descriptors, how they may be sourced, and what benefits they can provide.

The remainder of this first section discusses the benefits and shortcomings of the composition-data-driven approach before introducing ingredient and formulation descriptors alongside their utility and importance. Following this, six common methods are introduced for producing or retrieving additional information about ingredients and formulations that can be input for modeling.

5.1.1 Pros and Cons of the Composition-Data-Driven Approach

The composition-data-driven approach, which relies on utilizing material composition as ML inputs, can rapidly produce models predicting material properties. The subject-agnostic origins of compositional data make it easily usable and encourage

DOI: 10.1201/9781003385974-7

purely data-minded decisions on input structure, scaling techniques, and other processing (Rodrigues Jr. et al. 2021; Marschik et al. 2020). With large volumes of purely compositional data at hand, this can be a good first step to see how ML can be used, and, conveniently, there is no requisite knowledge of the underlying phenomena to input compositions alone. Finally, it may be the case that the composition-data-driven approach is the only[1] viable method with the information available. There is significant domain knowledge necessary to frame physics-informed models. The materials space is riddled with expensive equipment and data that are challenging to acquire experimentally or analytically. Expanding a dataset with such descriptors may be cost-prohibitive or veritably impossible, leaving composition the only data available for modeling.

Though compositional modeling has the benefits of simplicity and rapid development times, the approach can come with considerable downsides. The first and greatest drawbacks originate from the quality of the data itself. Most formulating groups will not have several thousand (or ideally more) unique formulations to input for ML to avoid the data challenges discussed in Chapter 4, such as an imbalance between the number of components and observations (Xu et al. 2023). Among the available formulations, there are other challenges, such as sparsity, which can slow model development for large feature spaces (high number of unique components) and reduce performance, and the disparity between a small number of components used many times while others are rarely used. As discussed in Chapter 3, this amounts to a challenge called the "Curse of Dimensionality," where the input vector is large compared to the number of observations available, leading to increased noise and model overfitting. Models trained on so many variables will have high complexity and require an exponential number of parameters or weights (dependent on the chosen model architecture). Finally, beyond data quality concerns, there is no inherent knowledge of the chemo-physical phenomena or constraints of individual component amounts. Resulting models are thus fundamentally limited – unable to generalize and be expanded to new data beyond the limits of the range of compositions provided as input.

5.1.2 Ingredient and Formulation Descriptors to Replace Composition Data

An alternative to composition-only modeling inputs is to use ingredient- or formulation-based descriptors that have been measured, calculated, or collected. This method intends to incorporate chemical and physical insights through a smaller number of descriptors from existing intermediate property data or models of relevant underlying phenomena.

> Domain knowledge is thus embedded into the descriptors and models, producing a more interpretable model that can simultaneously be simpler and more accurate.

1 Pure data augmentation and other data-centric AI techniques are beyond the scope of this text.

Modeling properties from descriptors has produced structure-property relationships for drug design (Shahlaei 2013; Wilson and Famini 1991; Yang and Huang 2006) and predicting material properties (Le et al. 2012; Dai et al. 2020).

Identifying useful descriptors requires considerable domain expertise. Thus, there is some upfront cost to selecting descriptors representing the composition and related to the modeled property. Once such descriptors are chosen, there is a second stage of comparing the benefit the descriptor may bring to the effort required to produce it.

Some descriptors are readily available through simple models from chemistry or physics. In contrast, others require tremendous effort to produce additional data at the benchtop, set up and solve analytical solutions, or scrape literature, for example.

It is beneficial to consult experts before data production to save time and resources.

After all the effort to identify and source novel descriptors replacing composition, the new models may be equivalent to or worse than compositional models. Some principal reasons for this occurring are as follows:

i. The chosen descriptors were irrelevant to the modeled property. This will reveal itself through certain feature selection processes, where the irrelevant descriptor will be dropped during the selection process. The best resolution for this is to spend more time with domain experts to learn from their insights and intuitions about features affecting modeled properties. One caveat is that some features may themselves be linearly uncorrelated to the performance property being modeled, but they participate in an interaction alongside another significant variable and thus their removal from the model can harm its performance. Model architecture selection will define whether this is a concern for the modeler.
ii. Descriptors fail to capture some or all underlying phenomena. This goes hand in hand with overrepresenting certain phenomena while minimizing others. When there are many descriptors relating to one underlying mechanism (while still sufficiently orthogonal) and few representing another, the many will overshadow the few and have a greater influence on model predictions. This is generally resolved by identifying additional descriptors for the few, which can be difficult or impossible, or balancing the number of descriptors for the many by their importance to predictions. This practice can induce collinearities between components and descriptors which must be resolved by feature selection methods to avoid increasing model complexity and error (Figure 5.1).
iii. There remain gaps in the descriptors. Finding a model or property capturing all the influences present for multicomponent formulations is usually difficult and, more frequently, impossible. The previous point describes the case where adding descriptors can account for phenomena; however, when no such descriptor exists, the unaccounted-for components may be input as a stand-in for the phenomena. This gap-filling approach is intended to incorporate phenomena specific to those components while no other data are available (Figure 5.2).

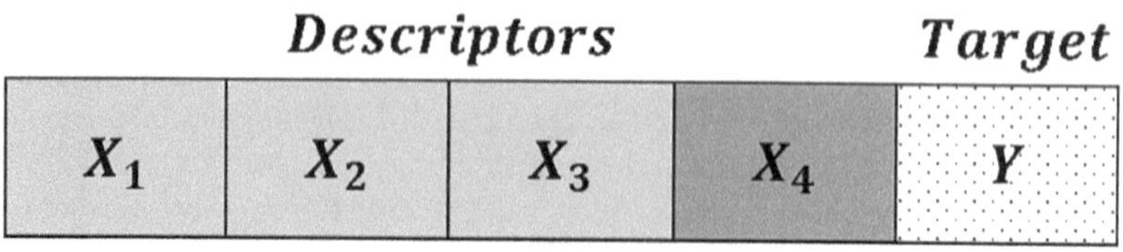

FIGURE 5.1 Overrepresentation of descriptors relating to one kind of component type and underrepresenting another.

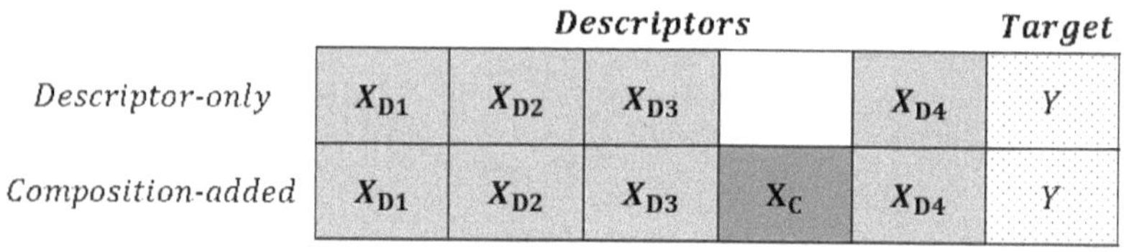

FIGURE 5.2 Inclusion of composition to fill gaps in descriptors can produce a more comprehensive model.

Keep these points in mind while reading each of the following methods and consider what gap the additional descriptors are resolving and if the addition may itself produce new gaps.

5.2 SUPPLEMENTS AND ALTERNATIVES TO COMPOSITIONAL DATA

Each of the following methods for gathering descriptors is best performed with the supervision and advice of subject matter experts relevant to the materials *and* the methods themselves; these persons can inform descriptor selection and accelerate the process of producing data. Additionally, it is important to carefully consider each descriptor's role in the data and whether there is already some other feature that may perform the same function or be closely related.

5.2.1 Direct Property Measurement

The first technique for expanding the data is the simplest – gather more information by measuring additional properties of each formulation. Here, a measured property is a feature of a formulated material or precursor (i.e., ingredient or intermediate material) that can be evaluated or observed in a laboratory, pilot plant, or manufacturing site by empirical or analytical techniques. This differs from properties evaluated by theory or simulation (discussed in the following sections of this chapter) and properties not directly describing the finished product itself (i.e., process conditions and metadata). These measured properties are generally a trusted method for expanding the information within a dataset as they do not depend upon assumptions or models to evaluate.

Deciding which experimental method to employ depends on the space and time scales of the property of interest. Recent material characterization and property measurement advances have increased accuracy and accelerated data production. For example, using a scanning tunneling electron microscope to observe polymerization dynamics (Zhan et al. 2022) brings greater detail to atom-scale interactions than was ever possible. This high-resolution examination comes with austere equipment costs and increased skill requirements and will produce data that is generally limited to atomic scale material properties; however, this was perfect for the two-dimensional polymers being studied. This technique may be grossly inadequate for other materials and properties examined, where more macroscopic methodologies can be required.

If you plan to undertake new data collection, working with the domain experts on your team to identify the most impactful measurements is critical.

The following chapter will further discuss advanced methods for generating descriptors from characterization methods with complex data outputs such as microscopy.

Beyond identifying the best equipment and techniques for collecting time- and space-scale appropriate data, looking for and aggregating data generated during routine procedures, where additional information is usually abundant in industry, can be beneficial. However, from the perspective of large-scale analysis, they may contain noise due to the variable and sometimes unknown context of their collection, for example, test method metadata, as discussed in the previous chapters. Nevertheless, they can still be valuable for some learning tasks depending on quality.

When combining data from multiple labs, designed experiments can reduce the noise and complexity associated with the data, improving the accuracy of the ML models and the insights they generate.

A quite accessible statistical method to improve the lab data quality is using Round-Robin protocols. Those protocols are designed to reduce the variance between laboratories by standardizing each testing procedure, measuring tools, and other conditions. As an illustrative example, a couple of international benchmarks for measuring the in-plane permeability of textile were published successively in 2011 (Arbter et al.) and 2014 (Vernet et al.). The Round-Robin method helped to reduce the variability from over one magnitude in the first benchmark down to about 25% in the second benchmark. However, variance will always exist between laboratories due to differences in the raw materials used, the accuracy of the measuring tools, and other factors; to fully understand the implications of these sources of variation on your modeling, you may need to engage a statistician in a full measurement systems analysis, which is outside the scope of this book. In the case that you cannot collect new data to standardize measurements between labs, you may still account for the different sources of data in the model, to be discussed in Chapter 7.

Efforts to integrate data from multiple labs can be further complicated by measurements from different instrument vendors using different data output formats. This

TABLE 5.1
Pros and Cons of Measured Property Descriptors for Ingredients and Formulations

Direct Property Measurement	
Directly measure or aggregate property descriptors of ingredients or formulated product	
Advantages	**Disadvantages**
• High trust and accuracy • Can be collected during normal routines	• Time and resource costs • Requires equipment and expertise • Limited precision

is one of the challenges in automating and standardizing digitalization from multiple machines and instruments to a common set of descriptor variables. Standardization of digital scientific data is the goal of the Allotrope Foundation (www.allotrope.org), an international consortium of scientific research-intensive industries. They aim to develop data models for analytical data, consisting of a standard data format, class libraries for interfacing with applications, and semantic capabilities to support standardized, structured metadata. Such initiatives facilitate the development of richer datasets where data, methods, and hardware components are seamlessly integrated even among disparate platforms.

Recently, high-throughput automation incorporating autonomous robotics and DoE- or AI-assisted experimental design capabilities have accelerated the production of experimental data (Pyzer-Knapp et al. 2022; Michael et al. 2008). From rudimentary automated screening capabilities with large volume assays for drug screening and development (Michael et al. 2008) to a high-tech combination of AI and robotics for autonomous, "smart" exploration of a design space (Pyzer-Knapp et al. 2020, 2022). As these technologies mature and data becomes increasingly available for training such capabilities, we will see an increasing role and dependence on such capabilities to accelerate experimental data production and analysis.

Table 5.1 summarizes the advantages and disadvantages of using measured property descriptors for formulation models.

5.2.2 Physical Process Descriptors

> Alongside incorporating physics, the addition of processing information, referred to as "process conditions" in Chapter 4, can tremendously benefit modeling.

Process descriptors refer to the whole experimental context (processing hardware equipment, operating procedure, process conditions, etc.) used in making a product with differentiated structure and properties from the starting ingredients. For example, in thermoplastic compounding, the process descriptors would refer to the critical parameters impacting the output, such as the melt temperature profile, screw profile, screw size, mixing time, and shear energy. External conditions such as ambient pressure, temperature, and relative humidity can considerably influence resultant material properties and thus are generally essential process conditions to collect. ML techniques have been used to optimize process conditions relative to process output and other at-line properties. A typical example of ML used to optimize a process is from Takada et al. (2021), who optimized a twin-screw extruder. Another example is the study of Jiang et al. (2023) who started from literature mining on "hot melt extrusion" and "amorphous solid dispersion" to build their dataset and applied multiple ML models to predict a drug's amorphization and chemical stability during the hot melt extrusion process.

Because the combinatorial size of processing conditions can be unreasonable to resolve experimentally, using a process model can lead to streamlined experimentation only to validate optimized conditions. Models can be computational simulations (an approach discussion in Section 5.4.2) or empirical. With Industry 4.0 and the digital manufacturing revolution, more process equipment can collect real-time data, including process settings and various at- and on-line sensor data. When limited to a single equipment, building its digital twin is a relatively simple task, and this exercise can highlight which descriptors may be meaningful to the formulation prediction problem.

However, when it comes to combining multiple pieces of equipment, even for the same conceptual process, the collection and inclusion of process descriptors face many similar challenges to experimentally derived ingredient descriptors discussed earlier. For example, many properties that sound alike between the process equipment cannot be taken as the same, and thus they cannot be aggregated as a single descriptor within the dataset. Continuing with the example of a twin-screw extruder, it may not be meaningful to use a single variable such as the Power Input across several extruders, as many technical details of each setup could impact the values (not of the same brand or model, extruder age, etc.).

Instead, one may prefer to rely on uniform physics-based process descriptors across devices such that the dimensionality of the process descriptors is reduced. Those descriptors are calculated, often dimensionless, and have a physical or technical meaning comparable across process/equipment/machine. For example, for a mixing device, one could express the process input in terms of average shear rate or total shear force work rather than impeller type and rpm.

Not only are such variables more interpretable to subject-matter experts, but they are more comparable between scales and equipment design.

See Leonova et al. (2014), who used calculated process descriptors and process model output for predicting blown film performance properties from material and process input. For blown film processing, the most common operating condition is expressed not in terms of absolute raw values (die diameter, die gap, line speed) but in relative quantities such as blow-up ratio (ratio of final bubble width over die diameter), draw down (ratio of final over initial film velocity), and output per inch length (die length normalized mass throughput). Many processes already have such normalized descriptors known by domain experts, and their advantage extends beyond engineering practicality to enhanced data science models.

In some scenarios, it may be beneficial to account for the order and timeline of processing steps. Indeed, some material reactions and material structure development may be sensitive to the sequence of introduction of the ingredients. An example is building food recipes (Galanis and Papakostas 2022) or chemical processes where reactions are sensitive to component concentrations or produce intermediates or by-products. Another highly sensitive application is pharmaceutical formulations where formulators may use so-called ordered mixing, a complex sequence involving different processing methods, to achieve the desired material structure (Saharan et al. 2008).

The ordering and timing information may come from continuous sources (e.g., delay times of ingredient introduction and residence times of each process steps), discrete sources (e.g., alternating batches and exchangeable mixing blocks elements of twin screw extruder), or unstructured and fuzzy sources (e.g., recipes, instructions from operating procedures such as "wait until liquid become transparent before introducing next component).

Those ordering and timing information must be encoded such that a machine learning method interprets them. Because it comes down to using the output of what has occurred before, machine learning methods with some "memory" property are preferred such as long short-term memory (LSTM) and transformer-based large language models (Hochreiter and Schmidhuber 1997).

Table 5.2 summarizes the advantages and disadvantages of using process descriptors for formulation models.

TABLE 5.2
Pros and Cons of Process Descriptors

Process Descriptors	
Incorporate descriptors of processing conditions (environmental, equipment, etc.) and operating procedures	
Advantages	**Disadvantages**
• Improve model generalizability • Additional optimization of process conditions • Readily supplements other data	• Requires additional measurements • Currently lacking in many datasets • Difficult to represent for some models (procedures, mainly)

5.2.3 Fundamental and Mechanistic Descriptors

Another method uses existing models describing chemical and physical material property phenomena. These fundamental and empirical mechanistic models relate ingredient, formulation, and/or process descriptors to material properties. This process is beneficial, as it is based on observations and existing physical laws which have a low financial burden. These are typically rapid methods, particularly where well-developed equations only require input variables to compute the properties.

Cheminformatic descriptors are among the most frequently used physics-based descriptors. These are calculated features describing atomic and molecular properties, such as molecular weight, polar surface area, and more, each calculated according to fundamental chemistry equations.

Cheminformatic descriptors can be used to describe "zero-dimensional" material properties, supplementing more macroscopic descriptors like process or experimental. They are most often applied to generate ingredient descriptors to enhance the feature space. The reader is directed toward several introductory resources on cheminformatics (Wishart 2016; LibreTexts 2024; Bender and Brown 2018), which also include references to public databases of calculated values for many molecules such as PubChem (Kim et al. 2023; PubChem 2024).

Mechanistic models can be applied to formulated systems in addition to individual molecular constituents. These models are, of course, highly specific to the product domain and application, but they can provide further insights into the behavior of the ingredients *in situ*, which can lead to improved model accuracy. Examples of such models include kinetics of polyurethane foam formation (Rao et al. 2017), formulation of pharmaceuticals (Mehta et al. 2019), processing of pharmaceutical tablets (Wu et al. 2005), and prediction of mechanical properties of concrete (Pal and Ramaswamy 2023).

Mechanistic descriptors, while potentially highly effective, can have considerable downsides. First, developing one can be costly if no model exists to calculate a desired descriptor. Novel models can require extensive data production and analysis or significant work to produce an analytical form, which may be insufficient anyway and justify using a purely descriptors-data-driven approach.

Mechanistic descriptors are only as good as the models used to calculate them.

Another shortcoming is that mechanistic descriptors are only as good as the models used to calculate them. For example, the commonly known ideal gas law is highly accurate under some conditions and for certain gases but becomes significantly inaccurate when those constraints are unmet. Some inaccuracies can be accounted for in the downstream machine learning, such as the case when the mechanistic model

TABLE 5.3
Pros and Cons of Mechanistic/Physics-Derived Descriptors

Physics-Based Descriptors	
Calculate descriptors from composition with equations of chemistry and physics	
Advantages	**Disadvantages**
• Rapid and inexpensive for existing models	• Costly if there is no existing model
• Based on first-principles or empirical data	• Only as accurate as the foundation of the model
• Controllable accuracy and sophistication	

is uniformly biased but important trends are preserved; more discussion on this approach is provided in Chapter 7.

Another example is group-contribution methods, where thermodynamic properties are accurately calculated for pure or simple materials but can neglect specific molecular interactions, requiring additional group-interaction parameters to resolve. Both examples provided earlier can be superb for some applications but will be inaccurate elsewhere, making their usage less effective. Finally, physics-based models necessitate a balance of accuracy and sophistication.

The time and expertise to develop a highly accurate model can outweigh the benefit received from the additional descriptors.

Table 5.3 summarizes the advantages and disadvantages of using mechanistic descriptors for formulation models.

5.2.4 Computational Simulations

The previously described sources of advanced descriptors depend upon analytical or empirical methods; these come with the benefit of capturing information directly and with high trust, leading to improved model accuracy and interpretability. However, they are simultaneously susceptible to the limitations and experimental error of the equipment or models used. Further, these can come with a high expense, among other restrictions, such as safety, equipment availability, and inability to study short-time or transient phenomena. Thus, producing large volumes of data can be challenging, given these limitations.

Depending on the application, finding material property databases, not to mention consistency in composition, experimental methods, and metadata within such sources, can be difficult or impossible. Recently, computational simulations have become a strong contender for producing such databases as they can rapidly generate large volumes of consistent, high-fidelity data.

Additionally, simulated data may be used to supplement descriptors or be used as the property to be modeled, depending on the application. Improved algorithms, decreased hardware costs, and novel techniques such as parallelization and AI-assisted simulations continue accelerating simulation times and expanding the quality and range of simulated properties. There are currently many open-source simulation suites for designing materials, performing simulations, and analyzing the outputs to calculate and produce property values.

Broadly, simulations utilize computers to numerically solve equations relevant to real-world physical and process models. Simulations are distinguishable from fundamental or mechanistic models because they require solutions by numerical algorithms, instead of being a calculable function. Simulation's fidelity, speed, and complexity depend upon the scale and necessary accuracy of the problem. For example, *ab initio* simulation tools resolve atom-scale features for small systems $(+-nm)$ but cannot reach long times or larger scales due to extensive computational costs. Finite element methods, utilizing coarse-grained models, can be used to examine significantly larger systems $(\geq \mu m)$, though they have decreased precision. Intermediate techniques, including Monte Carlo methods, molecular dynamics, and continuum modeling, face the same challenges as either, requiring a balance of scale, simulation time, and exactness As with all sources of descriptor data, cost versus accuracy is balanced. A high-accuracy simulation of a large system can require intractable time to perform, and the need might be equally satisfied by a coarser model with decreased accuracy but faster time.

One of the greatest advantages of simulations is the full control of the design space. This ability enables automated exploration, high-throughput simulation, and access to formulations that are challenging to produce under laboratory conditions.

With design parameters outlined, simulation tools can be combined with guided exploration algorithms, such as Bayesian optimization, to discover materials with improved or tailored properties.

If the modeled property is a simulated value, then the simulation-optimization scheme can be directly applied for directed material design and analysis can reveal descriptor-property relationships. Alternatively, if the simulated values are supplementing the descriptor set, they can be used as a control to optimally modify the target property.

Several recent developments have aimed to diminish computational simulations' scale and time limitations. Deep neural networks (DNN) have replaced the potential energy and forcefield models to accelerate molecular dynamics simulation, avoiding numerical solvers at each timestep (Wang et al. 2018). These DNN models were trained from *ab initio* simulations to represent the many-body interaction forces. They can be seamlessly integrated into MD software, such as LAMMPS, to reduce simulation times while maintaining high accuracy. Another approach is to dynamically adjust the fidelity of the simulation, improving granularity on points of interest while reducing it for areas with less desired physics. One example is adaptive time

TABLE 5.4
Pros and Cons of Computational/Simulated Descriptors

Computational Simulations	
Apply numerical solutions to model materials and calculate descriptors	
Advantages	**Disadvantages**
• Rapidly produce large volumes of data • Explore hard-to-reach design spaces • Generally inexpensive • Variable precision and speed • Many open-source software	• Can have high upfront hardware and software costs • Expertise needed for simulation setup and analysis

stepping, where an AI is trained to recognize redundant time steps and adjust the timestep sizes, and step jumps accordingly (Han et al. 2021). Combining simulation methods across scales is another such improvement, where a coarser model is applied to jump quickly forward in simulation time. A higher fidelity model is mapped upon it to correct the simulation and explore more high-speed or small-area effects (Olsen 2020). Each of these empowers the acceleration and utility of computational simulations for accurately calculating descriptors for modeling.

Table 5.4 summarizes the advantages and disadvantages of using computational descriptors for formulation models.

5.2.5 Data Mining

Each of the prior methods requires knowledge and technologies to generate original data. Instead of performing costly measurements or simulations, it is possible to extract valuable information from existing sources, such as academic journals, safety data sheets, and marketing information, among other sources, using data mining tools.

> Data mining employs algorithms to gather data from unstructured text, sound, or video sources and converts them so they may be digitally stored according to a structured schema.

These techniques can gather molecular structures, formulation data, processing conditions, measured properties, and more in the materials space. This data can then be used for guiding experimentalists, input for ML, and other such uses. Alongside data mining techniques are analytic tools, which summarize and describe trends in the collected data. Such techniques will not be discussed in this text.

Data mining methods apply algorithms to perform data parsing, preprocessing, and pattern recognition to split the raw source into packeted, utilizable forms for further analyses.

Currently, text mining is the leading data mining technique in the material design realm, primarily because much of the useful data is stored in text-based journal articles, books, safety sheets, or web pages.

Text mining parses a text-based document, such as the HTML code behind a webpage, a journal exported as a PDF, etc. It applies preprocessing and pattern recognition algorithms to collect desired data, including key properties, tables, and figures. This process is generally semi-automated, requiring some preparations to know what type of document is being read, where to collect information (e.g., find and record information between all occurrences of <table> and </table> within a webpage's HTML), and if additional processing (converting type, capitalization, and so on) is required before it can be stored.

With all the figures collected from a document, another data mining tool can be applied to convert graphical data into quantitative forms, such as tables. Digitizing tools can extract data from curves, axes, and labels of plots and convert those into computer-readable forms.

Kononova et al. (2021) prepared an excellent review of the text-mining process and how it has been employed in the materials space, including many case studies and links to large databases produced by the text-mining of scientific literature. Additionally, they reference several software specifically designed for extracting data from chemical texts.

Recently, the combination of data mining techniques has been explored as part of the "lab of the future" concept. AI assistants, computer vision, text and speech recognition, object tracking, etc., are simultaneously applied to record and appropriately identify and store information produced while performing a task. For example, preparing a benchtop sample can be enabled by a technology that uses cameras and microphones to automatically recognize which reagents are mixed and in what quantities, record measurements displayed on a scale, and hear verbal commentary. All this data can automatically be converted into a lab report, including the materials, procedures in the correct order, and measurement figures. This proposed system offers opportunities for rapid, low-cost descriptor production and extraction, including measured properties, process conditions, and metadata.

Supplementing AI in the laboratory, advanced databases and search tools are becoming ever more powerful and functioning to accelerate literature searching and collecting data from various sources. At the time of this writing, ChatGPT-3.5, a Large Language Model (LLM) produced by OpenAI (ChatGPT-3.5 2023), has become the figurehead in this area. The software takes in user-supplied prompts, such as "Please tell me about SMILES[2] notation" or "tldr:" followed by a link to a journal article or public web page, and it replies with a natural response summarizing the background, technique, and uses of SMILES and an abridged description of the linked media, respectively.

2 Simplified molecular-input line-entry system (Weininger 1988).

TABLE 5.5
Pros and Cons of Mining Descriptor Data

Data Mining	
Extract descriptors from existing sources (literature, web, images . . .)	
Advantages	**Disadvantages**
• Zero material or equipment costs • Benefit from external sources • Increasing amount of software for data mining in chemical space	• Source accumulation and data extraction process can require considerable upfront efforts • Chemical data is particularly difficult to extract (e.g., molecular structures) • Inconsistent methods for experiments or simulations can cause inhomogeneity

These supportive tools accelerate material design by reducing the time for literature review and data collection and have the potential to assist experimental design as well. However, concerns remain about the accuracy and truth of results, particularly for AI models such as Natural Language Processing (NLP) tools, which are a (highly effective) probabilistic model, producing text by statistics, not language rules or meaning.

Countless articles currently highlight failed examples of the tool to respond realistically, especially when references or specific details are requested. With this in mind, the software can still accelerate research efforts but must be used with caution.

Table 5.5 summarizes the advantages and disadvantages of using property descriptors mined from literature for formulation models.

5.2.6 Representation Models

Finally, the most abstract descriptors are representation models, which describe chemical information as encoded vectors, one-dimensional strings of characters, graphs, etc., so they may be easily communicated and, more importantly for modeling purposes, read by a computer program.

The likely most recognizable example of representative models in chemistry is the depiction of a molecular structure by a graph, showing (though sometimes not) atoms connected by lines symbolizing bonds with triangles and hashed lines indicating stereochemistry and free-floating dots for radicals.

This simplifies the concept of a three-dimensional molecule to a two-dimensional image while still capturing information essential to designing reaction pathways and understanding other cheminformatics. That is a significant amount of information stored within a small diagram. Other representation techniques take bolder steps to convert structures and process information into computer-readable forms. For example, text-based notations such as SMILES and BigSMILES can quickly communicate molecular and polymeric structures to a trained reader and computer programmer (Lin et al. 2019). The key benefit of representation modeling is conveying information in new ways with greater interpretability and utility, particularly for modeling.

To use a representation as a descriptor frequently requires modifying the approach to data preparation and ML while maintaining the overall intent of the models. Continuing with the earlier example, the drawing of a molecule (e.g., as a graph or an image) is generally incompatible with the one-dimensional input vector typical of many ML models. One resolution is to convert the molecular structure into the required schema by fingerprinting, where molecular features are reduced to a vector by a series of hashing and pattern recognition algorithms. Wigh et al. provide an excellent review of molecular representation methods (2022).

Alternatively, why not change the ML model to handle the molecular structure as-is? The Atomistic Line Graph Neural Network (ALIGNN) does just this by training a graph neural network on graph representations of a molecule's atoms and bonds, represented by nodes and edges, respectively (Choudhary and DeCost 2021).

A side-effect of using alternative representations is the introduction of new analysis and modeling capabilities. For example, converting molecular structures to text-based SMILES enables the application of natural language processing for sentiment analysis (Jastrzębski et al. 2016), where the sentiment can be drug activity or another property. Another example is variational autoencoding for generative molecular design (Alperstein et al. 2019). Conversion to data formats with much larger training data sets than materials projects can allow for use of foundational models borrowed from other domains. This will be further discussed in Chapter 7.

Representative models can also have shortcomings, like physics-based models and simulations, where the representation and model quality is only as high as the source data.

> A problem plaguing the automation of materials discovery is the machine generation of molecular structures through representations, such as SMILES or graphs, which are grammatically correct per the representation scheme but are not chemically or physically valid.

This can be a fault of validation methods for the representation itself or a side-effect of the ML method used. Kusner et al. (2017) demonstrated the failures of generative models to produce realistic SMILES strings, primarily missing the grammatical rules required to be considered a valid SMILES. By incorporating the grammar

TABLE 5.6
Pros and Cons of Alternative Representations as Descriptors

Representations	
Convert composition into representative forms (e.g., SMILES) to be input for machine learning	
Advantages	**Disadvantages**
• Some are interpretable for both humans and computers • Capture large volume of information in simple forms • Enables alternative modeling methods (e.g., graph neural networks)	• Representations can follow translation rules but be invalid • Abstraction may decrease interpretability

between the probabilistic model and input/output molecular structures, it was possible to ensure outputs were grammatically valid and enabled the model to focus solely on learning relationships within the data.

Table 5.6 summarizes the advantages and disadvantages of using molecular or material representations as descriptors for formulation models.

5.3 CONCLUSIONS

Descriptor engineering is an important stage in preparing data for ML modeling of material properties. In this chapter, several methods for descriptor engineering were presented, ranging from the immediate collection of data by additional experiments to the abstraction of molecular structure using representations. Replacing compositions with descriptors from these sources aims to incorporate chemistry, physics, and process knowledge into ML models, resulting in improved model performance, interpretability, and user trust. Each method has benefits and drawbacks, and it is important to consider the purpose of using potential descriptors for modeling before committing to the calculation or collection efforts.

REFERENCES

Alperstein, Zaccary, Artem Cherkasov, and Jason Tyler Rolfe. "All SMILES variational autoencoder." *Second Workshop on Machine Learning and the Physical Sciences*, NeurIPS (2019).

Arbter, R., et al. "Experimental determination of the permeability of textiles: a benchmark exercise." *Composites Part A: Applied Science and Manufacturing* 42, no. 9 (2011): 1157–1168. https://doi.org/10.1016/j.compositesa.2011.04.021.

Bender, Andrea, and Nathan Brown. *ChemMedChem* 13, no. 6 (2018, March): 467. https://doi.org/10.1002/cmdc.201800123.

ChatGPT-3.5. OpenAI (2023).

Choudhary, Kamal, and Brian DeCost. "Atomistic line graph neural network for improved materials property predictions." *NPJ Computational Materials* 7, no. 185 (2021). https://doi.org/10.1038/s41524-021-00650-1.

Dai, Dongbo, Qing Liu, Rui Hu, Xiao Wei, Guangtai Ding, Baoyu Xu, Tao Xu, Jincang Zhang, Yan Xu, and Huiran Zhang. "Method construction of structure-property relationships from data by machine learning assisted mining for materials design applications." *Materials & Design* 196 (2020): 109194. https://doi.org/10.1016/j.matdes.2020.109194.

Galanis, N. I., and G. A. Papakostas. "An update on cooking recipe generation with machine learning and natural language processing." *IEEE World Conference on Applied Intelligence and Computing (AIC)*, Sonbhadra, India (2022): 739–744. https://doi.org/10.1109/AIC55036.2022.9848929.

Han, Changnian, Peng Zhang, Danny Bluestein, Guojing Cong, and Yuefan Deng. "Artificial intelligence for accelerating time integrations in multiscale modeling." *Journal of Computational Physics* 427 (2021): 110053. https://doi.org/10.1016/j.jcp.2020.110053.

Himanen, Lauri, Amber Geurts, Adam Stuart Foster, and Patrick Rinke. "Data-driven materials science: status, challenges, and perspectives." *Advanced Science* 6, no. 21 (2019). https://doi.org/10.1002/advs.201900808.

Hochreiter, Sepp, and Jürgen Schmidhuber. "Long short-term memory." *Neural Computation* 9, no. 8 (1997). https://doi.org/10.1162/neco.1997.9.8.1735.

Jastrzębski, Stanisław, Damian Leśniak, and Wojciech Marian Czarnecki. "Learning to SMILE(S)." *International Conference on Learning Representations* (2016). ArXiv. http://arxiv.org/abs/1602.06289.

Jiang, Junhuang, Anqi Lu, Xiangyu Ma, Defang Ouyang, and Robert O. Williams III. "The applications of machine learning to predict the forming of chemically stable amorphous solid dispersions prepared by hot-melt extrusion." *International Journal of Pharmaceutics: X* 5 (2023): 100164. https://doi.org/10.1016/j.ijpx.2023.100164.

Kim, S., et al. "PubChem 2023 update." *Nucleic Acids Research* 51, no. D1 (2023): D1373–D1380. https://doi.org/10.1093/nar/gkac956.

Kononova, Olga, Tanjin He, Haoyan Huo, Amalie Trewartha, Elsa A. Olivetti, and Gerbrand Ceder. "Opportunities and challenges of text mining in materials research." *iScience* 24, no. 3 (2021). https://doi.org/10.1016/j.isci.2021.102155.

Kusner, Matt J., Brooks Paige, and José Miguel Hernández-Lobato. 2017. "Grammar variational autoencoder." *Proceedings* of *the 34th International Conference on Machine Learning*, Sydney (2017). https://doi.org/10.5555/3305381.3305582.

Le, Tu, V. Chandana Epa, Frank R. Burden, and David A. Winkler. "Quantitative structure-property relationship modeling of diverse materials properties." *Chemical Reviews* 112 (2012): 2889–2919. https://doi.org/10.1021/cr200066h.

Leonova, Tatyana, Stef van Eijndhoven, and C. F. Jaap den Doelder. "Physics-based film blowing modeling to enhance empirical film property predictions from polymer and processing parameters." *Antec*, Las Vegas (2014).

LibreTexts. "Cheminformatics." https://chem.libretexts.org/Courses/Intercollegiate_Courses/Cheminformatics (accessed January 18, 2024).

Lin, Tzyy-Shyang, Connor W. Coley, Hidenobu Mochigase, Haley K. Beech, Wencong Wang, Zi Wang, Eliot Woods, Stephen L. Craig, Jeremiah A. Johnson, Julia A. Kalow, Klavs F. Jensen, and Bradley D. Olsen. "BigSMILES: a structurally-based line notation for describing macromolecules." *ACS Central Sciences* 5, no. 9 (2019): 1523–1531. https://doi.org/10.1021/acscentsci.9b00476.

Marschik, Christian, Wolfgang Roland, Bernhard Löw-Baselli, and Georg Steinbichler. "Application of hybrid modeling in polymer processing." *Antec 2020 Papers* (2020).

Mehta, Chetan Hasmukh, Reema Narayan, and Usha Y. Nayak. "Computational modeling for formulation design." *Drug Discovery Today* 24, no. 3 (201 9, March): 781–788. https://doi.org/10.1016/j.drudis.2018.11.018.

Michael, Sam, Douglas Auld, Carleen Klumpp, Ajit Jadhav, Wei Zheng, Natasha Thorne, Christopher P. Austin, James Inglese, and Anton Simeonov. "A robotic platform for quantitative high-throughput screening." *Assay and Drug Development Technologies* 6, no. 5 (2008): 637–657. https://doi.org/10.1089/adt.2008.150.

Olsen, Jake Dylan. *The atomistic reconstruction of coarse-grained polymeric systems via machine learning techniques*. University of Oregon Scholars' Bank (2020).

Pal, Biswajit, and Ananth Ramaswamy. "A multi-physics-based approach to predict mechanical behavior of concrete element in a multi-scale framework." *Mechanics of Materials* 176 (2023): 104509. https://doi.org/10.1016/j.mechmat.2022.104509.

PubChem. "PubChem." https://pubchem.ncbi.nlm.nih.gov/ (accessed January 18, 2024).

Pyzer-Knapp, Edward O., Jed W. Pitera, Peter W. J. Staar, Seiji Takeda, Teodoro Laino, Daniel P. Sanders, James Sexton, John R. Smith, and Alessandro Curioni. "A mobile robotic chemist." *Nature* 583 (2020): 237–241.

Pyzer-Knapp, Edward O., Jed W. Pitera, Peter W. J. Staar, Seiji Takeda, Teodoro Laino, Daniel P. Sanders, James Sexton, John R. Smith, and Alessandro Curioni. "Accelerating materials discovery using artificial intelligence, high performance computing and robotics." *NPJ Computational Materials* 8, no. 1 (2022): 84.

Rao, Rekha, Lisa A. Mondy, Kevin N. Long, Mathew C. Celina, Nicholas Wyatt, Christine C. Roberts, Melissa M. Soehnel, and Victor E. Brunini. "The kinetics of polyurethane structural foam formation: foaming and polymerization." *AIChE Journal* 63 (2017): 2945–2957. https://doi.org/10.1002/aic.15680

Rodrigues Jr., Jose F., Larisa Florea, Maria C. F. de Oliveira, Dermot Diamond, and Osvaldo N. Oliveira Jr. "Big data and machine learning for materials science." *Discover Materials* 1, no. 12 (2021).

Saharan, Vikas Anand, Vipin Kukkar, Mahesh Kataria, Vandana Kharb, and Pratim Kumar Choudhury. "Ordered mixing: mechanism, process and applications in pharmaceutical formulations." *Asian Journal of Pharmaceutical Sciences* 3, no. 6 (2008): 240–259.

Shahlaei, Mohsen. 2013. "Descriptor selection methods in quantitative structure–activity relationship studies: a review study." *Chemical Reviews* 113, no. 10 (2013): 8093–8103. https://doi.org/10.1021/cr3004339.

Takada, Shingo, Toru Suzuki, Yoshihiro Takebayashi, Takumi Ono, and Satoshi Yoda. "Machine learning assisted optimization of blending process of polyphenylene sulfide with elastomer using high speed twin screw extruder." *Scientific Report s* 11 (2021): 24079. https://doi.org/10.1038/s41598-021-03513-3.

Vasudevan, Rama, Ghanshyam Pilania, and Prasanna V. Balachandran. "Machine learning for materials design and discovery." *Journal of Applied Physics* 129, no. 7 (2021). https://doi.org/10.1063/5.0043300.

Vasudevan, Rama K., Kamal Choudhary, Apurva Mehta, Ryan Smith, Gilad Kusne, Francesca Tavazza, Lukas Vlcek, Maxim Ziatdinov, Sergei V. Kalinin, and Jason Hattrick-Simpers. "Materials science in the artificial intelligence age: high-throughput library generation, machine learning, and a pathway from correlations to the underpinning physics." *MRS Communications* 9, no. 3 (2019): 821–838.

Vernet, N., et al. "Experimental determination of the permeability of engineering textiles: Benchmark II." *Composites Part A: Applied Science and Manufacturing* 61 (2014): 172–184. https://doi.org/10.1016/j.compositesa.2014.02.010.

Wang, Han, Linfeng Zhang, Jiequn Han, and Ee Weinan. "DeePMD-kit: a deep learning package for many-body potential energy representation and molecular dynamics." *Computer Physics Communications* 228 (2018): 178–184.

Weininger, David. "SMILES, a chemical language and information system. 1. Introduction to methodology and encoding rules." *Journal of Chemical Information and Computer Sciences* 28, no. 1 (1988): 31–36.

Wigh, Daniel S., Jonathan M. Goodman, and Alexei A. Lapkin. "A review of molecular representation in the age of machine learning." *WIREs Computational Molecular Science* 12, no. 5 (2022): e1603.

Wilson, Leland Y., and George R. Famini. "Using theoretical descriptors in quantitative structure-activity relationships: some toxicological indices." *Journal of Medicinal Chemistry* 34 (1991): 1668–1674. https://doi.org/10.1021/jm00109a021.

Wishart, David S. "Introduction to cheminformatics." *Current Protocols in Bioinformatics* 53 (2016, Marc h): 14.1.1–14.1.21. https://doi.org/10.1002/0471250953.bi1401s53

Wu, C.-Y., O. M. Ruddy, A. C. Bentham, B. C. Hancock, S. M. Best, and J. A. Elliott. "Modelling the mechanical behaviour of pharmaceutical powders during compaction." *Powder Technology* 152, no. 1 –3 (2005): 107–117. https://doi.org/10.1016/j.powtec.2005.01.010.

Xu, Pengcheng, Xiaobo Ji, and Wencong Lu. "Small data machine learning in materials science." *NPJ Computational Materials* 9, no. 42 (2023). https://doi.org/10.1038/s41524-023-01000-z.

Yang, Guang-Fu, and Xiaoqin Huang. "Development of quantitative structure-activity relationships and its application in rational drug design." *Current Pharmaceutical Design* 12, no. 35 (2006): 4601–4611.

Zhan, Gaolei, Zhen-Feng Cai, Karol Strutyński, Lihua Yu, Niklas Herrmann, Marta Martínez-Abadía, Manuel Melle-Franco, Aurelio Mateo-Alonso, and Steven De Feyter. "Observing polymerization in 2D dynamic covalent polymers." *Nature* 603 (2022): 835–840.

6 Machine Learning for Analysis of Structural Characterization

Arthi Jayaraman and Shizhao Lu

6.1 INTRODUCTION

Formulation science and engineering in the chemical industry is directed toward achieving the target physical properties that make the resulting commercial product attractive to the end-user. The target physical properties of the formulations and the end-user experience are controlled by the chemical composition and, in some cases, the spatial arrangements of those components, that is, the structure or morphology within the formulation. Here are a few examples of how the structure of the formulation impacts the eventual application: a) the state of the pigments' dispersion in solution within paints determines the ease of coating the surface with the paint and the eventual color one sees from the painted surface (Weiss 1997; Tiarks et al. 2003; Clough et al. 2019; Larson et al. 2022); b) in addition to the chemistry, shape, and size distribution of the drugs, the morphology of the drug carrier/encapsulation around the drug dictates the drug release profile (Hughes 2005; Mitragotri and Lahann 2012; Webber and Langer 2017); c) the phase separated/mixed state of the blends along with the extent of order/disorder within phase separated domains, orientational alignment of polymers, and presence/absence of solvent around the polymers during or after processing dictate the mechanical, electrical, and thermal properties of the blend (Barlow and Paul 1981; Chen and White 1993; Crosby and Lee 2007; Sen et al. 2015). These examples show why formulation design has to consider the understanding of the structure at equilibrium as well as evolution of the structure with time, conditions (e.g., temperature, pressure, humidity), and processing techniques (e.g., shearing, drying) that take the structure out of equilibrium if your problem requires that non-equilibrium treatment for accuracy.

As discussed in Chapter 4, ingredient properties, including structural characterization data, are commonly missing from a product development dataset. Recalling that ingredient properties heavily influence the model structure and unlock the ability to consider new ingredients in future formulations, it is valuable to locate or obtain this information before proceeding with building a predictive model.

 DOI: 10.1201/9781003385974-8

In cases where structural descriptions of the ingredient materials are not sufficient to capture the observed behavior of the final formulated material, then capturing structural descriptions of the formulated product may help. Using analytical data to create descriptors or modeling features from formulation morphology can then enable you to elucidate structure-property relationships between ingredient structures and formulation morphology, which could be further modeled to relate morphology to final performance.

We can use the techniques presented in this chapter to generate increasingly informative feature representations of the structure present in your system to tackle even the most complex prediction problems.

Structural characterization of formulations is generally done through a combination of microscopy (e.g., scanning electron microscopy, transmission electron microscopy, and cryo-transmission electron microscopy), spectroscopy (e.g., UV-vis, Fourier transform infrared or FTIR, infrared spectroscopy or IR, Raman spectroscopy, nuclear magnetic resonance or NMR), and scattering techniques (e.g., small angle neutron scattering, small angle X-ray scattering, wide-angle X-ray scattering, dynamic light scattering). Each of these techniques has its advantages and limitations in terms of equipment costs, accessibility (presence in house vs. national laboratories/characterization facilities), information (e.g., types of structural features and their length scales (Figure 6.1)), and expertise required (e.g., researcher with

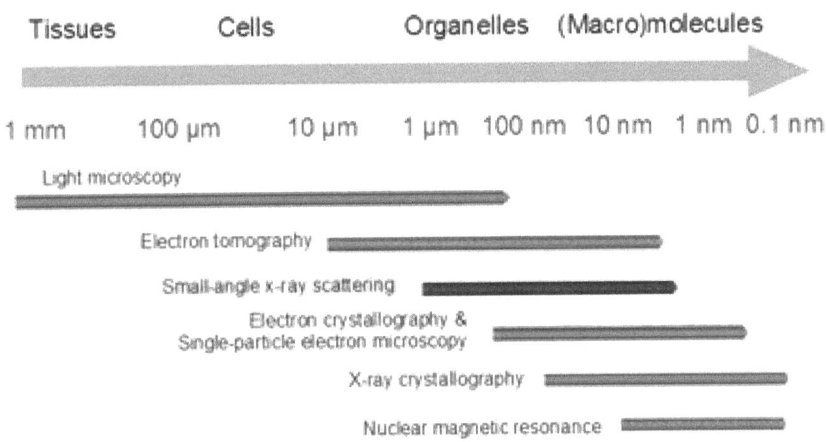

FIGURE 6.1 Length scales and characterization techniques for soft materials (natural and synthetic). This figure is adapted from https://en.wikipedia.org/wiki/File:Saxs_resolution4.png

minimal training or an expert technician trained specifically to operate the equipment). These techniques also differ in sample preparation, time, and effort needed for the collection of data and interpretation of data, as well as subjectivity and errors that are possible during data collection and interpretation. Several articles and reviews in the past decade (Kamerzell et al. 2011; Zölls et al. 2012; Letourneau and Volmer 2023; Sonje et al. 2023; Lyu et al. 2023) discuss key features of these commonly used structural characterization techniques for formulations. In this chapter, we share the many ways artificial intelligence (AI) [e.g., machine learning (ML) models, automation] can complement one or more of these structural characterization techniques and highlight the advantages, challenges, and potential pitfalls of combining AI with these characterization techniques.

AI can be used in many important ways for structural characterization of formulations – one is AI-driven automation of the hardware (e.g., measurement instruments, robotics, and control systems) that allows for simultaneous and high throughput characterization and the second is in use of ML models to automate and accelerate analyses and interpretation of the structural characterization data. AI-driven automation is necessary if the chemical industry wishes to move toward completely automated formulations design; however, that is not the focus of this chapter. Instead, this chapter is focused on ML-driven analyses and interpretation of structural characterization of formulations.

In this chapter, we describe the development, use, and value of ML methods for interpreting microscopy imaging and small-angle scattering data. To a smaller extent will be the discussion of ML models specifically to couple measurements of formulation structure from multiple techniques (e.g., scattering and microscopy). Within each of the following sections, we say when, why, and how one would want to use ML for interpretation and analysis of these two broad structural characterization techniques: microscopy imaging and small angle scattering.

6.2 MACHINE LEARNING METHODS FOR AUTOMATED ANALYSIS AND INTERPRETATION OF STRUCTURE FROM ELECTRON MICROSCOPY IMAGES

6.2.1 Types of Electron Microscopy Imaging Techniques

Material scientists have used different electron microscopy imaging techniques to characterize the morphology of the studied material. Scanning Electron Microscopy (SEM) uses reflected or back-scattered electrons to create an image of the studied material, which can provide information on the 3D shape and surface roughness of the studied material (Goodhew et al. 2000). SEM samples are relatively easy to prepare and do not have restrictions on sample thickness. The resolution of SEM can be as high as 0.5 nm. Transmission Electron Microscopy (TEM) uses electrons that go through the sample to image the studied material. TEM imaging provides information on the inner structure of the material such as bulk morphology, crystal structure, and defects. The resolution of TEM can be as high as 0.5 Å. While in TEM parallel electron beams are focused perpendicular to the sample plane, in Scanning Transmission Electron Microscopy (STEM), the beam is focused on a large angle and is converged onto a focal point. One can obtain the atomic arrangement, orientation, and crystalline structure of the material from STEM. The resolution of STEM is 0.5–2 Å. Cryo-TEM is a subclass of TEM that enables imaging (bio)material in a solvated environment by rapidly freezing the sample with liquid ethane or propane. Cryo-TEM reduces sample damage during preparation and from the electron beams and has gained attention as a valuable method for macromolecular structure determination.

Analysis of the results from these different electron microscopy techniques is usually a manual process with experts interpreting qualitative (e.g., shapes of morphologies) and quantitative (e.g., domain sizes or distribution of sizes) information of these images. Such interpretation is time-consuming and subjective.

Having ML serve as the tool for fast and objective analysis of these microscopy results is both desirable and necessary to bring in automation and reduce noise and errors for data-driven formulation design and characterization.

6.2.2 Unique Challenges of Image-Based Machine Learning for Electron Microscopy Images

As described in the previous chapter, before an ML model can be trained on any data, one must do feature engineering, which is the process of converting the raw data into features (i.e., any measurable input or output that will be predicted). This feature engineering can be done manually via programming guided by the subject matter expertise or by using ML approaches. Regardless of how it is done, feature engineering often requires domain expertise on the governing relationships of the input data and prediction target.

Deep learning is a subset of ML that does feature engineering automatically by simultaneously learning features/representations of the raw input data and perform prediction of the target data using the learned features as input.

Deep learning has shown immense potential in image learning, in general, with studies like the one by Krizhevsky et al. (Krizhevsky et al. 2017), which demonstrated excellent predictive capabilities with AlexNet (Krizhevsky et al. 2017) for the classes of crowd-source labeled objects from photographic images. Deep learning-based image recognition has also found application in autonomous driving with traffic photos (Grigorescu et al. 2020), medical diagnostics with medical scans (Shen et al. 2017), and material characterization with electron microscopy images (Ge et al. 2020; Treder et al. 2022). Unlike the photographic images of everyday objects, electron microscopy images in materials science have unique characteristics that require some special consideration when applying deep learning, as we discuss next.

6.2.3 Types of Image-Based Machine Learning Models

In this section, the reader will learn about two broad categories of image-based ML models – non-deep learning and deep learning. The size of the dataset will dictate which of these two approaches is best – if one has enough data (e.g., hundreds of thousands of data points) to use deep learning models or smaller data sets (e.g., hundreds of data points) that would make non-deep learning models a better choice.

Non-deep learning workflows typically have two parts: feature selection and classification. Feature selection can be done manually by domain experts or via non-manual selection by dimensionality reduction methods such as Principle Component Analysis (PCA) (Pearson 1901), t-distributed Stochastic Neighbor Embedding (t-SNE) (Van der Maaten and Hinton 2008), and Uniform Manifold Approximation and Projection (UMAP) (McInnes et al. 2018). For classification of the category that each image (turned into features) falls into, ML models such as linear classifier, non-linear classifier, k-Nearest Neighbor classifier (Wu et al. 2008), naïve Bayes classifier (Wu et al. 2008), support vector machine classifier (Guyon et al. 2002), decision tree classifier (Wu et al. 2008), and Gaussian process classifier (Williams and Rasmussen 1995), are often used. Ensemble methods such as max voting, weighted average, bagging, boosting, and stacking can be used to improve the classifier by combining multiple (same or different types) of classifiers together (Dietterich 2000). Random forest is an example of ensemble methods that combines multiple decision trees to reach a single result through max voting or averaging (Cutler et al. 2012).

Even though the typical dataset sizes for the previous non-deep learning methods can be small, there are a few drawbacks of non-deep learning methods for image learning. These include the need for prior domain knowledge to conduct feature engineering, lack of retention of spatial correlation of objects in the method in common dimensionality reduction methods, and inability to tackle tasks that require pixel-level prediction.

One commonly used deep learning approach is the Convolutional Neural Network (CNN), which is a type of artificial neural network utilized to learn, extract, and encode information from an image in an automated fashion (LeCun and Bengio 1995). The major advantage of CNN over the two-step non-deep learning methods is that CNN learns an optimal feature set to use for classification through training smaller processing units called filters. CNN typically consists of a hierarchical structure of convolutional layers sandwiched by an input layer and an output layer. Each convolutional layer consists of multiple filters each of which encodes pieces of information by conducting convolution on a small perception field of the entire image. The input image is sequentially encoded into smaller and smaller feature maps through the hierarchical convolutional layers. The last feature map is used as input for training the classifier, which is typically a couple of layers of fully connected networks after the hierarchical convolutional layers. As the convolutional networks progress deeper, the information learned in the feature maps goes from simple and local to abstract and global. The hierarchical convolutional architecture of CNN enables more generalizable learning of abstract and complex patterns in images and minimizes the risk of vanishing gradients and exploding gradients problems seen previously in deep fully connected neural networks. Some of the established CNN architectures include LeNet-5 (LeCun et al. 1998), AlexNet (Krizhevsky et al. 2017), VGG16 (Simonyan and Zisserman 2014), InceptionV3 (Szegedy et al. 2016), ResNet50 (He et al. 2016), Xception (Chollet 2017), Inception-ResNet-V2 (Szegedy et al. 2017), ResNeXt-50 (Xie et al. 2017), MobileNet (Howard et al. 2017), and EfficientNet (Tan and Le 2019). We direct the reader to a recent review (Alzubaidi et al. 2021) on concepts, architectures, and applications of CNN for more coverage on this topic.

While CNN works well for classification or regression tasks, it cannot yet be used for tasks that require pixel-level prediction. Pixel-level prediction may be needed for tasks that require the reconstruction of images as the output. For example, if one were to generate images like the one obtained from a bottleneck characterization, then a pixel-level prediction is going to be needed. In such cases, one would need to *encode* the image into a lower-dimensional representation and then *decode* the lower-dimensional representation back to an image. Figure 6.2 demonstrates four different architectures that incorporate CNN as an image encoder and include an inverse-CNN that can reconstruct or generate images from the encoded content. An autoencoder is a pair of neural networks (an encoder and a decoder) built to reconstruct the input fully unsupervised (i.e., without any manual labeling). The encoder converts the information from the input into a set of lower dimensional latent variables and the decoder deciphers information back to the original target space. Long et al. proposed an autoencoder called Fully Convolutional Network (FCN) with a CNN as the encoder part and a neural network consisting of transpose-convolutional layers as the decoder to reconstruct images (Figure 6.2(A)). (Long et al. 2015) A drawback of autoencoder is that the set of latent variables is discrete and tends to do poor reconstruction when

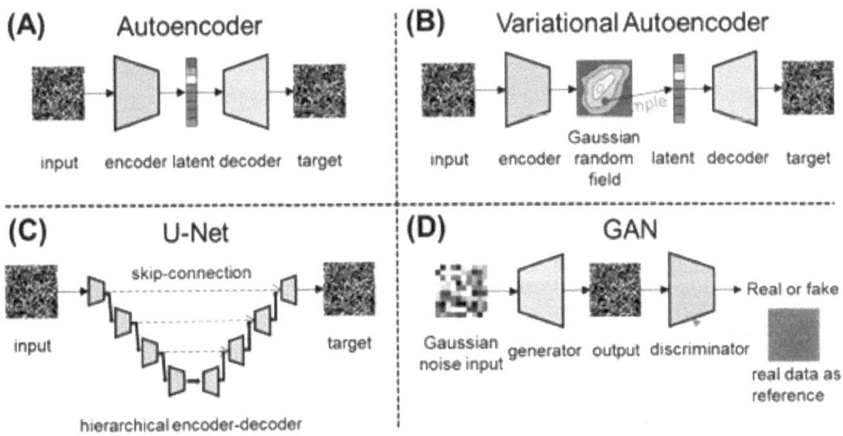

FIGURE 6.2 Different deep learning architectures for image reconstruction. **A)** Autoencoder, **B)** Variational Autoencoder, **C)** U-Net, **D)** Generative Adversarial Neural Network (GAN).

the test target is outside of the training set. To mitigate the discrete latent space problem, a modified version of autoencoder called variational autoencoder (Kingma and Welling 2013) projects the latent space to a continuous Gaussian random field and randomly samples a set of latent variables when reconstructing from the latent variables to the target space (Figure 6.2(B)). Although autoencoders and variational autoencoders work well when reconstructing 1-dimensional array data, it can be difficult to expect precise pixel-wise reconstruction of images due to the large number of pixels contained in each image. Ronneberger et al. proposed a modified version of FCN called U-Net by establishing skip connections between the same hierarchical layers in both the encoder and decoders, feeding more information about local features in the image from earlier encoder layers directly to later decoder layers (Figure 6.2(C)). (Ronneberger et al. 2015) U-Net was first developed to tackle the segmentation of biomedical images but can also be broadly applicable to other types of image segmentation problems.

Generative adversarial neural network (GAN) is an architecture proposed by Goodfellow et al. that trains a generator and a discriminator in an adversarial fashion with the purpose of generation of real-like images from Gaussian random noise signals (Figure 6.2(D)) (Goodfellow et al. 2020). The generator is trained to generate images, and the discriminator is trained to classify images as being real (i.e., true images from the sample domain) or fake (i.e., generated images). Training is conducted in a zero-sum way where the generator and the discriminator compete.

The specific tasks the aforementioned ML models can be used for in the context of analyzing image-based structural characterization of formulations are described in the next section.

6.2.4 Types of Electron Microscopy Image Analysis Tasks

In this section, we describe the types of tasks that one may want the ML models described in the previous section to perform. These tasks include classification, segmentation, or object detection to gain better understanding or characterization of the formulation's structure and resulting properties. For example, if one has time evolution images of atomic force microscopy (AFM) on a surface coated with paint or a film of solution undergoing a drying process, these images may need to be analyzed for extent of roughness (e.g., presence or absence of gradients in height identified as lines in the image) or emergence of defects (e.g., detecting bumps in the images against the smooth background). This means that segmentation or object detection tasks might be required. Another example – one may need to monitor the development of crystallinity in a deposited solution as the solvent evaporates; this may involve analyzing polarized light optical microscopy images for the emergence of birefringence. In such a case, the segmentation task to identify areas where there are high-intensity colors against the smooth dark background may be required. In another example, one may wish to identify the presence or absence of aggregates of particulate matter in a formulation; this would effectively be a binary classification problem.

In the next three sub-sections, we discuss ML modeling for these three tasks – classification, segmentation, and object detection – in more detail (Figure 6.3).

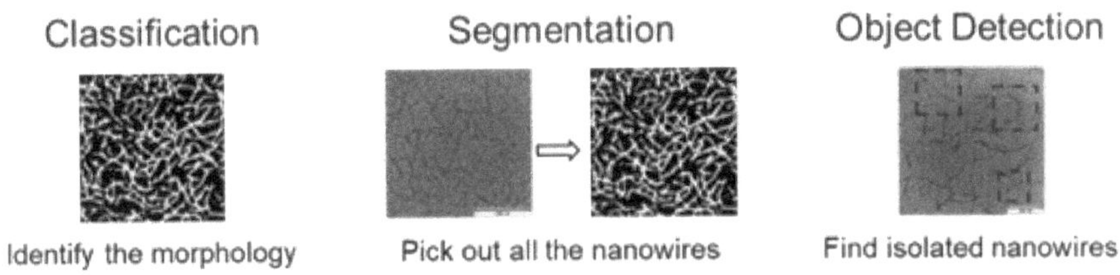

FIGURE 6.3 Schematic of different image analysis tasks for microscopy images, using the example of nanowires' images.

Classification

Image classification is usually defined as a supervised learning problem: establish a set of categories (types, phases, or morphologies of materials or presence or absence of particle aggregates in images) and train an ML model (a classifier) on labeled example images to be able to sort images into the established categories.

> Echoing the unique challenges we have mentioned previously, classification of objects, phases, and morphologies from microscopy images requires special consideration. Domain knowledge of the characterized material is necessary to produce good categories that are highly system and material dependent.

In the subsequent discussion, we list recent studies using both non-deep learning and deep learning methods to tackle classification of microscopy images of nanomaterials.

DeCost and Holm used Support Vector Machine classifier (SVM) with a bag of visual patches as features to classify microstructures of different metal alloys (DeCost and Holm 2015). Representative visual patches were selected with k-means clustering. They achieved 83% classification accuracy with a small dataset of ~100 images. Ma et al. used a series of machine learning methods: k-means clustering, feature extraction, and random forest classifier to classify the processing parameters of a SEM image of U-Mo alloy sample (Ma et al. 2020). Images are turned into area features and spatial features as input for their random forest classifier.

> Deep learning approaches automate or eliminate the feature extraction process commonly needed in non-deep learning approaches. However, usually deep learning approaches require large datasets (tens of thousands of images). Another common concern for deep learning is the interpretability of the model outputs.

Next, we mention studies that use transfer learning to mitigate the small-data dilemma as well as studies aiming to provide more interpretation of the model's performance and outputs.

Transfer learning is an approach to leverage a CNN model trained on large image datasets by using it as is (i.e., weights made untrainable) or only allowing a few layers to train when training on a different significantly smaller size dataset with images from the same or different domain as the larger image dataset.

Chowdhury et al. developed a transfer learning approach consisting of a feature extractor, a feature selector, and a classifier to classify different dendritic microstructures (Chowdhury et al. 2016). For the feature extractor, Chowdhury et al. compared a pretrained CNN with other featurization methods such as a visual bag of words, and a combination of texture, and shape statistics. For feature selector, they compared multiple dimensionality reduction methods such as PCA. For the classifier, they compared methods including SVM, random forest, neural networks, and voting ensemble classifier. Chowdhury et al. achieved ~90% classification for most feature extractors, feature selectors, and classifier combinations.

Modarres et al. used transfer learning of Inception-v3 CNN pretrained on ImageNet dataset (Deng et al. 2009) to classify SEM images belonging to different nanomaterial subcategories like particles, fibers, and tips (Modarres et al. 2017). Additional efforts were made to investigate the categories with lower classification accuracies. The low classification accuracy of some categories was explained by either similarity between images of distinct categories or images containing elements of multiple categories. Förster et al. used two LeNet-5 CNNs to classify the diameter and chiral indices of carbon nanotubes from high-resolution (HR) TEM images in a two-step process (Förster et al. 2020). The HRTEM images were obtained from carbon nanotube configurations sampled from molecular dynamics simulations. Luo et al. used a transfer learning approach with VGG-16 CNN pretrained on ImageNet dataset (Deng et al. 2009) to classify nanotube and nanocluster morphologies from TEM images (Luo et al. 2021). They showed improvement of their model's classification performance when training on a class-balanced dataset after implementing data augmentation on their original dataset. Kaufmann et al. used ResNet50 and Xception CNN architectures to classify crystal structures from SEM-derived diffraction pattern images (Kaufmann et al. 2020). They visualized feature importance of learned images that correlate closely to the different crystal planes.

The above examples of ML models developed for the classification task are primarily from the nanomaterials community. For soft polymer materials and/or formulations, there are far fewer such studies. This could be due to poor/inconsistent labeling of images and smaller data sets (say, 10–100 images for a system at different conditions). The study highlighted next tackles some of these issues and is relevant to researchers who may be interested in analyzing particulate aggregation in formulations or classifying the morphology of particles in polymer nanocomposites or complex fluids.

Lu et al. developed a semi-supervised transfer learning workflow to facilitate automatic and label-efficient classification of protein nanowire morphologies using small-size image datasets obtained from transmission electron microscopy characterization of protein nanowires (Lu et al. 2022). Semi-supervised learning refers to machine learning workflow consisting of an unsupervised learning part (i.e., no manual labels required, target label can be derived from the input data) and a supervised learning part.

Lu et al. first trained image encoders with less than a thousand generic microscopy images taken from CEM500K without any labels (i.e., self-supervised learning). They then transferred the trained image encoders to generate feature maps for the domain-specific microscopy images (i.e., TEM images of protein and peptide nanowires). They demonstrate that both classification tasks of the nanowire morphology (i.e., dispersed, network, bundle, and single nanowires) of these TEM images require less than ten labeled images per morphology class to achieve high classification performance. They further show that high-accuracy one-shot learning (i.e., labeling only one image per morphology class for training) is possible if one selects and labels good representative images for training based on the underlying distribution of the image dataset. The demonstration of high-accuracy one-shot learning is another step toward effectively label-free learning of microscopy images.

One may wonder why we would need automation of image analyses if we were dealing with small data sets.

> Automating analysis of microscopy images, regardless of data set size, is useful because it can eliminate researcher's interpretation and subjectivity.

To further eliminate subjectivity and work during label-intensive, supervised learning, it is useful to have ML models that are label-efficient (i.e., requires minimal manual labeling of images) and perform with high accuracy even when using a small dataset of images for training. These advantages of workflows like Lu et al.'s self-supervised transfer learning workflow – requiring a small data set of microscopy images and few labels for training – also make it accessible to researchers from around the world with only modest computational resources and image dataset availability.

Segmentation

Image segmentation is the task of delineating domains/phases of different physical or chemical features from an image based on pixel-level prediction of the domain/phases. If the researcher wants to differentiate between two domains, segmentation results in a binary image of the two domains. For example, in formulations, one may wish to distinguish domains of particulate matter against the background/matrix or wish to identify amorphous domain versus crystalline domain in semi-crystalline polymer films or liquid crystalline materials. Segmentation of more than two domains is also possible with the assignment of distinct colors for the different domains; however, this is relatively less commonly done. These segmented domains can then serve as a foundation for other algorithms that aim to identify individual particles and to sort particles by shape, size, and orientation. Such segmentation tasks on microscopy images can be useful when researchers link formulation design to the extent of crystallization or particulate aggregation/dispersion.

Rather than link the numerical data describing the formulation recipe and the raw image data from AFM/TEM/SEM to create such design-structure relationships, one would relate the formulation recipes to simplified images resulting from the segmentation task.

U-Net model architecture (Ronneberger et al. 2015) originally developed for medical image segmentation is widely applied for segmentation purposes of electron microscopy images in materials domains (Azimi et al. 2018; Ma et al. 2018; Madsen et al. 2018; Furat et al. 2019; Roberts et al. 2019; Strohmann et al. 2019; Tsopanidis et al. 2020; Hwang et al. 2020; Evsevleev et al. 2020; Yao et al. 2020). Evsevleev et al. used U-Net to achieve better delineation of segmentations of SiC particles and Al_2O_3 fibers from X-ray tomography 2D image slices (Evsevleev et al. 2020). Strohmann et al. used U-Net to perform segmentation on multiphase Al-Si alloys (Strohmann et al. 2019). In addition to the commonly used cross-entropy loss function, they also defined pixel-wise level weights for Al and Si pixels to better distinguish the two elements. While the aforementioned examples are from inorganic/metallurgy field, there are far fewer studies using segmentation-focused ML models for soft materials/formulations. One relevant study is by Yao et al., who used a U-Net-based workflow to study three aspects of colloidal nanoparticles: diffusion and interaction, reaction kinetics, and assembly dynamics, all of which are in real time with the liquid-phase TEM technique. (Yao et al. 2020). This study would be relevant to those who wish to quantify dynamic aspects of a formulation (reaction kinetics, particulate aggregation, opacity changes) with time evolution images from liquid-phase TEM techniques. Yao et al.'s U-Net-based workflow could efficiently identify the nanoparticle borders even for scenarios of noisy images, heterogeneous particle intensity, and where multiple particles overlap. They also tested U-Net models versus conventional thresholding methods on the tolerance against noise in microscopy image data and found that U-Net is much more robust for segmentation of nanoparticles from noisy microscopy images.

Object Detection

Object detection is a task to output bounding boxes of objects of interest from an image; these objects can be of the same type or forming the same domain.

One may want to do object detection to identify any particle aggregates, their sizes, and shapes, in a formulation imaged via microscopy.

For object detection, specialized architectures have been developed such as R-CNN (Girshick et al. 2014), Faster R-CNN (Ren et al. 2015), and You Only Look Once (YOLO) (Redmon et al. 2016). Each of these methods uses different ways to identify the regions of interest (or ROI) and further perform classification on the ROI as to whether the ROI resembles an image containing the object of interest.

Let us consider specific examples of materials problems where object detection has been used. Ziatdinov et al. used CNN with FCN to detect and classify different types of defects in high-quality, atomically resolved STEM images of two cases: graphene and monolayer MoSe2 with dopant (Ziatdinov et al. 2017). Combining chemistry restraints, they were able to distinguish silicon dopant defects with different coordination numbers (3-fold or 4-fold), which cannot be distinguished by the image-based deep learning model alone. Qu et al. applied YOLOv3 (Redmon and Farhadi 2018) to detect inorganic nanoparticles from TEM images of polymer nanocomposites (Qu et al. 2021). They used the radial number fluctuation of nanoparticles obtained from the detection model to establish a quantification metric for the classification of different nanoparticle morphologies seen in their TEM images.

The aforementioned studies summarize the recent advances in ML model development for various tasks – classification, object detection, and segmentation – that one may be interested in during the automated analysis of microscopy images. To complement microscopy images that are two-dimensional and limited to specific length scales, some labs also use scattering techniques to understand the ensemble average three-dimensional structure at various length scales. In the next section, we will dive into a subset of scattering techniques – small angle scattering – and how ML can enable automated analysis and interpretation of scattering results.

6.3 MACHINE LEARNING METHODS FOR ANALYZING SMALL ANGLE SCATTERING DATA

The structural characterization of mostly amorphous (i.e., non-crystalline or semi-crystalline polymers or oligomers) nanostructures within formulations relies heavily on interpretation of small-angle scattering (SAS) results. Understanding atomic crystalline arrangements requires wide-angle X-ray scattering (WAXS) based characterization. As much of the formulation industry cares about the mostly amorphous structures at larger length scales than atomic, this section will focus on SAS.

SAS measurements elucidate structures ranging from a few *nm* to the assembled domains scale (*nm* to microns) with the exact resolution depending on the instrument; for example, the NGB 30m SANS instrument at the National Institute of Standards and Technology (NIST) can capture structures whose sizes are ~1 *nm* to ~700 *nm*.

In SAS, X-ray or neutron beams are used on samples, and the elastic scattering of those beams is measured (Boles et al. 2016) (Figure 6.4). The output from SAXS and SANS experiments is the scattering intensity that depends on the chemistry of the components and the spatial correlations (relative packing, density) of the various components (Boles et al. 2016).

When analyzing data from these techniques, one must know the details of the scattering experiment. SAXS intensity is based on the electron density difference between the components of the system and the background (solvent). This makes SAXS more sensitive to the components of the system that have heavier elements

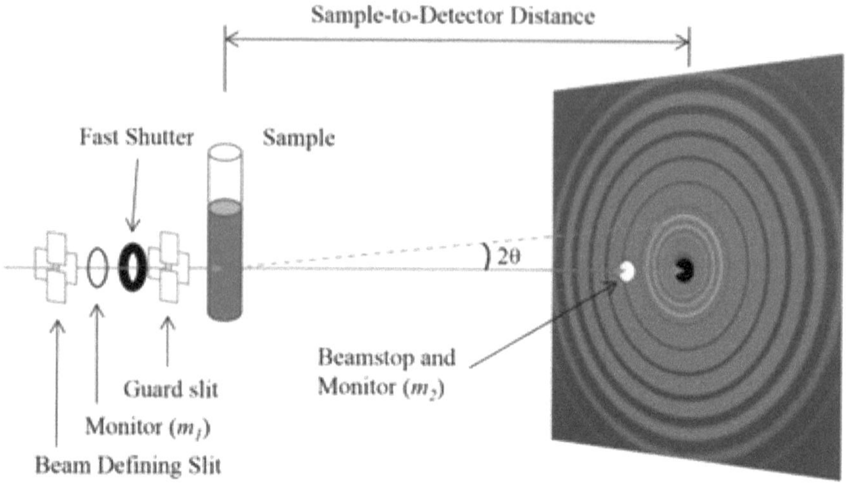

FIGURE 6.4 Schematic of small angle scattering experiment and its 2D scattering plot as output. This figure is adapted from https://pubs.acs.org/doi/10.1021/acs.chemrev.5b00690

(Mourdikoudis et al. 2018; Gosecka and Gosecki 2015). SANS intensity, on the other hand, is based on the neutron scattering length density of a material that can vary with the element's isotope, enabling the use of isotopes (e.g., deuteration) to tune the contrast between various components of the systems (Mourdikoudis et al. 2018; Gosecka and Gosecki 2015). Contrast-matching of solutions composed of multiple chemically distinct species is used to reduce or eliminate the scattering contrast between one of the species and the background (i.e., solvent) leading to the scattering primarily from the non-contrast-matched species, allowing for selective signal acquisition (Blin and Impéror-Clerc 2013; Ballauff 2011). Contrast-matching the solvent to primary particle in SAXS requires varying the electron density of the solvent by adding a contrast agent into the system (Garcia-Diez et al. 2015; Lombardo et al. 2020). In SANS contrast-matching, the solvent or nanoparticle(s) are deuterated appropriately to remove the scattering contrast (Ballauff 2011; Lombardo et al. 2020; Blin and Impéror-Clerc 2013). These details (e.g., SAXS or SANS, contrast matching, or lack thereof) become important in the labeling of the SAS data for analysis, manually or when using ML models.

The typical output from SAXS or SANS is a 2D scattering pattern. If the user expects some anisotropy in the structure of the formulation (e.g., domains in formulations can be extended and aligned in the direction of shear unlike the unsheared as cast formulation), it is best to work directly with the 2D scattering profile. The signatures of anisotropy can show up as asymmetric patterns in a 2D scattering profile; sometimes such asymmetry can be hard to identify manually, and ML models can be more precise. If the structure within the formulation is not expected to have any anisotropy or if the 2D scattering profile looks symmetric along all angles, then the 2D scattering pattern can be azimuthally averaged to obtain a 1D scattering profile – a curve of scattering intensity [I(q)] as a function of the wavevector (q) (Guinier et al. 1955; Glatter et al. 1982; Da Vela and Svergun 2020). Analysis of these 1D or 2D scattering patterns provides an interpretation of the structure at multiple scales within the sample used in the measurement.

6.3.1 Traditional Analysis of Small Angle Scattering Data

Traditionally, for formulations comprised of macromolecules, oligomers, particles (termed in general as "primary particles"), the SANS or SAXS experiments' output I(q) vs. q, contains structural information related to the form factor, P(q), (i.e., primary particle's size and shape) and the structure factor, S(q), (i.e., spatial arrangement of the primary particles) (Hammouda 2010; Pokorski and Hore 2019). Figure 6.5 shows two examples of "primary particle" and schematics of relevant structural information included in the P(q) and S(q). Example one depicts the case where the "primary particle" is a core-shell type micelle formed by amphiphilic macromolecules or oligomers. Example two depicts the case where the "primary particle" is a non-spherical nanoparticle.

One can analyze the I(q) vs. q profiles via shape-independent (Figure 6.6) or shape-dependent approaches to identify chain conformations (e.g., radius of gyration or Rg) within the assembly in the case of SANS, the microphase separated assembled structure (e.g., a micelle, lamellae in solid state), and macrophase separation (e.g., aggregation of micelles).

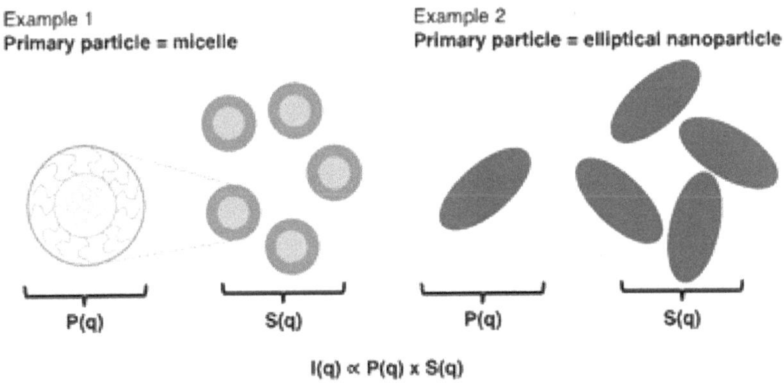

FIGURE 6.5 Two examples of "primary particle" and schematics of relevant structural information included in the P(q) (i.e., form of "primary particle") and S(q) (i.e., spatial arrangement of the "primary particl es"). Adapted from https://pubs.acs.org/doi/10.1021/jacsau.2c00697

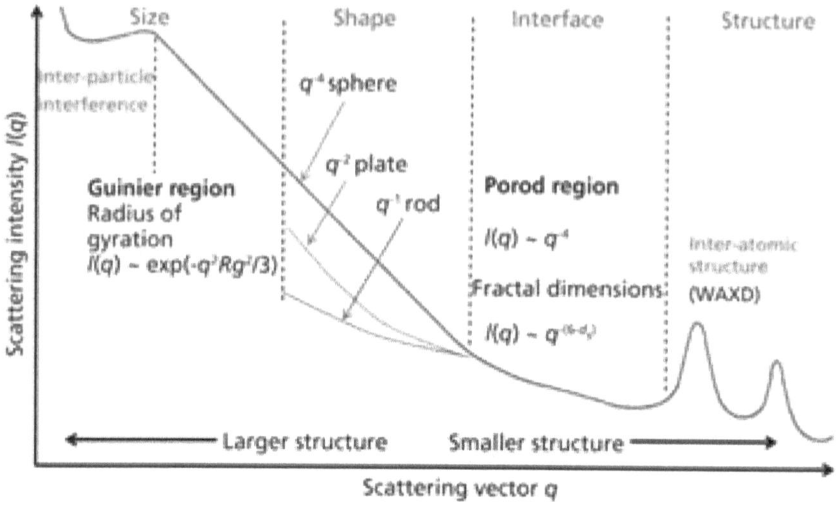

FIGURE 6.6 Regions of the scattering profile – Guinier region, Porod region. This figure is adapted from https://www.spectroscopyonline.com/view/recent-developments-small-angle-x-ray-scattering

Shape-dependent analyses rely heavily on analytical models derived for many standard nanostructures (e.g., spherical micelles, cylindrical micelles, hairy-rods, and core-shell-shell) as well as chain conformations within these structures (e.g., Gaussian chain conformations in the spherical micelle corona). Software packages [e.g., SASView https://www.sasview.org, SASfit (Breßler et al. 2015)] are then used to fit the I(q) vs. q data with these known analytical models to extract information about the assembled structure (e.g., shape, domain sizes, and chain radius of gyration) (Pedersen 1997b; Jeffries et al. 2021). To better understand the computational techniques described for AI-driven analyses, the reader may want to understand basics of shape-dependent analyses discussed next in the context of formulations at low and high concentrations. If the reader is an expert on this topic, then they can directly go to the following section on CREASE – Computational Reverse Engineering Analysis of Scattering Experiments.

At low concentrations of "primary particle," I(q) ~ P(q) as the structure S(q) ~1. In this condition, if the "primary particle" is a macromolecule then the structural information, one would obtain from analysis of I(q) vs. q could be the macromolecules' average radius of gyration and shape (e.g., cylindrical, or globular). If the "primary particle" is a macromolecular assembly (e.g., micelle), then the structural information would be the core and corona dimensions of the micelle or length and diameter of the cylindrical assembly or core radius and wall thickness of a vesicle. If the "primary particle" is a bare or coated nanoparticle the structural information could be the shape (e.g., rod, cylinder, sphere, ellipsoid) and relevant dimensions of the nanoparticle and coating. Analytical models are available for a range of conventional shapes, for example, spheres (Guinier et al. 1955), micelles (Pedersen 2000), and cylinders (Pedersen 1997a). In all cases, the relevant analytical models that are used to fit to the I(q) ~ P(q) were developed by making certain assumptions about the "primary particle" (Cooksey et al. 2017; Pedersen 1997b; Svaneborg and Pedersen 2000; Pedersen and Gerstenberg 1996; Pedersen and Gerstenberg 1996; Borbély 2000).

The choice of the analytical model warrants some prior knowledge of the structure or physical behavior for the given polymer(s) chemistry, but this information may not adequately describe structures formed by new polymer chemistries with novel architectures and non-equilibrium structures formed by new assembly techniques (Zhao et al. 2020; Early et al. 2021). Then the interpretation of the form of the "primary particle" based on these approximate model fits to the I(q) vs. q curves would be incorrect. For example in one recent study, (Beltran-Villegas et al. 2019) Beltran-Villegas et al. showed that a fit of the I(q) vs. q using an analytical model was not able to capture the correct core-corona micelle dimensions and aggregation number for a newly synthesized polymer. This was likely due to the polymer conformations not adopting Gaussian chain conformations, an approximation built into that analytical model. Therefore, for complex new polymer backbone chemistries and architectures that one may consider during design for new or improved formulations, appropriate analytical models may not exist. Therefore, there is a need for other scattering analysis approaches (discussed in the following sections) that do not rely on fitting with standard analytical scattering models that are irrelevant for such new polymer chemistries and/or architectures.

At high concentrations of "primary particles," SAXS and SANS are used to characterize structural information held in the P(q), including the macromolecular assembly (e.g., micelle) or particle features (e.g., rod, ellipsoid, sphere), as well as their spatial arrangement (e.g., assembly of the micelles or disordered/ordered aggregate of particles), which is held in the S(q). It is common to use the scattering profile at dilute conditions where the S(q) is ~1 (e.g., dilute macromolecular solution concentration or dispersion of particles at low concentration) to calculate P(q) as I(q) ~ P(q). The analysis of that P(q) provides an interpretation of the form of the "primary particle." This P(q) obtained from the system at low concentrations is then used during analysis of the I(q) vs. q for that system at higher concentrations to analyze S(q), assuming that the P(q) does not change with concentration. With the known P(q) of the "primary particle," analysis of the S(q) provides the spatial arrangement of interacting "primary particles." Analysis of I(q)/P(q) ~ S(q) is done using existing structure factor analytical models (Botet et al. 2020; Li et al. 2016). These structure factor models include hard sphere (Blum and Stell 1979; Salacuse and Stell 1982) and sticky hard sphere (Bharti et al. 2014; Baxter 1968) for characterizing the structure of "primary particles" in a fluid suspension. At higher packing fractions (above 0.4) when there is a formation of aggregates of "primary particle" and the fluid suspension assumption fails (Schmidt 1995; Zemb and Lindner 2002), more complex analytical models for aggregating particles are needed. These types of models require significant *a priori* knowledge of the system at hand (e.g., aggregate radius of gyration and aggregation number) (Larsen et al. 2020).

Let us revisit that commonly used assumption that the form factor P(q) observed at low concentration does not change with increasing concentration or other changes in system conditions to analyze S(q); this assumption is not always applicable. Such an assumption is valid only if the primary particles do not change shape or dimensions with changing conditions of interest (e.g., with increasing concentration, temperature, and salinity), which may not be the case for the formulation under consideration. For example, in Figure 6.5 when the "primary particle" is a core-shell type micelle formed by amphiphilic macromolecules or oligomers, one may not be able to assume that the P(q) found at low concentrations (when S(q) ~1) is also applicable at high concentration. The amphiphilic macromolecules could reorganize to form micelles of different dimensions or shapes as the concentration of the solution increases (Qi and Zhou 2019; Gast 1997). However, in the case where the "primary particle" is an inorganic (hard) nanoparticle, if the nanoparticle is not expected to evolve in shape and dimension with changing concentration, this assumption that the P(q) found at low concentration is applicable at higher concentration as well is correct. To handle universally all cases – ones where the "primary particle" form P(q) may change with changing conditions of the formulation and ones where P(q) remains constant with changing conditions – there is a need for a computational approach that simultaneously identifies and interprets the structural features of the system stored in the P(q) and S(q) at the conditions of interest.

The aforementioned sections present multiple reasons why we need alternate methods for analyzing SAS results that do not rely on traditional approaches (e.g., analytical models and manual fitting) to get correct interpretation.

Furthermore, in the context of this book – AI for formulations – for ML-driven analyses and automation in characterization analysis, we need computational approaches that can handle high-throughput characterization and interpret data with minimal manual intervention, for cases with and without *a priori* knowledge.

In the next couple of sections, we discuss one old (Reverse Monte Carlo) and one new (CREASE) computational method for analyzing SAS profiles; the latter is better suited to the AI-driven analysis of SAS profiles.

6.3.2 Reverse Monte Carlo Simulations

Reverse Monte Carlo (RMC) simulations (McGreevy 2001), a computational method that has been around for decades, reverse engineers the three-dimensional (3D) structure whose quantitative measure (e.g., radial distribution function or RDF, computed scattering profile, pore size structure, and percolation) matches the target quantitative measure. This is accomplished by evolving from some initial 3D structure of the species and iteratively making changes to the structure, accepting or rejecting each change based on whether the difference in quantitative measure (e.g., RDF) of the evolving structure and corresponding target measure decreases or increases, respectively (McGreevy and Pusztai 1988). Similarly, if the goal is to identify the 3D structure of particles whose computed scattering most closely matches the input (target) scattering, then one would conduct an RMC simulation where randomly selected particle moves are accepted or rejected based on how much better the computed scattering profile of that post-move structure versus pre-move structure matches the target experimental scattering intensity profile. The structures produced upon accepting MC particle moves evolve the system toward exhibiting a scattering intensity profile like the target scattering profile. RMC simulations are non-trivial to use for high-density and close-packed particle assemblies where the rate of structural changes can be slow (Tóth 1999). This slow rate of structural changes is exaggerated for macromolecular systems (e.g., polymer solutions, polymer blends, or polymer nanocomposites). Such issues lead to an associated high computational cost of running RMC simulations for the purpose of identifying structures whose computed scattering matches target experimental scattering. As such,

> RMCs are less feasible for automation and AI-driven formulation design and characterization.

6.3.3 Computational Reverse Engineering Analysis of Scattering Experiments (CREASE)

Jayaraman and coworkers have developed the "Computational Reverse-Engineering Analysis for Scattering Experiments" (CREASE) method (https://crease-ga.readthedocs.io/) as a way to address the growing needs for computational scattering analysis methods that are capable of analyzing high-throughput structural characterization results for amorphous soft materials with dispersity in arrangements as well as molecular sizes and for formulations both at high and low concentrations. CREASE avoids many of the issues described earlier with analytical model fits or RMC. This method is applicable to both conventional structures with existing analytical models and unconventional structures/chemistries that may not have good analytical models as well as handle analysis of P(q) and S(q) simultaneously. With advances in ML models and algorithms and CREASE presents a solution to the need for methods that can do an automated analysis of high throughput scattering profiles.

Figure 6.7 shows the general workflow in CREASE (as of summer 2023), where experimentally measured 1D or 2D scattering profiles are taken as input and

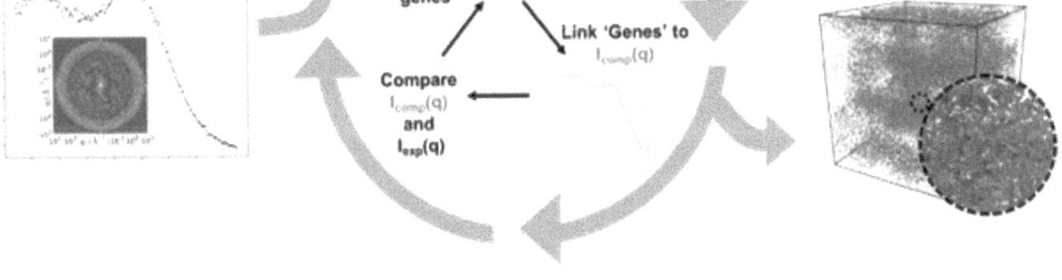

FIGURE 6.7 CREASE workflow. CREASE takes as input experimentally measured 1D or 2D scattering profiles and identifies as output the key structural features as well as representative 3D real space structures whose computed scattering profiles match the experimental scattering input.

CREASE through an internal optimization generates as output the key structural features as well as representative 3D real space structures whose computed scattering profiles matches the experimental scattering input.

CREASE's workflow is based on the *philosophy* that real space three-dimensional (amorphous) arrangements of constituents in soft materials can be reduced to a lower dimensional mathematical representation of key "structural features" and that the distributions of those structural features give rise to a computed scattering profile.

For example, for a system with core-corona spherical micelles at low concentrations (Beltran-Villegas et al. 2019), these structural features could be sizes of core and corona and probability distributions of those sizes. At higher concentrations, there would be additional structural features that describe the relative neighborhood of each micelle, for example through mathematical order parameters describing positional and orientational order (Heil et al. 2023). The user must first decide the types of structural features they are interested in; these features are mathematical descriptors of the fundamentally interesting structural information (e.g., sizes of domains, and shapes of domains) or parameters that the researcher knows will impact the formulations' eventual application (e.g., flexibility/semi-flexibility of the assembled domains). Once the user has decided on the structural features of interest, then CREASE runs an optimization loop where it iterates over sets of structural features, calculating the computed scattering profile, $I_{comp}(q)$ (more on this calculation below), for each set of structural features. In these iterations, the $I_{comp}(q)$ profiles and experimental (input) profile, $I_{exp}(q)$, are compared, and eventually this optimization converges toward the sets (note the plural!) of structural features whose $I_{comp}(q)$ profiles most closely matching the input $I_{exp}(q)$.

As of summer 2023, CREASE has been implemented in Python code (https://github.com/arthijayaraman-lab/crease_ga) using an accessible optimization method – genetic algorithm. CREASE's genetic algorithm (CREASE-GA) takes as input the SAXS and/or SANS scattering profile. It also uses the user's input on the types of structural features (i.e., "genes" of the "individuals" in GA) chosen through some knowledge about the general shape of the assembled structure from other imaging techniques and domain expertise. Then, CREASE-GA starts with an initial "generation" of multiple sets of structural features (i.e., multiple "individuals" with specific values of "genes") and iterates in the GA loop toward the optimal individuals whose genes give rise to computed scattering profile, $I_{comp}(q)$, that closely matches the input experimentally measured scattering profile, $I_{exp}(q)$. One important calculation in this loop is the $I_{comp}(q)$ for a given set ("individual") of structural features ("genes"); this has been done so far in one of two ways (Figure 6.8). One way (let us call it *Debye method*) is by creating for each set of genes its representative three-dimensional real space structures filled with point scatterers whose scattering length densities represent the constituents of the system and using Debye-equation on the scatterer positions to compute $I_{comp}(q)$. This algorithm can be computationally intensive even with computational advances that accelerate the speed of this algorithm

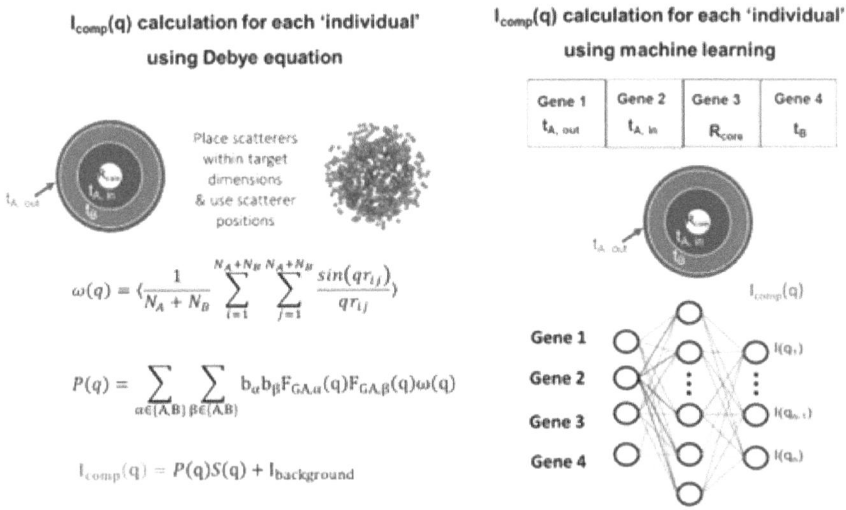

FIGURE 6.8 Two ways to calculate the "computed scattering profile" or $I_{comp}(q)$ for each "individual" in the genetic algorithm. Each individual has a unique set of genes (i.e., a set of structural features). Both the Debye equation method (left) and machine learning approach (right) connect the values of those structural features or genes to an $I_{comp}(q)$ vs. q profile. In this figure a vesicle with a hollow core, inner A-layer, B-layer, and an outer A-layer is used as an example; the dimensions related to those layers and core become the "genes" for this system. There can be additional genes depending on the user's interest. It is important to note that in the right (machine learning approach) the user may want to represent the structural features (genes) as normalized dimensions to make the trained machine learning model transferable [see for example Heil et al. (2022)].

in a code (Schmidt-Rohr 2007; Brisard and Levitz 2013; Olds and Duxbury 2014). Another way is by using a surrogate machine learning (ML) model that links the structural features directly to $I_{comp}(q)$; Jayaraman and coworkers have used neural networks (NN) trained on thousands of computed scattering profiles calculated from the Debye method for various sets of genes; this NN model would then serve as a surrogate model for the *Debye method.* The NN-CREASE-GA is orders of magnitude faster than the Debye method, after the initial reasonable time investment of ML model training. The user could use other more sophisticated ML models that can also accomplish the same purpose as the NN model in Figure 6.8. For example, Akepati et al. have replaced the NN surrogate ML model in CREASE with an XGBoost based surrogate model (Akepati et al. 2024).

Regardless of the type of surrogate model the researcher chooses to use, they are encouraged to write the code for the model on their own (or through their data science collaborators) to ensure they have an optimized code that works well for their selected structural features and keeps the computational cost to the minimum.

Depending on the users' system and structural features, they can use the published studies applying CREASE as a guide. These studies will show the user how they can make decisions about the structural features, understand the sensitivity of the scattering profiles to those structural features, create the surrogate ML model linking the structural features to the computed scattering profile, and set up the CREASE optimization loop.

- *Users interested in analyzing dimensions of assembled structures in polymer solutions at dilute concentrations:* CREASE has been applied to characterize structure using scattering profiles I(q) ~ P(q), that is, conditions where S(q) is ~1) for a variety of "primary particles" at low concentration – spherical and cylindrical micelles, (Beltran-Villegas et al. 2019; Wessels and Jayaraman 2021a, 2021b) vesicles, (Ye et al. 2021), and fibrils (Wu and Jayaraman 2022) bypassing the need for an analytical model. These publications can serve as a guide to users interested in understanding any structure that may form among the minority (macromolecular/oligomeric) components within formulations.
- *Users interested in analyzing amorphous assembled structure of primary particles (that do not change shape or size with varying conditions) in formulations at high concentrations regardless of extent of mixing/segregation:* CREASE has also been extended to analyze S(q) part of the scattering profiles from *concentrated* binary mixture of polydisperse spherical nanoparticles (i.e., P(q) is a sphere form factor) to determine the extent of segregation/mixing of the two types of nanoparticles and the precise mixture composition (Heil and Jayaraman 2021; Heil et al. 2022). These publications can serve as a guide to scientists who want to understand the structures that may form among the components present in high concentration within formulations; these components are also not expected to change in their form as concentrations or conditions vary.
- *Users interested in analyzing amorphous assembled structure of primary particles (particles, micelles) with unknown form [P(q)] and unknown structure [S(q)], simultaneously:* For systems where one does not know the P(q) or S(q) *a priori*, CREASE has been extended to simultaneously interpret structural information held in P(q) and S(q) and appropriately called "P(q) and S(q) CREASE" (Heil et al. 2023). In this study Heil et al. demonstrated the use of "P(q) and S(q) CREASE" method to interpret SANS scattering profiles of surfactant-coated silica nanoparticles at varying temperatures and concentrations; for these systems both the extent of silico nanoparticle aggregation [related to S(q)] and surfactant shell thickness characteristics [related to P(q)] are unknown and are expected to change with temperature and salt concentration at certain values.

For those considering using CREASE, there are unique advantages of CREASE-GA regardless of availability of appropriate analytical models for the system being characterized:

- *No requirement of a priori knowledge about constituent molecular conformations:* The physical description of the structures is using structural features and placement of scatterers. There are no polymer/oligomer/molecules in this GA step – only scatterers – so this overcomes issues one may have not knowing anything about molecular conformations.
- *Adaptable to the users' needs and to any morphologies:* Any structural feature of interest can be a gene and for the same system, two different users may be interested in different structural features (and thus, different genes). Some of those structural features may not be in any existing analytical model! Note: If and how well CREASE can identify a structural feature reliably from an input experimental scattering profile will depend on how much that structural feature affects the computed scattering profile and how that effect changes with the values of that structural feature. See for example recent work on analysis of scattering results from methylcellulose fibrils from Wu and Jayaraman (Wu and Jayaraman 2022). In that system, the length, Kuhn length, and diameter of fibrils are the structural features of interest; however, Wu and Jayaraman showed that KL could only identified well if it were within a certain range of values for the methylcellulose systems. Such sensitivity analysis is especially useful in deciding on the structural features.
- *Easy to use optimization and availability of many possible optimized solutions:* Genetic algorithm (GA) is the chosen optimization method because it is easy for others regardless of prior computational knowledge and experience to adopt. Furthermore, GA's output contains multiple numerical solutions whose computed scattering matches experimental scattering. This is useful as it informs us about the degeneracy of this solution; in other words, there can be many different structures whose computed scattering profile can match with experiments, so knowing this distribution from the converged individuals in the last couple of generations of GA is valuable.
- *Streamlined computational workflow*: The use of structural features (i.e., the "genes" in the genetic algorithm) instead of all coordinates of all relevant species as used in other computational methods (e.g., RMC simulations) enables faster optimization toward the best structures whose computed scattering matches the target scattering. Furthermore, the gene-based CREASE incorporates machine learning (ML) models that directly links the genes defining the structure to the computed scattering intensity profile, bypassing the need to create a 3D structure with nanoparticles to calculate the computed scattering intensity profile using the Debye (Debye 1915; Scardi et al. 2016) scattering equation (Wessels and Jayaraman 2021b).
- *Availability of 3D structures corresponding to the input scattering profile*: At the end, CREASE also gives as output representative real-space 3D structures – either as is because the system is made of particles whose positions can be generated from scatterer positions or via additional molecular modeling and simulation step to show chain conformations in the structures output from CREASE-GA. This can then be used as an input for other

calculations (e.g., resistor network model calculation for electrical conductivity (White et al. 2009) and finite-difference time-domain method for optical properties (Patil et al. 2021, 2022)).

- *High-throughput analysis of related systems:* Machine learning (ML) enhanced CREASE-GA can facilitate high-throughput analysis of related systems as long as the trained ML model, in particular the structural features that are inputs to the ML model, are valid for those related systems.
- *In silico hypothesis testing:* CREASE can be used to evaluate researcher's hypotheses about the structures. Multiple examples for this include Wessels and Jayaraman (2020, 2021b) and Heil et al. (2023)

6.3.4 Pairing Analysis of Scattering Data with Microscopy

While the previous two sections focused on microscopy and scattering in isolated discussions, researchers often use multiple complementary techniques (e.g., scattering, microscopy, spectroscopy) to obtain a holistic understanding of the structure within the formulation. Depending on the availability and accessibility of the different characterization techniques, each research facility or academic research lab may have access to high-throughput capability in one technique but face limitations (sample preparation, resolution) with other technique(s).

For establishing design-structure-property relationships, it would be valuable to have paired structural characterization from multiple techniques, regardless of ease or access.

We highlight here two studies that have accomplished such pairing or coupling – one study for inorganic nanoparticles and another study for organic macromolecular materials.

In one study, Yaman et al. (Yaman et al. 2023) demonstrated a dual Variational Autoencoders (VAE) ML workflow to couple the surface photonic response spectroscopy (SPR) profiles and SEM images of gold nanoparticles. They showed that by correlating the latent space of one VAE trained on SPR with the latent space of another VAE trained on SEM, they can correctly generate (or construct) an SPR profile from a corresponding SEM image or vice versa.

For soft (organic) macromolecular materials, Lu and Jayaraman have developed Pair-Variational Autoencoders (PairVAE) (Lu and Jayaraman 2023), which can generate structural characterization data from one technique taking as input data from another complementary characterization technique. Using paired SAXS and SEM data of newly observed block copolymer-assembled morphologies archived in an open access database (Doerk et al. 2023), Lu and Jayaraman trained the PairVAE. After successful training, the PairVAE was able to generate SEM images of the block copolymer morphology when it took as input that sample's corresponding SAXS 2D pattern, and vice versa. This method can be extended to other soft materials morphologies as well and serves as a valuable tool for easy interpretation of 2D SAXS patterns as well as an engine for generating ensembles of similar microscopy images to create a database for other downstream calculations of structure-property relationships. Most importantly, Lu and Jayaraman demonstrate that the PairVAE approach does not need much supervision during training and that it works well even with a small amount of data (e.g., ~ 50 pairs of characterization data).

The formulations community can benefit from the general idea of cross-reconstruction of complementary characterization data to foster more ways of generating data for design-structure-property relationships that accelerate formulation design and innovation.

REFERENCES

Akepati SVR, Gupta N, Jayaraman A. Computational Reverse Engineering Analysis of the Scattering Experiment Method for Interpretation of 2D Small-Angle Scattering Profiles (CREASE-2D). JACS Au. 2024 Mar 20;4(4):1570–1582. doi: 10.1021/jacsau.4c00068. PMID: 38665659; PMCID: PMC11040659.

Alzubaidi, Laith, Jinglan Zhang, Amjad J. Humaidi, Ayad Al-Dujaili, Ye Duan, Omran Al-Shamma, José Santamaría, Mohammed A. Fadhel, Muthana Al-Amidie, and Laith Farhan. 2021. "Review of deep learning: concepts, CNN architectures, challenges, applications, future directions." *Journal of Big Data* 8: 1–74.

Azimi, Seyed Majid, Dominik Britz, Michael Engstler, Mario Fritz, and Frank Mücklich. 2018. "Advanced steel microstructural classification by deep learning methods." *Scientific Reports* 8 (1): 1–14.

Ballauff, Matthias. 2011. "Analysis of polymer colloids by small-angle X-ray and neutron scattering: contrast variation." *Advanced Engineering Materials* 13 (8): 793–802.

Barlow, J.W., and D.R. Paul. 1981. "Polymer blends and alloys – a review of selected considerations." *Polymer Engineering & Science* 21 (15): 985–996. https://doi.org/10.1002/pen.760211502.

Baxter, R.J. 1968. "Percus–Yevick equation for hard spheres with surface adhesion." *The Journal of Chemical Physics* 49 (6): 2770–2774.

Beltran-Villegas, Daniel J., Michiel G. Wessels, Jee Young Lee, Yue Song, Karen L. Wooley, Darrin J. Pochan, and Arthi Jayaraman. 2019. "Computational reverse-engineering analysis for scattering experiments on amphiphilic block polymer solutions." *Journal of the American Chemical Society* 141 (37): 14916–14930. https://doi.org/10.1021/jacs.9b08028.

Bharti, Bhuvnesh, Jens Meissner, Sabine H.L. Klapp, and Gerhard H. Findenegg. 2014. "Bridging interactions of proteins with silica nanoparticles: the influence of pH, ionic strength and protein concentration." *Soft Matter* 10 (5): 718–728.

Blin, Jean-Luc, and Marianne Impéror-Clerc. 2013. "Mechanism of self-assembly in the synthesis of silica mesoporous materials: in situ studies by X-ray and neutron scattering." *Chemical Society Reviews* 42 (9): 4071–4082.

Blum, L., and G. Stell. 1979. "Polydisperse systems. I. Scattering function for polydisperse fluids of hard or permeable spheres." *The Journal of Chemical Physics* 71 (1): 42–46.

Boles, Michael A., Michael Engel, and Dmitri V. Talapin. 2016. "Self-assembly of colloidal nanocrystals: from intricate structures to functional materials." *Chemical Reviews* 116 (18): 11220–11289.

Borbély, Sándor. 2000. "Aggregate structure in aqueous solutions of Brij-35 nonionic surfactant studied by small-angle neutron scattering." *Langmuir* 16 (13): 5540–5545.

Botet, Robert, Sylvie Kwok, and Bernard Cabane. 2020. "Percus–Yevick structure factors made simple." *Journal of Applied Crystallography* 53 (6): 1570–1582.

Breßler, Ingo, Joachim Kohlbrecher, and Andreas F. Thünemann. 2015. "SASfit: a tool for small-angle scattering data analysis using a library of analytical expressions." *Journal of Applied Crystallography* 48 (5): 1587–1598.

Brisard, Sébastien, and Pierre Levitz. 2013. "Small-angle scattering of dense, polydisperse granular porous media: computation free of size effects." *Physical Review E* 87 (1): 013305. https://doi.org/10.1103/PhysRevE.87.013305.

Chen, Cheng Chung, and James L. White. 1993. "Compatibilizing agents in polymer blends: interfacial tension, phase morphology, and mechanical properties." *Polymer Engineering & Science* 33 (14): 923–930. https://doi.org/10.1002/pen.760331409.

Chollet, François. 2017. "Xception: deep learning with depthwise separable convolutions." *Proceedings of the IEEE Conference on Computer Vision and Pattern Recognition.*

Chowdhury, Aritra, Elizabeth Kautz, Bülent Yener, and Daniel Lewis. 2016. "Image driven machine learning methods for microstructure recognition." *Computational Materials Science* 123: 176–187.

Clough, Jess M., Elise Guimard, Cindy Rivet, Joris Sprakel, and Thomas E. Kodger. 2019. "Photonic paints: structural pigments combined with water-based polymeric film-formers for structurally colored coatings." *Advanced Optical Materials* 7 (15): 1900218. https://doi.org/10.1002/adom.201900218.

Cooksey, Tyler J., Avantika Singh, Kim Mai Le, Shu Wang, Elizabeth G. Kelley, Lilin He, Sameer Vajjala Kesava, Enrique D. Gomez, Bryce E. Kidd, and Louis A. Madsen. 2017. "Tuning biocompatible block copolymer micelles by varying solvent composition: core/corona structure and solvent uptake." *Macromolecules* 50 (11): 4322–4334.

Crosby, Alfred J., and Jong-Young Lee. 2007. "Polymer nanocomposites: the "Nano" effect on mechanical properties." *Polymer Reviews* 47 (2): 217–229. https://doi.org/10.1080/15583720701271278.

Cutler, Adele, D. Richard Cutler, and John R. Stevens. 2012. "Random forests." *Ensemble Machine Learning: Methods and Applications*: 157–175.

Da Vela, Stefano, and Dmitri I. Svergun. 2020. "Methods, development and applications of small-angle X-ray scattering to characterize biological macromolecules in solution." *Current Research in Structural Biology* 2: 164–170. https://doi.org/10.1016/j. crstbi.2020.08.004. https://www.sciencedirect.com/science/article/pii/S2665928X20300180.

Debye, Peter. 1915. "Zerstreuung von röntgenstrahlen." *Annalen der Physik* 351 (6): 809–823.

DeCost, Brian L., and Elizabeth A. Holm. 2015. "A computer vision approach for automated analysis and classification of microstructural image data." *Computational Materials Science* 110: 126–133.

Deng, Jia, Wei Dong, Richard Socher, Li-Jia Li, Kai Li, and Li Fei-Fei. 2009. "Imagenet: a large-scale hierarchical image database." *2009 IEEE Conference on Computer Vision and Pattern Recognition.*

Dietterich, Thomas G. 2000. "Ensemble methods in machine learning." *Multiple Classifier Systems: First International Workshop*, MCS 2000 Cagliari, Italy, June 21–23, 2000 Proceedings 1.

Doerk, Gregory S., Aaron Stein, Suwon Bae, Marcus M. Noack, Masafumi Fukuto, and Kevin G. Yager. 2023. "Autonomous discovery of emergent morphologies in directed self-assembly of block copolymer blends." *Science Advances* 9 (2): eadd3687.

Early, Julia T., Alison Block, Kevin G. Yager, and Timothy P. Lodge. 2021. "Molecular weight dependence of block copolymer micelle fragmentation kinetics." *Journal of the American Chemical Society* 143 (20): 7748–7758.

Evsevleev, Sergei, Sidnei Paciornik, and Giovanni Bruno. 2020. "Advanced deep learning-based 3D microstructural characterization of multiphase metal matrix composites." *Advanced Engineering Materials* 22 (4): 1901197.

Förster, Georg Daniel, Alice Castan, Annick Loiseau, Jaysen Nelayah, Damien Alloyeau, Frédéric Fossard, Christophe Bichara, and Hakim Amara. 2020. "A deep learning approach for determining the chiral indices of carbon nanotubes from high-resolution transmission electron microscopy images." *Carbon* 169: 465–474.

Furat, Orkun, Mingyan Wang, Matthias Neumann, Lukas Petrich, Matthias Weber, Carl E. Krill III, and Volker Schmidt. 2019. "Machine learning techniques for the segmentation of tomographic image data of functional materials." *Frontiers in Materials* 6: 145.

Garcia-Diez, Raul, Christian Gollwitzer, and Michael Krumrey. 2015. "Nanoparticle characterization by continuous contrast variation in small-angle X-ray scattering with a solvent density gradient." *Journal of Applied Crystallography* 48 (1): 20–28.

Gast, Alice P. 1997. "Polymeric micelles." *Current Opinion in Colloid & Interface Science* 2 (3): 258–263.

Ge, M., F. Su, Z. Zhao, and D. Su. 2020. "Deep learning analysis on microscopic imaging in materials science." *Mate rials Today Nano* 11. https://doi.org/10.1016/j.mtnano.2020.100087.

Girshick, Ross, Jeff Donahue, Trevor Darrell, and Jitendra Malik. 2014. "Rich feature hierarchies for accurate object detection and semantic segmentation." *Proceedings of the IEEE Conference on Computer Vision and Pattern Recognition.*

Glatter, Otto, Otto Kratky, and H.C. Kratky. 1982. *Small Angle X-ray Scattering*. Academic Press.

Goodfellow, Ian, Jean Pouget-Abadie, Mehdi Mirza, Bing Xu, David Warde-Farley, Sherjil Ozair, Aaron Courville, and Yoshua Bengio. 2020. "Generative adversarial networks." *Communications of the ACM* 63 (11): 139–144.

Goodhew, Peter J., John Humphreys, and Richard Beanland. 2000. *Electron Microscopy and Analysis*. CRC Press.

Gosecka, Monika, and Mateusz Gosecki. 2015. "Characterization methods of polymer core–shell particles." *Colloid and Polymer Science* 293 (10): 2719–2740.

Grigorescu, Sorin, Bogdan Trasnea, Tiberiu Cocias, and Gigel Macesanu. 2020. "A survey of deep learning techniques for autonomous driving." *Journal of Field Robotics* 37 (3): 362–386.

Guinier, André, Gérard Fournet, and Kenneth L. Yudowitch. 1955. "Small-angle scattering of X-rays." *Physics Today* 9 (8): 38–39.

Guyon, Isabelle, Jason Weston, Stephen Barnhill, and Vladimir Vapnik. 2002. "Gene selection for cancer classification using support vector machines." *Machine Learning* 46: 389–422.

Hammouda, Boualem. 2010. "SANS from polymers – review of the recent literature." *Journal of Macromolecular Science®, Part C: Polymer Reviews* 50 (1): 14–39.

He, Kaiming, Xiangyu Zhang, Shaoqing Ren, and Jian Sun. 2016. "Deep residual learning for image recognition." *Proceedings of the IEEE Conference on Computer Vision and Pattern Recognition.*

Heil, C.M., and A. Jayaraman. 2021. "Computational reverse-engineering analysis for scattering experiments of assembled binary mixture of nanoparticles." *ACS Materials Au* 1 (2): 140.

Heil, C.M., Yingzhen Ma, Bhuvnesh Bharti, and Arthi Jayaraman. 2023. "Computational reverse-engineering analysis for scattering experiments for form factor and structure factor determination ("P (q) and S (q) CREASE")." *JACS Au* 3 (3): 889–904.

Heil, C.M., Anvay Patil, Ali Dhinojwala, and Arthi Jayaraman. 2022. "Computational reverse-engineering analysis for scattering experiments (CREASE) with machine learning enhancement to determine structure of nanoparticle mixtures and solutions." *ACS Central Science* 8 (7): 996–1007. https://doi.org/10.1021/acscentsci.2c00382.

Howard, Andrew G., Menglong Zhu, Bo Chen, Dmitry Kalenichenko, Weijun Wang, Tobias Weyand, Marco Andreetto, and Hartwig Adam. 2017. "Mobilenets: efficient convolutional neural networks for mobile vision applications." *arXiv preprint arXiv:1704.04861.*

Hughes, Gareth A. 2005. "Nanostructure-mediated drug delivery." *Nanomedicine: Nanotechnology, Biology and Medicine* 1 (1): 22–30. https://doi.org/10.1016 /j.nano.2004.11.009. https://www.sciencedirect.com/science/article/pii/S1549963405000122.

Hwang, Heesu, Sung Min Choi, Jiwon Oh, Seung-Muk Bae, Jong-Ho Lee, Jae-Pyeong Ahn, Jeong-O. Lee, Ki-Seok An, Young Yoon, and Jin-Ha Hwang. 2020. "Integrated application of semantic segmentation-assisted deep learning to quantitative multi-phased microstructural analysis in composite materials: case study of cathode composite materials of solid oxide fuel cells." *Journal of Power Sources* 471: 228458.

Jeffries, Cy M., Jan Ilavsky, Anne Martel, Stephan Hinrichs, Andreas Meyer, Jan Skov Pedersen, Anna V. Sokolova, and Dmitri I. Svergun. 2021. "Small-angle X-ray and neutron scattering." *Nature Reviews Methods Primers* 1 (1): 1–39.

Kamerzell, Tim J., Reza Esfandiary, Sangeeta B. Joshi, C. Russell Middaugh, and David B. Volkin. 2011. "Protein–excipient interactions: mechanisms and biophysical characterization applied to protein formulation development." *Advanced Drug Delivery Reviews* 63 (13): 1118–1159. https://doi.org/10.1016 /j.addr.2011.07.006. https://www.sciencedirect.com/science/article/pii/S0169409X11002006.

Kaufmann, Kevin, Chaoyi Zhu, Alexander S. Rosengarten, Daniel Maryanovsky, Tyler J. Harrington, Eduardo Marin, and Kenneth S. Vecchio. 2020. "Crystal symmetry determination in electron diffraction using machine learning." *Science* 367 (6477): 564–568.

Kingma, Diederik P., and Max Welling. 2013. "Auto-encoding variational bayes." *arXiv preprint arXiv:1312.6114.*

Krizhevsky, Alex, Ilya Sutskever, and Geoffrey E. Hinton. 2017. "ImageNet classification with deep convolutional neural networks." *Communications of the ACM* 60 (6): 84–90. https://doi.org/10.1145/3065386.

Larsen, Andreas Haahr, Jan Skov Pedersen, and Lise Arleth. 2020. "Assessment of structure factors for analysis of small-angle scattering data from desired or undesired aggregates." *Journal of Applied Crystallography* 53 (4): 991–1005.

Larson, Ronald G., Antony K. Van Dyk, Tirtha Chatterjee, and Valeriy V. Ginzburg. 2022. "Associative thickeners for waterborne paints: structure, characterization, rheology, and modeling." *Progress in Polymer Science* 129: 101546. https://doi.org/10.1016/j. progp olymsci.2022.101546. https://www.sciencedirect.com/science/article/pii/ S0079670022000442.

LeCun, Yann, and Yoshua Bengio. 1995. "Convolutional networks for images, speech, and time series." *The Handbook of Brain Theory and Neural Networks* 3361 (10): 1995.

LeCun, Yann, Léon Bottou, Yoshua Bengio, and Patrick Haffner. 1998. "Gradient-based learning applied to document recognition." *Proceedings of the IEEE* 86 (11): 2278–2324.

Letourneau, Dane R., and Dietrich A. Volmer. 2023. "Mass spectrometry-based methods for the advanced characterization and structural analysis of lignin: a review." *Mass Spectrometry Reviews* 42 (1): 144–188. https://doi.org/10.1002/mas.21716.

Li, Tao, Andrew J. Senesi, and Byeongdu Lee. 2016. "Small angle X-ray scattering for nanoparticle research." *Chemical Reviews* 116 (18): 11128–11180.

Lombardo, Domenico, Pietro Calandra, and Mikhail A. Kiselev. 2020. "Structural Characterization of biomaterials by means of small angle X-rays and neutron scattering (SAXS and SANS), and light scattering experiments." *Molecules* 25 (23): 5624.

Long, Jonathan, Evan Shelhamer, and Trevor Darrell. 2015. "Fully convolutional networks for semantic segmentation." *Proceedings of the IEEE Conference on Computer Vision and Pattern Recognition.*

Lu, Shizhao, and Arthi Jayaraman. 2023. "Pair-variational autoencoders (PairVAE) for linking and cross-reconstruction of characterization data from complementary structural characterization techniques." *JACS Au* 3 (9): 2510–2521.

Lu, Shizhao, Brian Montz, Todd Emrick, and Arthi Jayaraman. 2022. "Semi-supervised machine learning workflow for analysis of nanowire morphologies from transmission electron microscopy images." *Digital Discovery* 1 (6): 816–833.

Luo, Qixiang, Elizabeth A. Holm, and Chen Wang. 2021. "A transfer learning approach for improved classification of carbon nanomaterials from TEM images." *Nanoscale Advances* 3 (1): 206–213.

Lyu, Zhiheng, Lehan Yao, Wenxiang Chen, Falon C. Kalutantirige, and Qian Chen. 2023. "Electron microscopy studies of soft nanomaterials." *Chemical Reviews* 123 (7): 4051–4145.

Ma, Boyuan, Xiaojuan Ban, Haiyou Huang, Yulian Chen, Wanbo Liu, and Yonghong Zhi. 2018. "Deep learning-based image segmentation for al-la alloy microscopic images." *Symmetry* 10 (4): 107.

Ma, Wufei, Elizabeth J. Kautz, Arun Baskaran, Aritra Chowdhury, Vineet Joshi, Bulent Yener, and Daniel J. Lewis. 2020. "Image-driven discriminative and generative machine learning algorithms for establishing microstructure–processing relationships." *Journal of Applied Physics* 128 (13): 134901.

Madsen, Jacob, Pei Liu, Jens Kling, Jakob Birkedal Wagner, Thomas Willum Hansen, Ole Winther, and Jakob Schiøtz. 2018. "A deep learning approach to identify local structures in atomic-resolution transmission electron microscopy images." *Advanced Theory and Simulations* 1 (8): 1800037.

McGreevy, R.L. 2001. "Reverse Monte Carlo modelling." *Journal of Physics: Condensed Matter* 13 (46): R877.

McGreevy, R.L., and L. Pusztai. 1988. "Reverse Monte Carlo simulation: a new technique for the determination of disordered structures." *Molecular Simulation* 1 (6): 359–367.

McInnes, Leland, John Healy, and James Melville. 2018. "Umap: uniform manifold approximation and projection for dimension reduction." *arXiv preprint arXiv:1802.03426*.

Mitragotri, Samir, and Joerg Lahann. 2012. "Materials for drug delivery: innovative solutions to address complex biological hurdles." *Advanced Materials* 24 (28): 3717–3723. https://doi.org/10.1002/adma.201202080.

Modarres, Mohammad Hadi, Rossella Aversa, Stefano Cozzini, Regina Ciancio, Angelo Leto, and Giuseppe Piero Brandino. 2017. "Neural network for nanoscience scanning electron microscope image recognition." *Scientific Reports* 7 (1): 1–12.

Mourdikoudis, Stefanos, Roger M. Pallares, and Nguyen T.K. Thanh. 2018. "Characterization techniques for nanoparticles: comparison and complementarity upon studying nanoparticle properties." *Nanoscale* 10 (27): 12871–12934.

Olds, Daniel P., and Phillip M. Duxbury. 2014. "Efficient algorithms for calculating small-angle scattering from large model structures." *Journal of Applied Crystallography* 47 (3): 1077–1086. https://doi.org/10.1107/S1600576714005925.

Patil, Anvay, Christian M. Heil, Bram Vanthournout, Markus Bleuel, Saranshu Singla, Ziying Hu, Nathan C. Gianneschi, Matthew D. Shawkey, Sunil K. Sinha, and Arthi Jayaraman. 2021. "Structural color production in melanin-based disordered colloidal nanoparticle assemblies in spherical confinement." *Advanced Optical Materials*: 2102162.

Patil, Anvay, Christian M. Heil, Bram Vanthournout, Saranshu Singla, Ziying Hu, Jan Ilavsky, Nathan C. Gianneschi, Matthew D. Shawkey, Sunil K. Sinha, Arthi Jayaraman, and Ali Dhinojwala. 2022. "Modeling structural colors from disordered one-component colloidal nanoparticle-based Supraballs using combined experimental and simulation techniques." *ACS Materials Lette rs* 4 (9): 1848–1854. https://doi.org/10.1021/acsmaterialslett.2c00524

Pearson, Karl. 1901. "LIII. On lines and planes of closest fit to systems of points in space." *The London, Edinburgh, and Dublin Philosophical Magazine and Journal of Science* 2 (11): 559–572.

Pedersen, J.S. 1997a. "Analysis of small-angle scattering data from colloids and polymer solutions: modeling and least-squares fitting." *Advances in Colloid and Interface Science* 70 (0): 171.

Pedersen, J.S. 1997b. "Analysis of small-angle scattering data from colloids and polymer solutions: modeling and least-squares fitting." *Advances in Colloid and Interface Science* 70: 171–210.

Pedersen, J.S. 2000. "Form factors of block copolymer micelles with spherical, ellipsoidal and cylindrical cores." *Journal of Applied Crystallography* 33 (3-1): 637–640. https://doi.org/10.1107/S0021889899012248.

Pedersen, J.S., and Michael C. Gerstenberg. 1996. "Scattering form factor of block copolymer micelles." *Macromolecules* 29 (4): 1363–1365.

Pokorski, Jonathan K., and Michael J.A. Hore. 2019. "Structural characterization of protein-polymer conjugates for biomedical applications with small-angle scattering." *Current Opinion in Colloid & Interface Science*. https://doi.org/10.1016/ j.cocis.2019.08.001. http://www.sciencedirect.com/science/article/pii/S1359029419300020.

Qi, Meiwei, and Yongfeng Zhou. 2019. "Multimicelle aggregate mechanism for spherical multimolecular micelles: from theories, characteristics and properties to applications." *Materials Chemistry Frontiers* 3 (10): 1994–2009.

Qu, Eric Zhonghang, Andrew Matthew Jimenez, Sanat K. Kumar, and Kai Zhang. 2021. "Quantifying nanoparticle assembly states in a polymer matrix through deep learning." *Macromolecules* 54 (7): 3034–3040.

Redmon, Joseph, Santosh Divvala, Ross Girshick, and Ali Farhadi. 2016. "You only look once: unified, real-time object detection." *Proceedings of the IEEE Conference on Computer Vision and Pattern Recognition.*

Redmon, Joseph, and Ali Farhadi. 2018. "Yolov3: an incremental improvement." *arXiv preprint arXiv:1804.02767.*

Ren, Shaoqing, Kaiming He, Ross Girshick, and Jian Sun. 2015. "Faster r-cnn: towards real-time object detection with region proposal networks." *Advances in Neural Information Processing Systems* 28.

Roberts, Graham, Simon Y. Haile, Rajat Sainju, Danny J. Edwards, Brian Hutchinson, and Yuanyuan Zhu. 2019. "Deep learning for semantic segmentation of defects in advanced STEM images of steels." *Scientific Reports* 9 (1): 12744.

Ronneberger, Olaf, Philipp Fischer, and Thomas Brox. 2015. "U-net: convolutional networks for biomedical image segmentation." *International Conference on Medical Image Computing and Computer-Assisted Intervention.*

Salacuse, J.J., and G. Stell. 1982. "Polydisperse systems: statistical thermodynamics, with applications to several models including hard and permeable spheres." *The Journal of Chemical Physics* 77 (7): 3714–3725.

Scardi, Paolo, Simon J.L. Billinge, Reinhard Neder, and Antonio Cervellino. 2016. "Celebrating 100 years of the Debye scattering equation." *Acta Crystallographica. Section A, Foundations and Advances (Online)* 72 (BNL-112736–2016-JA).

Schmidt, P.W. 1995. *Modern Aspects of Small-Angle Scattering*. NATO Science Series C. Springer, 1–55.

Schmidt-Rohr, Klaus. 2007. "Simulation of small-angle scattering curves by numerical Fourier transformation." *Journal of Applied Crystallography* 40 (1): 16–25. https://doi.org/10.1107/S002188980604550X.

Sen, Sanghamitra, Shradha Patil, and Dimitris S. Argyropoulos. 2015. "Thermal properties of lignin in copolymers, blends, and composites: a review." *Green Chemistry* 17 (11): 4862–4887. https://doi.org/10.1039/C5GC01066G.

Shen, Dinggang, Guorong Wu, and Heung-Il Suk. 2017. "Deep learning in medical image analysis." *Annual Review of Biomedical Engineering* 19: 221.

Simonyan, Karen, and Andrew Zisserman. 2014. "Very deep convolutional networks for large-scale image recognition." *arXiv preprint arXiv:1409.1556*.

Sonje, Jayesh, Seema Thakral, Susan Krueger, and Raj Suryanarayanan. 2023. "Enabling efficient design of biological formulations through advanced characterization." *Phar maceutical Research*. https://doi.org/10.1007/s11095-023-03495-z.

Strohmann, Tobias, Katrin Bugelnig, Eric Breitbarth, Fabian Wilde, Thomas Steffens, Holger Germann, and Guillermo Requena. 2019. "Semantic segmentation of synchrotron tomography of multiphase Al-Si alloys using a convolutional neural network with a pixel-wise weighted loss function." *Scientific Reports* 9 (1): 19611.

Svaneborg, Carsten, and Jan Skov Pedersen. 2000. "A Monte Carlo study on the effect of excluded volume interactions on the scattering from block copolymer micelles." *The Journal of Chemical Physics* 112 (21): 9661–9670.

Szegedy, Christian, Sergey Ioffe, Vincent Vanhoucke, and Alexander Alemi. 2017. "Inception-v4, inception-resnet and the impact of residual connections on learning." *Proceedings of the AAAI Conference on Artificial Intelligence*.

Szegedy, Christian, Vincent Vanhoucke, Sergey Ioffe, Jon Shlens, and Zbigniew Wojna. 2016. "Rethinking the inception architecture for computer vision." *Proceedings of the IEEE Conference on Computer Vision and Pattern Recognition*.

Tan, Mingxing, and Quoc Le. 2019. "Efficientnet: rethinking model scaling for convolutional neural networks." *International Conference on Machine Learning*.

Tiarks, F., T. Frechen, S. Kirsch, J. Leuninger, M. Melan, A. Pfau, F. Richter, B. Schuler, and C.L. Zhao. 2003. "Formulation effects on the distribution of pigment particles in paints." *Progress in Organic Coatings* 48 (2): 140–152. https://doi.org/10.1016/S0 300-9440(03)00095-X. https://www.sciencedirect.com/science/article/pii/S030094400300095X.

Tóth, Gergely. 1999. "Simultaneous Monte Carlo determination of particle size distribution and pair-correlation function of spherical colloids from a diffraction experiment." *Langmuir* 15 (20): 6718–6723.

Treder, Kevin P., Chen Huang, Judy S. Kim, and Angus I. Kirkland. 2022. "Applications of deep learning in electron microscopy." *Microscopy* 71 (Supplement_1): i100–i115.

Tsopanidis, Stylianos, Raúl Herrero Moreno, and Shmuel Osovski. 2020. "Toward quantitative fractography using convolutional neural networks." *Engineering Fracture Mechanics* 231: 106992.

Van der Maaten, Laurens, and Geoffrey Hinton. 2008. "Visualizing data using t-SNE." *Journal of Machine Learning Research* 9 (11).

Webber, Matthew J., and Robert Langer. 2017. "Drug delivery by supramolecular design." *Chemical Society Reviews* 46 (21): 6600–6620. https://doi.org/10.1039/C7CS00391A.

Weiss, Keith D. 1997. "Paint and coatings: a mature industry in transition." *Progress in Polymer Science* 22 (2): 203–245. https://doi.org/10.1016/S0 079-6700(96)00019-6. https://www.sciencedirect.com/science/article/pii/S0079670096000196.

Wessels, M.G., and A. Jayaraman. 2020. "Self-assembly of amphiphilic polymers of varying architectures near attractive surfaces." *Soft Matter* 16 (3): 623–633. https://doi.org/10.1039/c9sm02104c. https://www.ncbi.nlm.nih.gov/pubmed/31808757.

Wessels, M.G., and A. Jayaraman. 2021a. "Computational reverse-engineering analysis of scattering experiments (CREASE) on amphiphilic block polymer solutions: cylindrical and fibrillar assembly." *Macromolecules* 54 (2): 783–796.

Wessels, M.G., and A. Jayaraman. 2021b. "Machine learning enhanced computational reverse engineering analysis for scattering experiments (crease) to determine structures in amphiphilic polymer solutions." *ACS Polymers Au* 1 (3): 153–164.

White, Sadie I., Brian A. DiDonna, Minfang Mu, Tom C. Lubensky, and Karen I. Winey. 2009. "Simulations and electrical conductivity of percolated networks of finite rods with various degrees of axial alignment." *Physical Review B* 79 (2): 024301.

Williams, Christopher, and Carl Rasmussen. 1995. "Gaussian processes for regression." *Advances in Neural Information Processing Systems* 8.

Wu, Xindong, Vipin Kumar, J. Ross Quinlan, Joydeep Ghosh, Qiang Yang, Hiroshi Motoda, Geoffrey J. McLachlan, Angus Ng, Bing Liu, and Philip S. Yu. 2008. "Top 10 algorithms in data mining." *Knowledge and Information Systems* 14: 1–37.

Wu, Zijie, and A. Jayaraman. 2022. "Machine learning-enhanced computational reverse-engineering analysis for scattering experiments (CREASE) for analyzing fibrillar structures in polymer solutions." *Macromolecules* 55 (24): 11076–11091. https://doi.org/10.1021/acs.macromol.2c02165.

Xie, Saining, Ross Girshick, Piotr Dollár, Zhuowen Tu, and Kaiming He. 2017. "Aggregated residual transformations for deep neural networks." *Proceedings of the IEEE Conference on Computer Vision and Pattern Recognition*.

Yaman, Muammer Y., Sergei V. Kalinin, Kathryn N. Guye, David S. Ginger, and Maxim Ziatdinov. 2023. "Learning and predicting photonic responses of plasmonic nanoparticle assemblies via dual variational autoencoders." *Small* n/a (n/a): 2205893. https://doi.org/10.1002/smll.202205893. https://onlinelibrary.wiley.com/doi/abs/10.1002/smll.202205893.

Yao, Lehan, Zihao Ou, Binbin Luo, Cong Xu, and Qian Chen. 2020. "Machine learning to reveal nanoparticle dynamics from liquid-phase TEM videos." *ACS Central Science* 6 (8): 1421–1430.

Ye, Ziyu, Zijie Wu, and Arthi Jayaraman. 2021. "Computational reverse engineering analysis for scattering experiments (CREASE) on vesicles assembled from amphiphilic macromolecular solutions." *JACS Au* 1 (11): 1925–1936.

Zemb, Thomas, and Peter Lindner. 2002. *Neutrons, X-rays and Light: Scattering Methods Applied to Soft Condensed Matter*. North-Holland.

Zhao, Dan, En Wang, and Timothy P. Lodge. 2020. "Hybridization of a bimodal distribution of copolymer micelles." *Macromolecules* 53 (18): 7705–7716.

Ziatdinov, Maxim, Ondrej Dyck, Artem Maksov, Xufan Li, Xiahan Sang, Kai Xiao, Raymond R. Unocic, Rama Vasudevan, Stephen Jesse, and Sergei V. Kalinin. 2017. "Deep learning of atomically resolved scanning transmission electron microscopy images: chemical identification and tracking local transformations." *ACS Nano* 11 (12): 12742–12752.

Zölls, Sarah, Ruedeeporn Tantipolphan, Michael Wiggenhorn, Gerhard Winter, Wim Jiskoot, Wolfgang Friess, and Andrea Hawe. 2012. "Particles in therapeutic protein formulations, part 1: overview of analytical methods." *Journal of Pharmaceutical Sciences* 101 (3): 914–935. https://doi.or g/10.1002/jps.23001. https://www.sciencedirect.com/science/article/pii/S0022354915316865.

Section 3

Predictive Modeling

Alix Schmidt and Kristin Wallace

By the time you have your data set ready for modeling, the team is likely to be tired of data wrangling and ready to see some results. Although you may have produced several models during data preparation to gauge the effectiveness of the data set and improvements made during feature engineering, most of the fun is yet to come. As discussed in Chapter 3, the goal of the modeling exercise is to find the model form and parameters that best represent the underlying structure of the data, often without knowing that underlying structure *a priori*. This is where model experimentation comes in: by testing a variety of machine learning algorithms suited to various data patterns and qualities, the team is able to identify the best-performing model and learn about the system. Additionally, there are many decisions to be made about how to map your formulation design business problem to a model output in the most effective way.

Section 3 is all about predictive modeling to solve product formulation challenges. We begin with an introduction to predictive modeling in Chapter 7. Here the authors present concepts of modeling such as the bias-variance trade-off, how to think about the advantages and disadvantages of various algorithms, and the wide variety of business problems that can be solved with models. Chapter 7 also contains a survey of relevant machine learning algorithms, from standard boilerplate methods to some more advanced techniques, where application to formulation design is cutting edge. Chapter 8 goes in depth into one technique in particular: partial least squares (PLS) regression. This method is tried and true – well suited to the formulation problem

DOI: 10.1201/9781003385974-9

and with a significant body of literature to support the reader in further study. The chapter includes a detailed explanation of the algorithm as well as recommended software packages for its application.

Chapter 9 discusses an often-overlooked challenge, that of gaining trust in the model. What can teams do to understand how the model works, if it agrees with the prevailing subject matter knowledge, and what are the boundaries past which the model should *not* be used? This discussion on model validation is valuable to all readers, technical and non-technical alike.

7 Machine Learning Techniques for Predicting Properties of Formulations

Maxwell Hutchinson, Erin Antono and Sean Paradiso

7.1 INTRODUCTION

In this chapter, we present several effective and well-established techniques for predicting the properties of formulations given their formulation ratios, descriptors, and process conditions. Our goal is to identify high-value, low-risk opportunities for the application of these techniques to a variety of problems that are important to formulation development and production. Importantly, as presented in Chapter 3, a significant portion of this challenge is not solely in choosing and fitting a machine learning algorithm but in defining your prediction goal, understanding the modeling work process, understanding your data, and then planning your data preparation, model fitting, and model validation work. As you will see in this chapter, careful consideration of these components can help you understand which prediction tasks are relevant to your team, what are the bounds of what is possible with your data, and which techniques you may need to incorporate to be successful.

We begin this explanation by highlighting various formulation prediction use cases to define the contexts in which the following machine learning methods are used. We then define a general formulation prediction problem and discuss how it varies depending on the use case. We conclude the introduction with some words of warning about realistic expectations for the performance of machine learning techniques, particularly early in a project.

Following the introduction, we get to the modeling – Section 7.2 describes the most common and effective prediction methods: Linear Models, Random Forests, Gaussian Processes, and Neural Networks. These methods, along with multi-variate analysis and related techniques, are the core building blocks for effective modeling of formulations.

Following that, Section 7.3 describes several advanced techniques that can be employed when a simpler model is unable to provide adequate accuracy, such as when data is scarce. Although unsupervised learning methods, for example, auto-encoders,

DOI: 10.1201/9781003385974-10

have been applied to problems where labeled training data is scarce (Sevgen et al. 2021), they are out of scope for this chapter.

We conclude the chapter with a walk-through of an example problem in Section 7.4.

7.1.1 Prediction Use Cases

This chapter is focused on the prediction of quality properties of formulations, given their formulation ratios, processing conditions, and other properties. Such predictions can be applied to many different business and research problems, as described in the first chapter of this book. Here, we review three of the most common applications of formulation property prediction: inverse design, ingredient substitution, and imputation. This will set the stage for the definition of the general formulation prediction problem, which is used in the rest of the chapter.

Inverse Design

> The most common use of formulation property prediction is inverse design: given a set of target quality properties, design a formulation that meets those targets (Ling, Hutchinson, Antono, Paradiso, et al. 2017).

Property prediction is used to evaluate the performance of candidate recipes more quickly and less expensively than in the laboratory. The property prediction method must be paired with another method to generate candidate recipes, for example, a list, a grid, and an optimizer.

> Inverse design problems are almost exclusively multi-objective; a core challenge in inverse design is often balancing two opposing performance trends.

Inverse design problems can vary greatly in their difficulty, for example, tuning a known composition to generate a 5% lift in a quality property versus designing a completely new formulation with a goal of doubling performance.

Inverse design is the primary subject of Chapters 10 and 11. Given its primary importance as a use case, we will highlight the connection between the property prediction method and the rest of the inverse design problem in this chapter. For example, the reliable prediction of uncertainty is key to evaluating performance of potential recipes, particularly difficult ones that no individual candidate is likely to achieve.

Ingredient Substitution

Of course, not all formulation design is novel or radical – often, we are looking to simply modify an existing formulation by substituting ingredients. This could be motivated by environmental, health, and safety hazards (and corresponding regulations); ingredient availability; and/or cost reduction.

When searching for substitute ingredients, an effective initial approach is to filter candidate ingredients based on ingredient properties.

This filtering should consider both properties relevant to ingredient performance and function, as well as potential hazards. Computational screening is attractive due to the large number of possible ingredients. This screening can be supported through ingredient property databases or ingredient property prediction if ingredient properties are not available. Machine learning models are particularly useful for properties such as environmental hazards where traditional models are less accurate (Syeda et al. 2022).

In some cases, one may also need to consider the interactions between a new ingredient and the rest of the formulation (Jhamb et al. 2019). This requires the prediction of formulation properties with the candidate substitution ingredients. Scientists may also need to take on more involved reformulation to accommodate ingredient substitution (Harten et al. 2020). This reformulation problem is essentially the inverse design problem, where you would use property prediction to evaluate a candidate recipe and pair it with a method to generate candidates.

When considering machine learning models for property prediction in the ingredient substitution use case, there are a few factors that make the modeling problem a bit more tractable than a typical inverse design problem. First, the search space is typically constrained to the neighborhood of the original formulation and there is likely to be training data that is similar to the candidate formulations. You may also be able to leverage more domain knowledge in your model building if you are familiar with trends and governing equations in the neighborhood of the existing formulation. Finally, the property prediction will not be extrapolative, as the target performance will be similar to the original formulation.

Sparse Characterization and Imputation

The final prediction use case is for data imputation. Consider the example dataset for four materials in Table 7.1, which illustrates several cases of missing property information. Everything is known about the first formulation. The second is missing a process condition. The second, third, and fourth formulations are each missing a different quality property, and the fourth is also of unknown composition.

As discussed in Chapter 4, in experimental datasets, both process conditions and quality properties are not always uniformly populated for all formulations. This can be for a variety of reasons: characterization cost/difficulty, changing standards, formulations developed for different use cases, etc. The missing data can impede or outright prevent the application of a wide variety of data-driven methods ranging from business intelligence reporting to outlier detection. In the extreme case of Formulation 4, the formulation itself is unknown, so only the quality properties B_4 and C_4 are available to use to impute A_4.

TABLE 7.1
Example Sparse Dataset

Formulation ID	Composition Vector	Process Condition *a*	Property A	Property B	Property C
1	X_1	α_1	A_1	B_1	C_1
2	X_2		A_2	B_2	
3	X_3	α_3	A_3		C_3
4				B_4	C_4

The sparse characterization or imputation use case leverages property prediction to fill in missing characterization data.

7.1.2 Prediction Task Components

With those three use cases in mind, we will now review the general formulation property prediction problem, which will be used as the framework for the rest of the chapter. Figure 7.1 depicts a simple formation prediction problem procedure. The formulation data consists of formulation ratios, processing conditions, quality properties, and optionally ingredient properties. The processing conditions and formulation ratios are featurized into process and mixture features, respectively. Those are fed into the machine learning model, which is trained to learn the quality properties from them. When performing a prediction, the formulation ratios and processing conditions are specified and the quality properties are predicted.

Training Data and Feature Engineering

Machine learning is a data-driven method – the quantity and quality of the data is the single most important factor for the success of a machine learning project.

Typical data sources for formulations project are described extensively in Chapter 4 and reviewed briefly here.

Formulation Ratios

The formulation ratios are quantities, in some basis, of the ingredients to the formulation. They could be represented as a large vector, with each position corresponding to the quantity of a specific ingredient. An alternative is to use a sparse representation with two coordinated vectors, the first containing the names of the ingredients

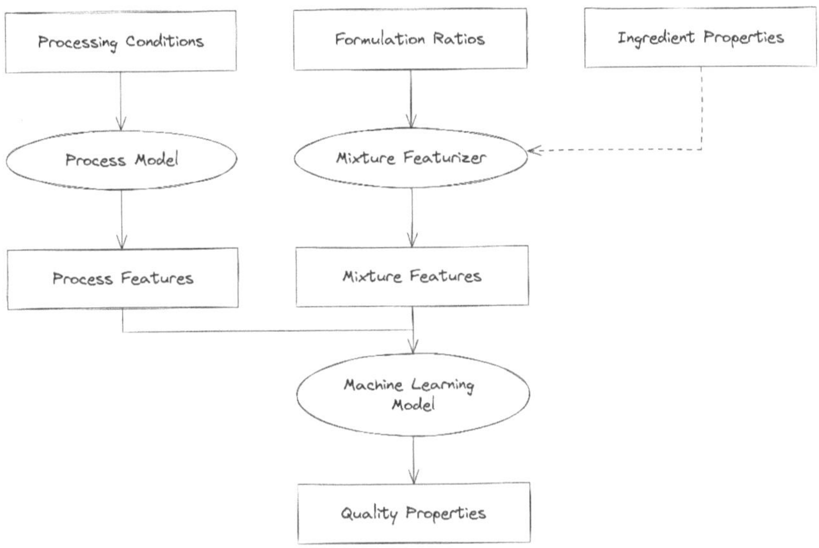

FIGURE 7.1 General formulation prediction diagram.

that are present in the formulation and the second containing the quantities of those ingredients. The quantities can usually be normalized such that they sum to one.

When the number of possible ingredients is large, the signal in the formulation ratios can become very dilute, resulting in a weak model. This makes sense intuitively: in order to predict the impact of an ingredient based on its name and quantity alone, it must have been observed in a variety of formulations with similar ingredients to the case being predicted.

With formulations ratios alone, predictions with rare ingredients will be much less accurate, and new ingredients are impossible to predict.

This problem can be addressed by incorporating additional data about the ingredients themselves, that is, ingredient properties.

Ingredient Properties

Ingredient properties describe the ingredients used in the formulation. Common examples include molecular weight, density, viscosity, and vapor pressure. They can be nominal, that is, as specified by the manufacturer, or measured, depending on what kinds of variations are the most important in the modeling task. Often, different ingredients will have different sets of ingredient properties.

The most relevant and available properties are dependent on the function of the ingredient within the formulation.

Ingredient properties and formulation ratios are typically combined into mixture properties using a mixture model.

Process Conditions

Representing the formulation process can be quite challenging, since processing often includes multiple steps with different conditions for each step. Ideally, there are a few very important parameters that capture the majority of the variation of the process, for example, total cure time or mixing temperature. When parameters vary over time, a simple piecewise decomposition into phases with duration and constant conditions can be effective.

> When the behavior of the formulation depends on more fine grain details of the process, it might not be a good candidate for property prediction in the sense used in this chapter.

Complex and/or subtle dependence on the details of the process will require far greater quantities of data to model, as the model will need to discover which aspects of the process are important.

Quality Properties

> Quality properties are the properties of interest that are to be predicted.

Common examples include physical properties like yield strength or hardness, thermal properties like thermal conductivity or heat capacity, and electromagnetic properties like electrical conductivity or magnetic permeability. It may be necessary to predict these properties under several operating conditions, for example, a range of temperatures, humidities, or system ages, depending on what business objectives and success criteria your team defined at the start of the project. It is also common to predict several related properties at the same time, for example, the yield strength and density.

Process Model/Process Featurizer

The process model transforms the processing conditions into features that are expected to be predictive of the quality properties, that is, it featurizes the processing conditions. The simplest process model returns the processing conditions unmodified, and many process models include the raw processing conditions as features. Some additional features may be normalizations or non-dimensionalizations of the processing conditions, for example, the flow rate divided by the nozzle area or the mixing Reynolds number. Others may be simple arithmetic combinations of processing conditions, especially aggregations across process steps, for example, the total mixing time in a multi-step mixing process.

Process models serve two purposes: first, they concentrate more predictive signal in a smaller number of features (recall the curse of dimensionality), and, second, they provide a way to unify data from multiple data sources.

Process models are described more thoroughly in Chapter 5, Section 5.2.2.

Mixture Model/Mixture Featurizer

Like the process model, the mixture model transforms the formulation ratios into a constant-length feature vector that is expected to be predictive of the quality properties, that is, it featurizes the mixture defined by the formulation ratios. The simplest mixture model returns the quantity of each possible ingredient as a vector, generally a highly sparse vector. If the ingredients can be grouped into classes, then the aggregate quantities of each class may also be included, for example, the total quantity of solvents. If ingredient property data is also available, then mixture properties can be computed as weighted generalized means of the ingredient properties by the formulation ratios, for example, the average molecular weight of the formulation. If there are also ingredient groups, then mixture properties can be computed per group in addition to over the entire formulation, for example, the average vapor pressure of the solvents.

Even when ingredient properties and/or groups are available, it may make sense to include the formulation ratios of some ingredients, particularly for important ingredients like water.

Mixture featurization is described more thoroughly in Chapter 4, Section 4.4.2, and Chapter 5, Section 5.1.2.

Models

The machine learning model itself takes the features and predicts the output properties.

The model may be able to predict all of the output properties in a single coherent step, but it is also common for the model to be composite, that is, composed of several models that together predict all of the output properties.

The simplest case is usually a composite model where each quality property is predicted by its own model independently. A more advanced example is a sequential model, where the quality properties are predicted in a sequence with all prior quality properties being used as an input to the following quality properties. These three examples can be seen in Figure 7.2.

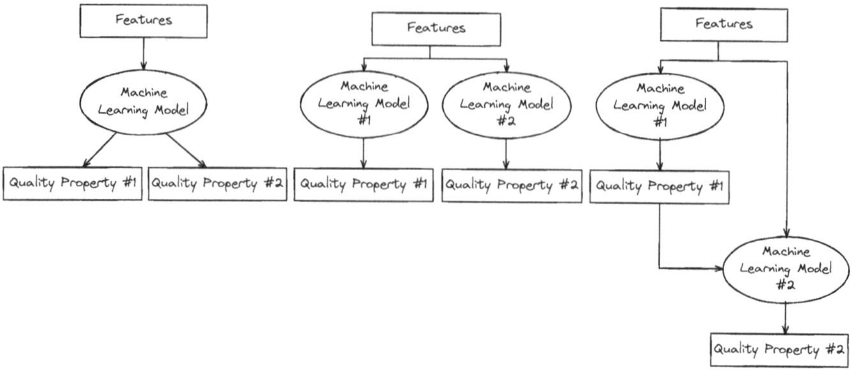

FIGURE 7.2 Three model structures for the prediction of two quality properties.

Training Data

In order to train a model, you need training data. In the simplest case, the training data looks exactly like the predictions you'd like to make, but with the quality properties already known. In this context, those known properties are called "labels."

> In product formulation projects, the amount of labeled training data is often very small, or there might not even be any at the start of the project.

In the absence of labeled training data, one can use unlabeled data for unsupervised learning methods, which are out of the scope of this text.

In many cases, there may be some training data that describe the ingredients or formulations but are not the specific quality properties of interest. For example, the properties of ingredients are valuable data, but they have neither the same inputs nor generally the same properties as the quality properties being predicted for the formulation. Formulation data that is labeled for related quality properties can also be highly valuable for transfer learning, which is discussed in Section 7.3.

When assembling training data from multiple sources, it is particularly important to consider the data that is left implied (i.e., un-recorded). For example, every stress-strain measurement in one dataset might have been performed on one temperature and humidity while another dataset might have the same measurements performed at a different temperature and humidity. Changes in equipment or experimental procedure may also have an impact on results. As discussed in Chapter 3, these variations can contribute significantly to the experimental noise in the data and pose challenges for modeling.

> The decision of whether and how to combine datasets must be assessed with scientific scrutiny to determine whether the data are comparable.

If there are insufficient details to make this assessment, some data-centric options are possible: One technique is to add a categorical descriptor indicating the data source to test for data compatibility. If there are important features omitted from the data that vary across the data points, then this source descriptor may capture that variation and indicate a data issue. Note that the categorical descriptor can also capture sampling bias, that is, systematic differences in the types of formulations present in the two datasets, which would not be a concern for combining the datasets; in this case, the formulation properties do vary in a statistically detectable way, but that difference is "intentional" in the type of material being formulated rather than an unintentional consequence of missing metadata. A second option would be to treat the two datasets as having two different kinds of labels, that is, "Quality Property from Dataset A" and "Quality Property from Dataset B," and then use a transfer learning technique described in Section 7.3.1 to combine them at the model level.

Model Validation

Before using a model to make predictions, one must validate that model.

> In the field of machine learning, model validation typically involves hiding some of the labels to produce smaller dataset, training the model on that dataset, and making predictions for the hidden labels in a process called cross-validation. If the predictions are accurate, then it suggests that predictions on similar unknown data will also be accurate.

Additional validation can be performed using human experts when applying machine learning in organizations with extensive in-house expertise, as is common for formulation projects. Model validation of both sorts is described extensively in Chapter 9.

Predicted Uncertainty

Model validation techniques tend to work with the data that you have. However, property prediction is often applied to formulations that do not yet exist, especially in the inverse design use case. Those predictions may differ significantly from the held-out data used in cross-validation, resulting in different distributions of errors. For example, consider a model trained only on formulations containing more than 80% water. Cross-validation would also only consider formulations containing more than 80% water. In the process of inverse design, two formulations with only 70% and only 60% water, respectively, could be generated and predicted. We might generally expect the error of the predictions of the 70% water formulation to be greater than the error in cross-validation, and for the error in the 60% water formulation to be even greater.

Fortunately, there are approaches for estimating errors that look specifically at the formulation on which the prediction is being made. We refer to these methods as "predicted uncertainty," as the uncertainty in the prediction is also being predicted. This is closely related to the fields of uncertainty quantification and more generally to statistical learning.

You can think of these methods as adding an error bar to the prediction, though the prediction can be of a richer object like a multi-variate distribution. Importantly, different predictions will have different error bars. These error bars can serve a similar purpose to model validation in that they help characterize the confidence one should have in the prediction.

They can also be used to understand which areas of the input space are well understood versus which have not been evaluated yet. This can be used when designing new formulations to encourage exploration in the input space, as discussed in Chapter 11.

7.1.3 Setting Expectations

In this section, we discuss common challenges in the useful application of machine learning to formulation property prediction and offer some words of warning to prospective practitioners with the hope of managing expectations.

Model Accuracy

Model predictions will not be perfectly accurate. The performance of machine learning models is fundamentally limited by the capacity of the model to fit the response surface, the data that the model has to learn from, and how challenging the response surface is to fit.

In practical formulation modeling, the data will nearly always be the limiting factor. Formulation datasets tend to be extremely small (10s to 1000s of samples) in comparison to other machine learning applications due to the high cost of collecting physical experimentation data. Experimental data is also subject to noise from sample creation and data collection processes, which further limits the accuracy of models.

Practitioners should *expect* their model to have limited accuracy, especially if they have a low amount of data, if they suspect the data to be noisy, or if they suspect that the response surface is challenging to fit (e.g., subject-matter experts know there are non-linear relationships at play). This often correlates with an expert's intuition for how challenging it would be to predict the property. In the authors' experience, it is not unusual for the initial model to have a coefficient of determination (R^2) below 0.5. In other words, the model is able to explain less than half of the variation in the predicted value.

There are two common approaches for dealing with such low-accuracy models. The first is to adopt a data strategy, begin to collect more data, and wait until the data supports more accurate models. The second is to carefully select projects for which the initial model quality is not a primary determinant of project success, for example, inverse design in a mostly unexplored region of formulation space. Recall the discussion in Chapter 3 regarding understanding the "baseline" accuracy to beat in order to have a valuable model; in this example case, because domain experts may have a low prediction accuracy in the unexplored region (i.e., they are not sure what will happen), even a low accuracy model is providing guiding information. These two approaches are not mutually exclusive; it is our opinion that organizations should establish data strategies while beginning work on appropriate projects. The types of projects that are ready to kick-off will expand as more data is collected. In this way, the initial model can create value for "low-hanging fruit" opportunities while further data collection occurs, and in the future an improved model can tackle more challenging tasks.

The details of a successful data strategy are broadly outside the scope of this chapter. See Chapter 1 for a high-level overview of strategies for digital transformation and Chapter 4 for a review of approaches for managing data specifically.

Projects that include additional data collection in their scopes are generally the least sensitive to initial data quality. In fact, property prediction coupled with an exploratory design goal is an excellent way to generate new high-quality datasets. Conversely, lower accuracy models cannot replace experimentation; don't simply rely on predictions; always plan to verify them experimentally.

Effective workflows often include a human in the loop. For example, when performing an inverse design, you can use the model to evaluate a large number of possibilities and rank/prioritize promising experiments, and then have a human select the set of experiments to actually run.

Local Model Accuracy

Models will not be uniformly accurate across the input space. Just as the model accuracy summarized by statistics like RMSE (root mean squared error) and R^2 are impacted by the quantity/quality of data and the complexity of the response surface, the local model accuracy is impacted by the local data density and local complexity of the response surface. For example, if there is a phase transition in the system but the training data includes only points on one side of it, then the model will be inaccurate on the other side. For formulations, this can also happen when making predictions with a new or rarely used ingredient, or an ingredient that behaves very differently from other past ingredients. Predictions that are outside or beyond the region of the training data, a phenomenon called covariant shift, are almost always inaccurate.

We again emphasize the importance of ensuring that your target design space for creating business value has overlapped with your training data space.

Poor local model accuracy in a region of interest presents all the same challenges as overall model accuracy, even when the model is reported to be accurate overall. However, it can be challenging to recognize this problem ahead of time. For example, we have found it to be easy to accidentally make predictions far from the training data, especially with very high-dimensional problems, for example, lots of possible ingredients. Similarly, regions where the quality property is varying rapidly may have worse fits and not be identifiable a priori.

The best way to recognize local model accuracy issues ahead of time is to use a property prediction method that predicts uncertainty. Predictions with high uncertainty indicate lower model accuracy, while predictions with low uncertainty should generally correspond to more accurate predictions. The pros and cons of specific predicted uncertainty methods are discussed more in Sections 7.2 and 7.3.

Another approach is to lean on expertise.

A domain expert can often quickly recognize a formulation as being very different from anything else in the training data.

Experts can sometimes even recognize challenging areas of the formulation space where multiple factors conspire to complicate the property prediction task. Just as before, it can be especially useful to run formulation candidates by experts before investing in synthesizing and measuring them, especially early in a project.

Model Granularity

A model's sensitivity to changes in inputs is also limited by data. The performance of the model near data will always be better than the performance far away from data. If you have data on coarse input changes/only a few levels (e.g., only tried each ingredient once), expect that your model will not be too effective at predicting the response to small changes in formulation ratios. This is particularly true of non-smooth models, for example, random forests, as described in Section 7.2.4. Fine-tuning a formulation generally requires the target degrees of freedom to be sampled at a higher granularity/resolution. It is more important to have a finer grid than to only vary a single degree of freedom at a time.

Output Extrapolation

Because our business goals often involve creating better-performing materials, we often want to predict extreme values that lie outside the range of observed quality properties. Some model types, for example, random forests, are explicitly unable to do this, because their predictions are mixtures of observed values. Other model types will be able to do this but may "blow up" or otherwise behave unexpectedly in extreme regions.

Generally, any prediction that exceeds a known value should be treated with skepticism.

Rather than try to identify formulations that will exhibit extreme values, property prediction with predicted uncertainty can be used to estimate how likely a formulation is to be extreme. The formulations with high likelihood of extreme behavior will typically have relatively large or small predicted value and also very high uncertainty, indicating that they lie in an unexplored region of formulation space.

Model Speed

The speed at which a digital tool can provide a result determines how your users will interact with it day-to-day. A key benefit of ML models is that predictions can be made quickly. They will be extremely fast compared to collecting data through experimental methods and also most physics-based computation methods.

However, ML models aren't instantaneous, particularly when predicting uncertainty. Typically, predictions for a single formulation can be performed quickly enough to fit into an interactive workflow (i.e., the user is changing values and re-running to probe the model in real-time), while batches of predictions will incur enough latency to break interactivity.

Workflows that involve evaluating millions or more of possible formulations, like inverse design, will often take hours or even days. Fortunately, the small size of formulation datasets usually means the model itself will be small and projects can at least kick-off on personal computers. As projects mature, it is often beneficial to take them to the cloud so that they can be regularly backed up, monitored, and executed asynchronously.

7.2 COMMON PREDICTION METHODS

With the data fully featurized and understood, we may turn our attention to the central component of the formulation prediction project: the learning algorithm. In this section, we review several common prediction methods that are usually the best place to start when developing a formulation property model. For some problems, these methods may even be usable "off the shelf," that is, using popular code packages with relatively little customization. Unfortunately, the challenges common to formulation property prediction often prevent these methods from providing adequate performance on their own. The subsequent section discusses advanced techniques for improving model performance when these methods are insufficient.

Those advanced techniques build upon, rather than replace, these common methods, so it is generally advisable to start simple even if you do not expect the performance to be adequate.

Due to the importance of the inverse design use case, we will particularly note the ability of the model to predict uncertainty. As discussed more thoroughly in Chapter 11, the most effective methods for inverse design balance exploration and exploitation, with the exploratory signal coming from the predicted uncertainty. Therefore, methods that are good at predicting uncertainty are generally good at formulation property prediction for inverse design.

Practically, these methods can all be found in the popular scikit-learn (Pedregosa et al. 2011) Python package. Applicable packages for each method are also available for the R language. For the Java Virtual Machine, the Statistical Machine Intelligence and Learning Engine (SMILE) (Li 2014) package has many core methods, and the Lolo package developed by this Chapter's authors is particularly well suited for predicting uncertainty (Folie and Hutchinson 2023) and inverse design (Ling, Hutchinson, Antono, Paradiso, et al. 2017). In any case, the core prediction methods must be preceded by domain and problem-specific setup and followed by analysis specific to the use case, as described in Chapter 3. The MAST-ML (Jacobs et al. 2020) and matminer (Ward et al. 2018) packages provide a framework for problem setup, though they have been developed with more of a materials science focus.

7.2.1 Linear Methods

Linear models are a family of algorithms widely used for both classification and regression problems where the response is modeled as a linear combination of the input features.

> It is easy to dismiss linear models as inadequate or dated, given the exotic estimators available today, but we believe they are important to review here for two reasons: establishing a simple baseline and interpretability.

In the context of formulations, we have found it useful to establish modeling performance baselines using linear regression for scalar responses (logistic regression for classification). To get a baseline, one can model the quality property as a linear function of the formulation ratios x and mixture properties ϕ_j:

$$f(x) = \sum_i \alpha_i x_i + \sum_j \beta_j \phi_j(x) \tag{7.1}$$

The model weights α_i and β_j describe the effect of the formulation ratios and mixture properties, respectively. Additional terms may be added to Equation 7.1 for any additional blocks of data you have, such as process conditions or other descriptors of the system. Ideally, many of the weights are nearly zero while a few have larger magnitudes, which can suggest terms that can be discarded and others to focus on. Regularization can encourage this splitting of input features into important versus unimportant ones. For example, l_1 -regularization (also known as LASSO regression (Tibshirani 1996)) adds the term $\lambda_\alpha \sum_i |\alpha_i| + \lambda_\beta \sum_j |\beta_j|$ to the squared error, which can have a sparsifying effect on the estimated model weights. Other common types of regularization are l_2 (also known as Ridge Regression or Tinkanov Regularization (Wieringen 2015)) and elastic net regularization (Zou and Hastie 2005), which combines the two. We recommend starting with l_1 regularization.

The code libraries mentioned will have options for applying regularization and adapting the parameter estimation method from the regression case (numerical quality variables) to the classification case (categorical quality variables, see also Support Vector Machines).

It is recommended as a general best practice to standardize all variables excluding formulation ratios to a common mean and variance – typically zero mean and unit variance – in order to remove any biases stemming from varying units of measure.

Conceptually, this ensures that a variable such as viscosity with mean 10,000 won't be given millions of times more weight in the regression than refractive index with mean 0.0015. Formulation ratios are excluded from this procedure since they are inherently collinear and variable-wise transformation would break the sum-to-100% relationship, which gives their values meaning. Instead, they should be scaled such that they sum to 1 (and not 100). After completing the linear model fitting, you will have your first model for predicting formulation performance.

There are two major drawbacks when applying linear methods to the formulations property prediction:

1) Predicting uncertainty: linear regression makes only point predictions.
2) Fitting complex functions: linear regression specifies the functional form upfront and cannot adapt it when it does not fit the data. Linear regression lacks a valuable property known as *asymptotic consistency* – the ability to fit an arbitrary function given enough data.

The following section on Bayesian linear regression attempts to address the challenge of predicting uncertainty. Modern machine learning techniques are broadly focused on producing asymptotically consistent or at least low-bias estimators. The Gaussian Process Regression, Random Forests, and Neural Networks each use different approaches to reduce bias, that is, fit arbitrarily complex functions.

A final challenge with linear regression and related methods is their potentially high cost of fitting. They scale with the cube of the amount of training data, making them impractical for very large datasets. The good news is that formulation data sets are seldom large enough for training time to become an issue.

Bayesian Linear Regression

Bayesian probability theory provides a rigorous and still relatively straightforward way to predict the uncertainty of linear model predictions. In the Bayesian linear regression formalism, we learn a distribution over plausible models that fit the data by treating each coefficient in the linear model as a random variable (Bishop 2006). To predict a new formulation, you can then draw samples of the response for a given input formulation by repeatedly drawing samples for each coefficient and using those coefficients along with the input features (representing a concrete linear model) to calculate the response.

In practice, Bayesian linear regression is much less frequently used than its more powerful cousin, Gaussian Processes, which not only provides a mechanism for predicting uncertainty but also can model more complex functions.

7.2.2 Gaussian Process Regression

Gaussian Processes (GPs), and particularly Gaussian Process Regression (GPR), are a popular model choice in cases when predictive uncertainty quantification is important, for example, inverse design with Bayesian Optimization (Amar et al. 2019; Cao et al. 2020). Unlike traditional regression models that assume a fixed functional form, GPR models the relationship between variables as a distribution over functions. A key feature of GPR, which is both an advantage and potentially a drawback, comes from the ability to choose and tune the kernel function used to model the covariance between data points (Williams and Rasmussen 1995). In principle, this choice provides modelers the ability to control overfitting and smoothness based on their understanding of the problem. In practice, this can become a cumbersome and error-prone process that can actually yield quite poor models if done ineffectively, especially in high dimensional problems, that is, the curse of dimensionality. Still, GPR remains a workhorse for data-driven optimization (Zhang and Xu 2021). The uncertainty behavior is intuitive (the greater the distance between points, the larger the uncertainty estimate) and the smoothness can be controlled through the choice of kernel.

When employing Gaussian Processes, expect to iterate several times on the choice of kernel, inputs and their scale, and the particulars of the hyperparameter optimization process.

Because of this high level of model choice, we strongly recommend using a three-tiered validation scheme when developing Gaussian Processes: first, set aside approximately 20% of the training data as a holdout set that will only be used for validation at the end. Then, iterate on the definition of the model using k-fold cross-validation on the remaining 80% of the data. Only when you are satisfied with the model performance via cross-validation should you validate on the initial hold-out set.

Since most formulations problems tend to feature large input spaces, we recommend combining GPR with dimensionality reduction techniques such as Principal Component Analysis (PCA) (a singular value decomposition where only the first few modes are retained) or a kernel that performs feature selection (Liu et al. 2019). You can also use domain expertise to rank the features by expected order of impact and then optimize over the number to include. You may also want to try different scalings of the inputs, since most kernels are defined with a uniform scale across all the inputs, that is, isotropic kernels. For the kernel, we recommend starting with Matérn kernel plus white noise, though many different kernels have been employed successfully in the literature.

Note that Gaussian Process Regression is closely related to a simpler method known as Kernel Ridge Regression (KRR). KRR employs the kernel trick and l_2 regularization, just like Gaussian Process Regression with assumed noise, that is, a "white kernel." However, KRR omits the calculation of the variance and only provides a point estimate. This also affects the fitting procedure – any kernel hyperparameters in KRR must be optimized outside of the fitting procedure versus in GPR where the kernel hyperparameters may be fit concurrently with the other model coefficients.

7.2.3 Support Vector Machines

Support Vector Machines (SVMs) are classification methods that are closed related to linear and kernel regression (Cortes and Vapnik 1995). A support vector machine tries to define a hyperplane through the domain such that examples of one class are on one side and examples of the other class are on the other side, with the maximum distance between examples and the plane itself. SVMs commonly employ the same "Kernel Trick" that Gaussian Processes and Kernel Ridge Regression do, which turns them into flexible non-linear classifiers.

SVMs, like many classifiers, produce a score that indicates how much like the predicted class the input is; for SVMs, it is the distance from the hyperplane. A follow-up technique called Platt scaling can be used to convert these into probability estimates (Platt 1999).

7.2.4 Random Forests

> Random forests are extremely flexible and robust machine learning models that are a good starting point for low-bias modeling.

Random forests are ensemble models built from decision trees (Breiman 2001). Each decision tree in the ensemble is trained on a different random weighted subset of the training data, which reduces the risk of over-fitting and facilitates statistical analyses of the predictions, which will be discussed later in this section. The decision trees themselves operate by partitioning the input space via a series of binary decisions, or *splits*. The series of splits culminates with some number of final endpoints, or *leaves*. The output value for each leaf is defined by applying an aggregation to the training data within leaf. Most decision trees used for regression aggregate with the mean, and those used for classification aggregate with the mode. To make a prediction, a new sample point traverses the tree, following a path determined by the values of its input variables, until it arrives at a leaf. This is repeated for all the trees in the ensemble, and the prediction is made by combining the results of all trees in some way (e.g., the mean or mode).

The trees are trained by recursively identifying the most informative binary decisions along a single axis – that is, which variable and how to split it – that minimizes the variation of the output property in each of the two resulting groups. For regression problems, the variance is typically used as a measure of the variation in a group. The type of decision depends on the feature, that is, independent variable, that is being considered. Numeric features are partitioned using a condition like $x > c$, where c is the midpoint between adjacent values of x in the training data. Categorical features are partitioned by defining a subset of the categories that correspond to one side of the decision, with the rest going to the other side.

Random forests are so flexible because their underlying decision trees work for both regression and classification, and they can handle both numeric and categorical features. They are robust because the predictions are made by aggregating output property values, so predictions do not stray far from the training data. Nonetheless, they are low-bias: as the amount of training data increases, if the depth of the tree is allowed to grow as well, then the partitions will get smaller and smaller, improving the fit of the model to the shape of the data. The cost of these properties is that random forests, which inherently create a piece-wise constant representation of a surface, require a relatively large amount of data to learn smooth polynomial terms relative to models with smooth components. This is offset somewhat by random forests' ability to easily fit discontinuities, which are common in under-sampled signals.

Most software implementations of random forests will have many parameters. Compared to Gaussian Processes, the performance of a random forest will generally depend less on the specific choice of those parameters, and the defaults will likely perform quite well. The most important are the number of trees, the depth of the trees, and the number of features considered at each split.

We recommend using at least 100 trees, though more is always better (but more expensive), growing the trees to full depth, and considering every feature for every split.

If you know that there is significant measurement error and enough training data that averaging measurements might reduce the error as a repeated measurement might, then setting a small minimum leaf size may help.

Random forests, being an ensemble method, provide several good ways to predict uncertainty. For example, a simple prediction interval can be calculated by computing the variance of the prediction over the trees in the ensemble. This estimate can be improved somewhat via re-calibration (Palmer et al. 2022; Ling, Hutchinson, Antono, Paradiso, et al. 2017). There are also methods for calculating confidence intervals (Wager et al. 2014), and for calculating intervals that capture the correlation of multiple predictions (Folie and Hutchinson 2023), which is particularly valuable for inverse design where those correlations are critical. These methods do come at additional cost, ranging from minimal for prediction intervals to quite significant for confidence intervals. Therefore, start with a simple and inexpensive method and refine it as needed.

7.2.5 Artificial Neural Networks

Artificial neural networks (ANNs) are parameterized functions composed of a network of artificial neurons that apply a non-linear function to the weighted sum of their inputs. In large networks, the repeated application of weighted sums and non-linear functions means that ANNs have very flexible functional forms, that is, the ANN does not bias the model. ANNs and, specifically, deep learning techniques have seen massive and widespread adoption over the past decade (LeCun et al. 2015). This is due primarily to the performance of these models on massive data sets, where the availability of data compensates for the low bias of the estimator itself.

When faced with small data sets, the low bias of neural networks can lead to poor generalization ability (overfitting). Nonetheless, there are strategies that can be employed to make neural networks work well even with limited data. One approach is to use transfer learning, where pre-trained neural networks that have been trained on large data sets to learn useful domain representations are utilized as a starting point. This technique is discussed in Section 7.2.3.

As with linear methods, regularization techniques are crucial when training neural networks on small data sets. Techniques such as weight decay, dropout, and early stopping can help prevent overfitting and improve generalization. Weight decay adds a penalty term to the loss function, discouraging large parameter values, while dropout refers to the practice of randomly setting a fraction of the neurons to zero during training, implicitly penalizing models that rely too heavily on specific features. Finally, even with dropout, some models benefit from "early stopping" or watching and periodically evaluating validation loss during training and stopping when it starts to degrade.

The regularization techniques described earlier, especially the practice of early stopping, are part of the reason we do not typically advocate for using neural networks as a default estimator for formulation modeling.

The complex training process of neural networks requires more expertise and presents more hazards than that of Random Forests and Linear Methods, so we have found them to be generally less reliable.

Bayesian Neural Networks

Another, more theoretical, challenge with neural networks is the fact that the typical mean squared error loss used for regression is multimodal. For a given dataset, there may be many different realizations of the neural network weights that yield similarly low loss values, yet make meaningfully different predictions, especially when interpolating between observed points in the input domain.

This can be a fairly uncomfortable observation for some, especially given how common it is to simply train a model once and then use it in production. However, in recent years, techniques have begun to spring up that embrace this multimodal reality by adopting a Bayesian perspective. By treating the neural network weights as random variables, it has been shown (Wilson and Izmailov 2020) that sampling the neural network response yields an approximation to the Bayesian model averaging over the accessible local minima you may have fallen into on any given training run. In practice, this means that with a bit more computational effort you can train a very flexible ensemble of models that each fit your data well – then sample from them to produce a robust estimated mean and variance for the test point.

Although we find this approach quite compelling, we note that there are few if any good examples of using it in the literature for formulation problems. This may be in part because the software available is less mature than alternatives discussed earlier. Given the pace of software development in the deep learning space; however, this may soon change.

7.2.6 Imputation: Filling in the Blanks

Another approach for predicting unknown quality properties, and even filling in missing process conditions and other features, is to impute them from known values (Schmitt et al. 2015). There are two modes of imputation: imputation of missing inputs/features, which is much more common, and imputation of quality properties. Imputation of missing input features is distinct from other property prediction scenarios in that the imputed value is typically not of direct interest. Consequently, the error in the imputed feature does not matter outside of the error it induces in other predicted quality properties. For example, if the imputed baking temperature of 350 °F differed significantly from the actual 425 °F but the result in either case was delicious, then the error in the imputation does not directly matter. Imputation of missing quality properties, on the other hand, is very similar to other property prediction tasks but has access to other measured quality properties. For example, if every highly rated cake had a high sweetness, then we could impute a high sweetness for a cake with a high rating, even though that rating itself is a quality property.

The simplest way to impute the value of a missing feature is to plug in a constant value. The mean value is a common choice, that is, imputation by the mean.

> In product formulation problems, it is often preferable to use domain expertise to select the value rather than to use the mean.

Imputation by a constant value is typically inappropriate for quality properties where variation in the quality property values is a core aspect of the problem.

A slightly more sophisticated approach is to impute the mean on a local region of the domain rather than across the entire domain. The k-nearest neighbors (kNN) model collects the nearest neighbors using a distance metric defined on the data that is present and then calculates the mean or weighted mean of the missing value across those neighbors. The data used to compute the distance can include other known quality properties in addition to formulation ratios or descriptors. You can think of kNN as first identifying the most similar formulations and then imputing the mean over them. The kNN method is parameterized by the value of k and the choice of distance metric. For imputation, the special case $k=1$ has the added advantage of always producing an imputed value that is present in the data. For $k>1$, this behavior can be recovered by using the mode rather than the mean. Even so, for imputation we recommend a small value of k, for example, 5. For the distance metric, a variation of the Euclidean distance works well: the root mean square column-wise distance, taken to include only columns that are present in both formulations. This can additionally be scaled to emphasize certain columns that are expected to be more strongly correlated with the missing value, for example, scaling by the correlation between the known data for the missing feature or property and the other feature or property.

We can continue to model the dependence between the known data by imputing via a random forest. Recall that decision trees work by partitioning the domain via binary decisions and then predict the mean (or mode) in each of the parts of the partition, that is, the leaves of the tree. This is conceptually very similar to imputing via the mean on a local region. In the case of random forests, the regions are defined during model training, in contrast to the prediction-centered domains used in kNN. Like in kNN, random forests can even be defined to accept missing values in their training data and prediction inputs, though missing value support is less common. Another advantage of imputation via a random forest is access to the predicted uncertainty, as described earlier.

Thus far, we have focused on imputing a single unknown feature or quality property. In principle, any property prediction technique can be used to impute multiple missing properties when applied to the data repeatedly. This approach is called Imputation by Chained Equations (Van Buuren and Oudshoorn 2000). In addition to handling multivariable sparse data, this technique can take advantage of models that can predict uncertainty to iteratively refine a prediction of the uncertainty in the imputation.

Another approach of note is called matrix completion. If you think of the data as a matrix with missing values, then matrix completion is the process of finding the complete matrix that matches the known data and satisfies some other property. In the context of imputing missing data, that property is usually that the matrix is low rank – that is, that it has the fewest degrees of freedom. Unfortunately, the most popular application of matrix completion, recommender systems, uses Non-Negative Matrix Factorization (NMF). NMF is not generally suitable for formulations problems because the feature values may be negative.

Often the best approach to dealing with missing values, features in particular, is not to impute them at all and instead adopt modeling techniques that can handle missing data.

As previously discussed, the kNN, decision tree, and random forest models can be formulated to tolerate missing features explicitly. With a bit of care, many other methods can be adapted as well. An important note for this approach is to clearly distinguish in the data between the case where the data was missing from the case where the data is not applicable. For example, a recipe for a cake that includes a baking step but does not specify the oven temperature is missing the baking temperature, whereas a recipe for a salad that has no baking step has a not-applicable baking temperature.

7.2.7 Additional Methods

The machine learning methods described so far are not exhaustive. Being a comparatively young field, there are both new machine learning methods and variants of existing methods introduced in the literature every year. There is enough interest in applying machine learning to engineering and the physical sciences that many of these new methods will be exercised in a field similar enough to product formulations to get some feeling for the way they could work in a product formulations project.

Among existing additional methods, the partial least squares (PLS) and XGBoost methods stand out as deserving of some attention. PLS is an extension of linear regression that linearly transforms both the inputs and outputs into a latent space before performing a linear regression.

> PLS is particularly powerful for problems with multiple quality properties and is discussed at length in Chapter 8.

XGBoost is a particular implementation of gradient-boosted trees, an alternative ensembling approach to that of random forests (Chen and Guestrin 2016).

> Gradient boosting helps the ensemble of trees to fit more complex functional forms, at the cost of robustness and the ability to predict uncertainty.

If you have a large quantity of training data and are trying to fit the most accurate model, XGBoost may be worth trying. However, given that product formulations tend to not have very much data and the common inverse design use case benefits so much from predicted uncertainty, we have found few opportunities to take advantage of XGBoost's strengths.

7.3 ADVANCED PREDICTION APPROACHES

Often, a first pass using a common method described in the previous section does not result in a model with adequate performance. This section outlines several more complex approaches to improve the performance of the model. The most common issue with model quality is a lack of relevant training data. Therefore, many techniques aim to reduce the training data requirements by either (a) transferring additional information into the model (transfer learning or inductive transfer), or (b) introducing additional structure to the model (inductive bias).

The team should consider whether spending time adding to model complexity or collecting new data (or both) is the most effective path forward.

Furthermore, there are some common machine learning assumptions that may not hold for a formulation property prediction problem. By removing those assumptions, performance can be improved so long as there is adequate data to constrain the more complex resulting model.

7.3.1 Transfer Learning

Transfer learning refers to a broad class of techniques that transfer information from one prediction task to another, including multi-task learning, graphical modeling, and pre-training (Weiss et al. 2016).

Transfer learning is especially appealing in formulations and materials, and chemicals more broadly, due to the abundance of some kinds of computational data and the narrow scope of many new projects.

Information can be transferred from the abundant computational or reference data to related but distinct tasks in a narrow project, improving the performance of models for the related tasks. In this subsection, we will review several transfer learning techniques.

Transfer learning is particularly powerful because the abundant data need not be accurate, so long as its relationship with the primary target is consistent. Computational simulations, like those discussed in Chapter 5 Section 5.2.4, typically have systematic errors for various properties but still contain strong signals in those properties so long as the calculations are performed using a consistent set of parameters.

Multi-Task Learning

In multi-task learning, a single model is trained on two different prediction tasks concurrently, with most of the model being shared between the two tasks. For example, a multi-task random forest model would share the decision trees between the two tasks but have the value predicted in the leaf of each tree be based on the training data for only that task. This can be seen schematically in Figure 7.3, which depicts a single decision tree out of a multi-task random forest that predicts $y(x)$ and $z(x)$ using the same decisions.

Multi-task learning works best when the data volume for the tasks are similar; when there is a large imbalance, the model tends to ignore the task with less data (Hutchinson et al. 2017).

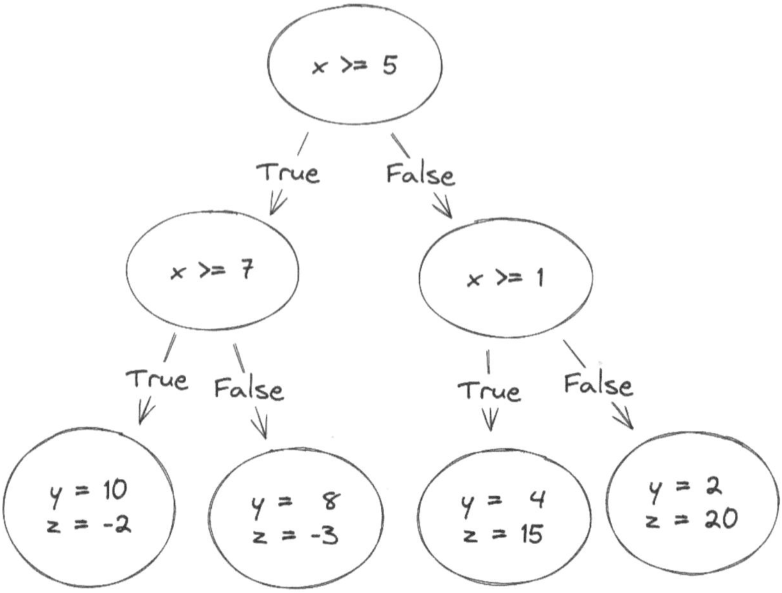

FIGURE 7.3 Multi-task decision tree that predicts both $y(x)$ and $z(x)$.

Multi-task learning also requires the two tasks to be very closely related over the entire domain of the model. If the tasks diverge, the model will lose accuracy as it is forced to try to fit both signals. For these reasons, multi-task learning is not a reliable general approach and multi-task models must be carefully validated. However, in the formulation domain, subject-matter experts are often able to identify quality properties that are related and share common underlying mechanisms. If this is the case for your prediction problem, multi-task learning can be an attractive pursuit.

Graphical Modeling

Graphical modeling involves dividing the prediction task into intermediate terms for which there is either (a) additional data or (b) the expectation of a simpler functional form, and then connecting them in a graph structure, that is, a network.

> The goal of graphical modeling is to find a set of relationships and associated training data where the amount of data matches or exceeds the amount needed to fit the relationship.

Large amounts of training data will support fitting more complex relationships while smaller datasets should be paired with simpler relationships. Effective graphical modeling therefore requires knowledge of the available training data and some intuition for which relationships are simpler versus more complex.

For example, consider the functions $y(x) = exp(x) - 1$ and $z(x) = 4 \times exp(x) + x + 1$, which could arise from similar phenomena. The exponential function can be hard to fit, particularly when combined with another linear term. If we substitute *y(x)* into *z(x)*, then we get $z(x, y) = 4y + x + 5$, which lacks the exponential. Therefore, *z(x,y)* is the easiest to fit, *y(x)*is harder, and *z(x)* is the hardest. If we have as much or more *(x,y)* data than *(x,z)* data, then we can use the abundant *(x,y)* data to learn a relatively accurate model of *y(x)*. Then we can use the model to turn *(x,z)* data into *(x,y(x),z)* data and use machine learning to fit the much simpler *z(x,y)*. This can be seen schematically in Figure 7.4.

The potential inclusion of additional data makes this a transfer learning technique, but the process of decomposing the problem can also incorporate domain knowledge, that is, an inductive bias. When the dependence of the target property on those intermediate models is simpler than the dependence on the inputs, and the intermediate models can be trained to be sufficiently accurate, then the machine learning task becomes easier. The graphical model can be made more robust by concatenating the features with the intermediates and using both as inputs to the latter

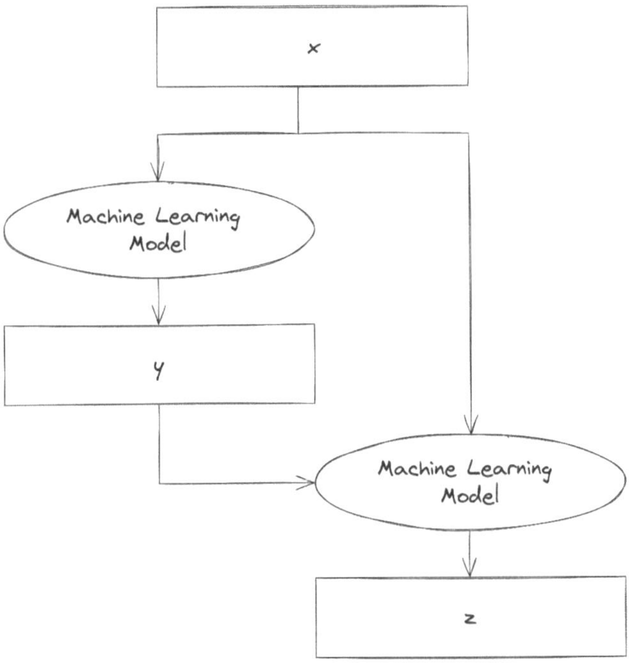

FIGURE 7.4 Graphical model for *y(x)* and *z(x,y)*. This graph follows the third example structure from Figure 7.2.

machine learning models. Then, the model can decide which of the inputs to use based on which produces the best fit, that is, it could learn $z(x,y) = exp(x) + x + 1$ or $z(x,y) = 4y + x + 5$, or some combination of the two. Using this approach, graphical modeling is quite robust (Hutchinson et al. 2017).

Another use case for graphical modeling is to decompose a more complex relationship into simpler ones using a definition, law, or highly accurate empirical relationship. For example, using the definition of a figure of merit in constituent quality properties may result in a simpler relationship to model, or vice versa. In a similar vein, non-dimensionalization of the quality properties, either with known or predicted values, can simplify the task. In that respect, graphical modeling allows for a generalization of feature engineering that can be applied to quality properties in addition to features with the ultimate goal of creating the least complex modeling problem. A similar approach has been used to learn models of mechanical properties of polyurethanes fit on underlying physical variables, which are in turn computed based on a physical model of the experimental degrees of freedom (Menon et al. 2019).

Unfortunately, most off-the-shelf machine learning tools do not support this sort of Graphical Modeling at the time of this writing. The closest analog in scikit-learn, for example, is the "Pipeline" class, which can chain together and concatenate pre-processing steps but not multiple regression or classification models. The Citrine Platform supports graphical modeling via the "GraphPredictor" class ("Citrine Python Client Documentation" n.d.).

Building general support for graphical modeling is quite challenging, but individual graphical models can be constructed and evaluated with careful scripting. The two main steps are (a) deciding in what order to fit the models and (b) propagating the predictions and predicted uncertainty through the graph. For (a), a good starting point is to always fit the trainable model with the largest available training data. For (b), a safe choice is the Monte Carlo method for uncertainty propagation following the NIST Uncertainty Machine (Lafarge and Possolo 2015), which is much better at handling the often large predicted uncertainties than a linear approximation.

Pre-Trained Neural Networks and Fine Tuning

Another way to transfer information from a task with abundant data to another task with scarce data is to first train the model on the abundant data and then to adjust the model for use in the task with scarce data. The first training is called "pre-training," and the resulting model is a pre-trained model. Then the two approaches for repurposing the model are (Ling, Hutchinson, Antono, DeCost, et al. 2017; Devlin et al. 2018):

1) Feature-based approach: To use a layer of the pre-trained neural net as features for a new model to predict the specific task's labels
2) Fine-tuning: To re-train the neural network on the new task with the pre-trained parameters as an initial condition

> Given that product-formulation problems, and materials and chemical problems more broadly, are specialized compared to more general language and image recognition problems, a feature-based approach using a relatively early layer of the network can be both more efficient and effective than a later layer or fine tuning (Ling, Hutchinson, Antono, DeCost, et al. 2017).

With this approach, the most important decision to make is which layer of the network to use as the features, with earlier layers corresponding to simpler and more foundational features and later layers being more specialized to the pre-training task. The more similar the specific task to the pre-training task, the more performant the later layers will be as features.

7.3.2 Correlated Uncertainty Predictions

Formulations projects often require the prediction of multiple properties. Assuming independence of those properties is a reasonable and common simplifying approximation (Zhang and Xu 2021). However, it is typically a poor assumption, particularly when designing a formulation with anti-correlated properties like strength and density. For example, consider the two anti-correlated quality properties in Figure 7.5. Assuming the predicted distributions are independent places the target within the outermost confidence level, while property accounting for correlations puts the target outside of it. Generally, failing to account for correlations and anti-correlations

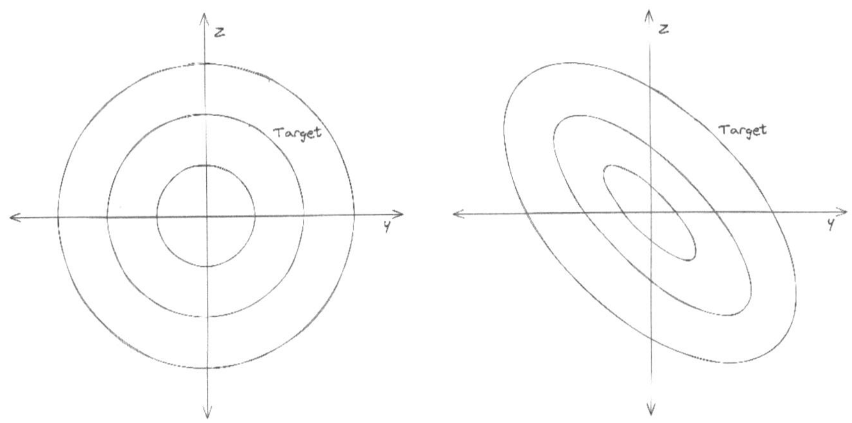

FIGURE 7.5 Predicted probability distributions for two quality properties, y and z, that are assumed to be independent (left) or with correlations (right). The target is in the same position on both axes but falls outside the distribution when correlations are modeled.

can result in extremely optimistic or pessimistic assessments of the likelihood that a formulation achieves multiple property targets.

At a minimum, the practitioner should exercise additional caution and solicit expert feedback when making decisions based on un-modeled correlations.

Fortunately, there has been recent progress in modeling correlations explicitly using ensemble-based methods for predicted uncertainty (Folie and Hutchinson 2023). Proper treatment of correlations has been shown to accelerate materials discovery by 2x-8x for correlated objectives, compared to the independence assumption. On the other hand, simpler treatments of correlations, like computing a correlation matrix based on the training data, are unreliable and can actually increase the number of trials needed to achieve correlated targets.

7.3.3 Graph and Message Passing Neural Networks

Message Passing Neural Networks or, more recently, Graph Neural Networks are neural network architectures specialized to operate on graph-structured data. In the context of formulation design, as mentioned in Chapter 5, graphs can be used to represent the structures of the constituent polymeric binders, solvents, and molecular additives.

Most of the chemistry literature on graph neural networks focuses on the latter: training property predictors on small molecules or monomers to predict physical or optoelectronic properties (St. John et al. 2019; Hasebe 2021; Gurnani et al. 2022), so we will explain this approach here. As with any supervised learning technique, our goal is to massage molecule inputs into fixed-length feature vectors we can use as inputs to a classification or regression model. Of course, there is a rich history of walking molecules and producing aggregate graph-based features (such as 1- and 2-point correlations of atomic properties – see (Landrum 2010)), and MPNNs are a modern approach that can learn flexible kernels and aggregations.

In practice, each input molecule is encoded as an undirected graph with atoms as nodes and bonds as edges, attaching elemental and bond properties as appropriate (note that Hydrogen atoms are typically excluded). For a good reference on useful atom and bond features, see Hasebe (2021). Then, the message passing algorithm is applied iteratively to update each node's property vector – often implemented as a matrix-vector multiply, where matrix elements are learned during training and represent how to weight neighboring atoms based on bond type and atomic property. The more iterations that are applied, the greater the distance information is allowed to propagate across the network. Finally, to produce a fixed-length vector for each (arbitrary-sized graph) input, some aggregation or summarization is required. This last layer is called a "readout" layer, producing a global fixed-length vector representation of the graph by aggregating over the elements in each node's property vector (now more appropriately termed feature or embedding vector after the message passing algorithm has been applied). A straight sum, simple or weighted average are all sensible choices.

Finally, given the contemporary excitement around Transformer models and their underlying attention mechanisms (Vaswani et al. 2017), which power large language models like GPT, we note that attention is an alternative aggregation mechanism that can be applied to a readout layer (Veličković et al. 2018). It is not hard to imagine attention being used productively to capture long-range interactions, known to be important in polymers and proteins, although we do not have direct experience in their use.

7.4 WORKED EXAMPLE: BAKING A CAKE

In this section, we work through a concrete, synthetic example to demonstrate how one might approach a simple formulation property prediction problem. Our example is an at-home formulation problem: the baking of a cake, inspired by our favorite example problem at Citrine Informatics ("GEMD Python Client" n.d.). Our task will be to optimize the recipe of a cake, that is, inverse design. This problem is intended to be realistic but also illustrative. It should provide a blueprint for a first attempt at a similar real-world problem, but your mileage may vary.

7.4.1 Predicting Multiple Quality Properties from Formulation Ratios and Ingredient Properties

A common task in product formulation is to predict the quality properties of a formulation based on the formulation ratios and the properties of those ingredients. Consider the example of Cake Formulations in Table 7.2 and quality properties in Table 7.3.

A simple model would take the formulation ratios as inputs to four separate regression models for each of the four quality properties. For this example, we will use random forests. With some initial inspection or exploratory data analysis, we can see immediately that this approach will not quite work. This model neglects processing conditions, and several of the formulations have identical formulation ratios but result in different quality properties. However, let us assume that the impact of the processing conditions is secondary, either because it is small or because the processing conditions have been chosen to result in good outcomes given the formulation ratios.

Given enough training data, these models trained only on formulation ratios will work. However, we can supplement them with ingredient property data in order to provide more signal to the models. We can divide the ingredients into dry and wet ingredients. In this case, we might imagine that the gluten content of the dry ingredients and the water content of the wet ingredients might influence the texture, water content, and rise of the final cake. We can calculate mixture features as mass-weighted generalized means of these properties using the following formula:

$$f_{\alpha}^{p} = \left(\sum_{i} q_i v_i^{p}\right)^{1/p} \quad (2)$$

where q_i is the mass fraction of the ith ingredient, α_i is the property of the ith ingredient, and p is the power of the mean, for example, 1 for a normal mean. These mixture features can be appended to or replace the quantities of the ingredients as

TABLE 7.2
Cake for Formulations for the Worked Examples, on a Percent Mass Basis

ID	Flour	Sugar	Baking Powder	Milk	Eggs	Butter
0	32.11	23.36	0.73	17.52	11.68	14.60
1	29.82	24.84	0.62	17.39	14.91	12.42
2	25.54	22.99	0.38	17.88	15.33	17.88
3	30.40	24.86	0.55	16.57	11.05	16.57
4	31.45	20.97	0.39	18.35	15.73	13.11
5	32.21	21.48	0.67	13.42	16.11	16.11
6	31.42	20.94	0.52	18.32	15.71	13.09
7	28.03	22.93	0.64	17.83	12.74	17.83
8	32.21	21.48	0.67	16.11	16.11	13.42
9	26.70	24.27	0.49	16.99	14.56	16.99
10	27.66	24.90	0.41	16.60	13.83	16.60
11	26.93	21.53	0.40	18.84	13.46	18.84
12	29.19	23.36	0.73	20.44	11.68	14.60
13	27.41	24.91	0.37	14.94	14.94	17.43
14	28.83	23.59	0.39	18.35	13.11	15.73
15	31.41	23.56	0.52	15.71	13.09	15.71

TABLE 7.3
Cake Quality Properties

ID	Water Content (%)	Density (g/ml)	Yield Stress (kPa)	Taste Rating (0–10)
0	30.78	0.81	4.90	8.10
1	27.81	0.78	5.29	7.34
2	12.10	0.59	9.05	5.05
3	30.79	0.82	4.90	8.15
4	20.80	0.71	6.49	6.53
5	23.65	0.73	5.94	6.95
6	20.60	0.70	6.54	6.30
7	17.27	0.65	7.34	5.92
8	31.80	0.82	4.79	8.25
9	29.27	0.80	5.09	7.81
10	12.57	0.59	8.86	5.17
11	29.53	0.81	5.06	7.57
12	20.53	0.71	6.55	6.34
13	24.47	0.74	5.80	6.93
14	21.50	0.71	6.35	6.62
15	16.58	0.64	7.52	5.82

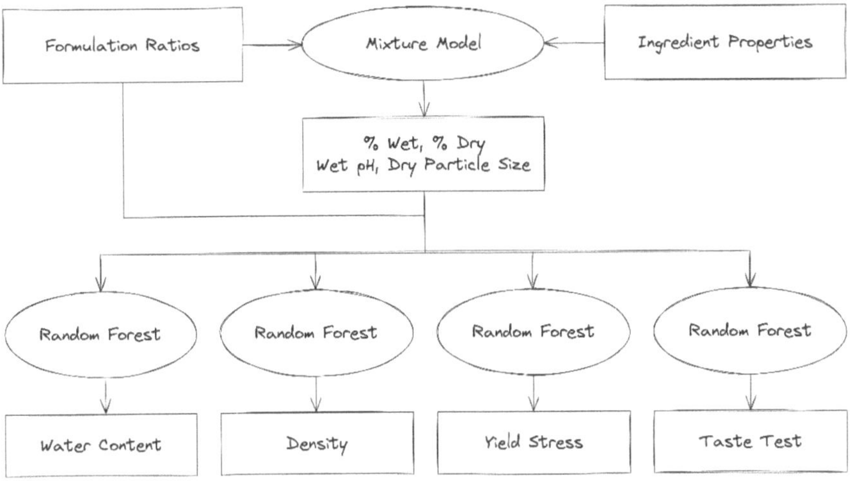

FIGURE 7.6 Schematic of a model for cake quality properties composed of four separate random forest models for each of the four cake quality properties.

inputs to the random forests. Generally, if there are lots of ingredients and ingredient properties, then the mixture features can replace the formulation ratios; otherwise, both mixture features and formulation ratios should be used together. A schematic of the model can be seen in Figure 7.6.

7.4.2 Predicting Quality Properties Using Process Parameters

When baking a cake, the mixing and baking steps are likely to have an effect on the quality of the cake. We can supplement the example data with processing conditions for both the baking and mixing steps, as seen in Table 7.4.

Note that the case where the dry mixing step is omitted is handled via either a zero value for either the speed or time. That can be further standardized via a simple process model that zeros both properties if either are zeroed, which improves consistency and provides signal in both the speed and duration.

This process model can calculate the total mixing time, time-averaged mixing speed, and maximum mixing speed. These derived features improve the flexibility of the model. For example, if the speed was changed during either mixing step, for example, starting the wet mixing at a lower speed and increasing it after a period of time, then these features could still be used to compare the formulation process to the earlier simpler process. Likewise, the baking process can be described by the time-integrated baking temperature.

In addition to flexibility, the process model can incorporate domain information. For example, if we suspected the drying of the cake during baking to be important, then we could compute a time-integrated water vapor pressure at the current oven temperature, which would capture that high temperatures can evaporate off water much faster.

The updated model can be seen in Figure 7.7.

TABLE 7.4
Cake Process Conditions

ID	Dry Mixing Speed (rpm)	Dry Mixing Time (m)	Wet Mixing Speed (rpm)	Wet Mixing Time (m)	Baking Temperature (deg F)	Baking Time (m)
0	100.00	1.21	125.69	2.38	389.67	43.55
1	100.00	1.99	174.05	3.27	363.15	45.23
2	100.00	1.45	104.75	2.40	392.77	59.16
3	100.00	1.12	178.48	3.49	367.92	43.66
4	100.00	2.09	148.88	4.83	386.20	50.08
5	100.00	2.94	142.72	5.76	362.49	47.79
6	100.00	1.30	108.13	3.69	351.61	50.20
7	100.00	2.02	194.94	3.35	377.85	53.13
8	100.00	1.71	164.81	2.66	393.76	43.10
9	100.00	2.63	163.37	5.06	389.56	44.53
10	100.00	2.74	116.15	4.47	372.67	58.60
11	100.00	1.51	127.37	5.39	394.10	44.24
12	100.00	1.91	151.50	2.26	354.23	50.27
13	100.00	0.00	140.60	5.86	364.47	47.30
14	100.00	0.00	136.63	3.42	376.34	49.59
15	100.00	0.00	197.71	5.77	370.11	53.72

The dry mixing step includes only the dry ingredients. The wet mixing occurs after the wet ingredients are added.

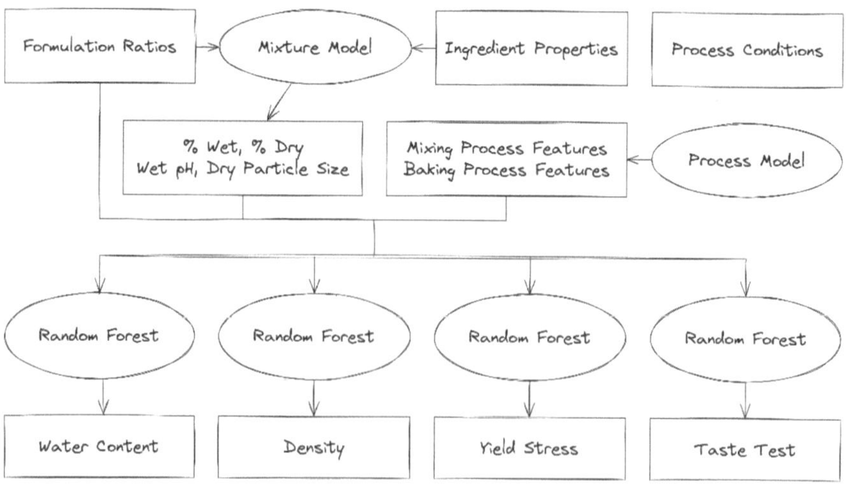

FIGURE 7.7 Schematic for the cake model including process conditions and a process model.

7.4.3 Incorporating Domain Knowledge via Graphical Modeling

The four cake quality properties are clearly related. The water content and density are positively correlated. The taste rating seems to be positively correlated with water content and anti-correlated with yield stress (Figure 7.8). This makes sense – a tasty cake is soft and moist.

The first thing we can do is feed the water content, density, and yield stress predictions into the model for taste rating. Not only can this improve the accuracy of the taste rating model, but it can also improve the consistency of the predictions and couple their predicted uncertainty.

Next, consider the density model. If we assume that water is the only constituent that leaves the cake during processing, then we can define the density with respect to the density of the ingredients, the quantity of lost water, and volume of gasses produced and captured within during the rise, for example, steam and CO_2. The density of the ingredients and ingredient water content can be computed as mixture features, and the quantity of lost water can be calculated from the water content of the final cake. That leaves us with an expression for the density as a function of the volume of gas captured during the rise. We can invert this expression to calculate the volume of gas captured as a new quality property.

The volume of gas captured has some advantages over the density. It may be less correlated with the water content and baking time and is probably strongly correlated with the pH of the wet ingredients and the quantity of baking powder. We can compare the accuracy of the volume of gas captured model and the density model to decide which is the easier relationship to predict. Let's assume that the volume of gas captured is a more accurate model and has well-calibrated predicted uncertainty.

	Water Content	Yield Stress	Taste Test
Density	0.52	0.25	-0.14
Water Content		-0.39	0.77
Yield Stress			-0.67

FIGURE 7.8 Correlation between the four quality properties of the cake.

We can stitch the volume of gas captured model back into what has become a graphical model, as seen in Figure 7.9. As previously discussed, graphical models of this sort require careful training and prediction procedures. In this example there are equal quantities of data for each of the models, so the order in which they are trained does not matter as much. However, due to the very low quantity of data, we can expect the accuracy of the models for the volume of gas captured, final water content, and yield stress to be poor. Therefore, we apply a trick to prevent the taste test model from relying on the accuracy of those inputs: we train the taste test model on the *predicted* quality properties. The full training procedure is as follows:

1) Apply the process model to the process conditions to calculate the mixing and baking features.
2) Load the ingredient properties data into the mixture model, and apply it to the formulation ratios to calculate the mixture properties, including the density of the ingredients and their water content.

FIGURE 7.9 Schematic for a graphical model that relates the density to the volume of captured gas and includes the density, water content, and yield stress as inputs to the model for the taste test.

3) Use the inverse of the density function to calculate the volume of captured gas quality property.
4) Train three independent random forests for the volume of captured gas, final water content, and yield stress quality property.
5) Hide the labels for the volume of captured gas, density, final water content, and yield stress and predict those quality properties using the models.
6) Including the *predicted* density, final water content, and yield stress as inputs, train a random forest for the taste test quality property.

Then, to perform a prediction on a new formulation:

1) Apply the process model and mixture model to calculate the mixture features, mixing features, and baking features.
2) Apply the three random forest models to calculate the volume of captured gas, water content, and yield stress.

If also predicting the uncertainty, sample the predicted distributions of captured gas, water content, and yield stress.

1) Apply the density function to the samples to calculate the density.
2) Apply the final random forest to each sample to calculate the taste test.
3) Compute the mean, variance, and covariance of the samples of the quality properties.

7.5 CONCLUSIONS

There are a variety of use cases for machine learning techniques for predicting properties of formulations. The use case strongly informs the available data and institutional knowledge, as well as the performance requirements for the machine learning model. For example, the common inverse design use case typically begins with a very small amount of data and greatly benefits from a machine learning method that can predict uncertainty but does not require an accurate machine learning model at the onset.

There is an even greater variety of machine learning techniques to consider. There is a place for each of linear methods, Gaussian Processes, Random Forests, and Neural Networks in a data-driven formulations program, though certain models are better suited for specific kinds of projects. The ability to predict uncertainty and produce reasonable predictions in both high and low data limits leads the authors to generally recommend starting with Random Forests, which are used in the example problem. Another property prediction technique, partial least squares (PLS), is discussed extensively in the next chapter.

Sometimes a standard method does not provide a sufficiently accurate model, particularly when data is scarce. Transfer learning methods can help to pull more data into the scope of the project, even when it does not perfectly line up with the prediction task. Graphical modeling can further incorporate domain knowledge to supplement small datasets.

Machine learning has produced many methods that can be used to predict properties of formulations. They are not magic, and one should typically expect a simple model on a small dataset to have modest accuracy. However, with the right expectations and a strong understanding of the prediction use case, machine learning can be a valuable addition to product formulation development.

REFERENCES

Amar, Yehia, Artur M. Schweidtmann, Paul Deutsch, Liwei Cao, and Alexei Lapkin. 2019. "Machine Learning and Molecular Descriptors Enable Rational Solvent Selection in Asymmetric Catalysis." *Chemical Science* 10 (27): 6697–6706. Royal Society of Chemistry.

Bishop, Christopher M. 2006. *Pattern Recognition and Machine Learning*. New York: Springer.

Breiman, Leo. 2001. "Random Forests." *Machine Learning* 45: 5–32. Springer.

Cao, Liwei, Danilo Russo, Werner Mauer, Huan Huan Gao, and Alexei A. Lapkin. 2020. "Machine Learning-Aided Process Design for Formulated Products." *Computer Aided Chemical Engineering* 48: 1789–1794. Elsevier.

Chen, T., and C. Guestrin. 2016, August. "Xgboost: A Scalable Tree Boosting System." *Proceedings of the 22nd ACM SIGKDD International Conference on Knowledge Discovery and Data Mining* (pp. 785–794).

"Citrine Python Client Documentation." n.d. https://citrineinformatics.github.io/citrine-python/index.html.

Cortes, C., and V. Vapnik. 1995. "Support-Vector Networks." *Machine Lea rning* 20: 273–297. https://doi.org/10.1007/BF00994018

Devlin, Jacob, Ming-Wei Chang, Kenton Lee, and Kristina Toutanova. 2018. "Bert: Pre-Training of Deep Bidirectional Transformers for Language Understanding." *arXiv Preprint arXiv:1810.04805.*

Folie, Brendan, and Maxwell Hutchinson. 2023. "Multivariate Prediction Intervals for Bagged Models." *Machine Learning: Science and Technology* 4 (1): 015022. IOP Publishing.

"GEMD Python Client." n.d. https://github.com/CitrineInformatics/gemd-python/.

Gurnani, Rishi, Christopher Kuenneth, Aubrey Toland, and Rampi Ramprasad. 2022. "Polymer Informatics at-Scale with Multitask Graph Neural Networks." *arXiv Preprint arXiv:2209.13557.*

Harten, Paul, Todd Martin, Michael Gonzalez, and Douglas Young. 2020. "The Software Tool to Find Greener Solvent Replacements, Paris III." *Environmental Progress & Sustainable Energy* 39 (1): 13331. John Wiley & Sons, Inc.

Hasebe, Tatsuya. 2021. "Knowledge-Embedded Message-Passing Neural Networks: Improving Molecular Property Prediction with Human Knowledge." *ACS Omega* 6 (42): 27955–27967. ACS Publications.

Hutchinson, Maxwell L., Erin Antono, Brenna M. Gibbons, Sean Paradiso, Julia Ling, and Bryce Meredig. 2017. "Overcoming Data Scarcity with Transfer Learning." *arXiv Preprint arXiv:1711.05099.*

Jacobs, Ryan, Tam Mayeshiba, Ben Afflerbach, Luke Miles, Max Williams, Matthew Turner, Raphael Finkel, and Dane Morgan. 2020. "The Materials Simulation Toolkit for Machine Learning (Mast-Ml): An Automated Open Source Toolkit to Accelerate Data-Driven Materials Research." *Computational Materials Science* 176: 109544. Elsevier.

Jhamb, Spardha, Xiaodong Liang, Rafiqul Gani, and Georgios Kontogeorgis. 2019, March. "Systematic Model-Based Methodology for Substitution of Hazardous Chemicals." *ACS Sustainable Chemistry & Engineering* 7. https://doi.org/10.1021/acssuschemeng.8b06064.

Lafarge, Thomas, and Antonio Possolo. 2015. "The Nist Uncertainty Machine." *NCSLI Measure* 10 (3): 20–27. Taylor & Francis.

Landrum, Greg. 2010. "RDKit." https://www.rdkit.org.
LeCun, Yann, Yoshua Bengio, and Geoffrey Hinton. 2015. "Deep Learning." *Nature* 521 (7553): 436–444. Nature Publishing Group UK London.
Li, Hai feng. 2014. "Smile." https://haifengl.github.io.
Ling, Julia, Maxwell Hutchinson, Erin Antono, Brian DeCost, Elizabeth A. Holm, and Bryce Meredig. 2017. "Building Data-Driven Models with Microstructural Images: Generalization and Interpretability." *arXiv Preprint arXiv:1711.00404*.
Ling, Julia, Maxwell Hutchinson, Erin Antono, Sean Paradiso, and Bryce Meredig. 2017. "High-Dimensional Materials and Process Optimization Using Data-Driven Experimental Design with Well-Calibrated Uncertainty Estimates." *Integrating Materials and Manufacturing Innovation* 6 (3): 207–217. Springer International Publishing.
Liu, Kailong, Yi Li, Xiaosong Hu, Mattin Lucu, and Widanalage Dhammika Widanage. 2019. "Gaussian Process Regression with Automatic Relevance Determination Kernel for Calendar Aging Prediction of Lithium-Ion Batteries." *IEEE Transactions on Industrial Informatics* 16 (6): 3767–3777. IEEE.
Menon, Aditya, James A. Thompson-Colón, and Newell R. Washburn. 2019. "Hierarchical Machine Learning Model for Mechanical Property Predictions of Polyurethane Elastomers from Small Datasets." *Frontiers in Materials* 6: 87. Frontiers Media SA.
Palmer, Glenn, Siqi Du, Alexander Politowicz, Joshua Paul Emory, Xiyu Yang, Anupraas Gautam, Grishma Gupta, Zhelong Li, Ryan Jacobs, and Dane Morgan. 2022. "Calibration After Bootstrap for Accurate Uncertainty Quantification in Regression Models." *npj Computational Materials* 8 (1): 115. Nature Publishing Group UK London.
Pedregosa, Fabian, Gaël Varoquaux, Alexandre Gramfort, Vincent Michel, Bertrand Thirion, Olivier Grisel, Mathieu Blondel, et al. 2011. "Scikit-Learn: Machine Learning in Python." *The Journal of Machine Learning Research* 12: 2825–2830. JMLR.org.
Platt, John. 1999. "Probabilistic Outputs for Support Vector Machines and Comparisons to Regularized Likelihood Methods." *Advances in Large Margin Classifiers* 10 (3): 61–74.
Schmitt, Peter, Jonas Mandel, and Mickael Guedj. 2015. "A Comparison of Six Methods for Missing Data Imputation." *Journal of Biometrics & Biostatistics* 6 (1): 1. OMICS Publishing Group.
Sevgen, Emre, Edward Kim, Brendan Folie, Ventura Rivera, Jason Koeller, Emily Rosenthal, Andrea Jacobs, and Julia Ling. 2021. "Toward Predictive Chemical Deformulation Enabled by Deep Generative Neural Networks." *Industrial & Engineering Chemistry Research* 60 (39): 14176–14184. ACS Publications.
St. John, Peter C., Caleb Phillips, Travis W. Kemper, A. Nolan Wilson, Yanfei Guan, Michael F. Crowley, Mark R. Nimlos, and Ross E. Larsen. 2019. "Message-Passing Neural Networks for High-Throughput Polymer Screening." *The Journal of Chemical Physics* 150 (23): 234111. AIP Publishing LLC.
Syeda, Sultana R., Easir A. Khan, Orakotch Padungwatanaroj, Nichakorn Kuprasertwong, and Anjan K. Tula. 2022. "A Perspective on Hazardous Chemical Substitution in Consumer Products." *Current Opinion in Chemical Enginee ring* 36: 100748. https://doi.org/10.1016/j.coche.2021.100748.
Tibshirani, Robert. 1996. "Regression Shrinkage and Selection via the Lasso." *Journal of the Royal Statistical Society Series B: Statistical Methodology* 58 (1): 267–288. Oxford University Press.
Van Buuren, Stef, and Catharina G.M. Oudshoorn. 2000. *Multivariate Imputation by Chained Equations*. Leiden: TNO.
Vaswani, Ashish, Noam Shazeer, Niki Parmar, Jakob Uszkoreit, Llion Jones, Aidan N. Gomez, Łukasz Kaiser, and Illia Polosukhin. 2017. "Attention Is All You Need." *Advances in Neural Information Processing Systems* 30.
Veličković, Petar, Guillem Cucurull, Arantxa Casanova, Adriana Romero, Pietro Liò, and Yoshua Bengio. 2018. "Graph Attention Networks." *arXiv Preprint arXiv:1710.10903v3*.

Wager, Stefan, Trevor Hastie, and Bradley Efron. 2014. "Confidence Intervals for Random Forests: The Jackknife and the Infinitesimal Jackknife." *The Journal of Machine Learning Research* 15 (1): 1625–1651. JMLR.org.

Ward, Logan, Alexander Dunn, Alireza Faghaninia, Nils E.R. Zimmermann, Saurabh Bajaj, Qi Wang, Joseph Montoya, et al. 2018. "Matminer: An Open Source Toolkit for Materials Data Mining." *Computational Materials Science* 152: 60–69. Elsevier.

Weiss, Karl, Taghi M. Khoshgoftaar, and DingDing Wang. 2016. "A Survey of Transfer Learning." *Journal of Big Data* 3 (1): 1–40. SpringerOpen.

Wieringen, Wessel N. van. 2015. "Lecture Notes on Ridge Regression." *arXiv Preprint arXiv:1509.09169.*

Williams, Christopher, and Carl Rasmussen. 1995. "Gaussian Processes for Regression." *Advances in Neural Information Processing Systems* 8.

Wilson, Andrew G., and Pavel Izmailov. 2020. "Bayesian Deep Learning and a Probabilistic Perspective of Generalization." *Advances in Neural Information Processing Systems* 33: 4697–4708.

Zhang, Yun, and Xiaojie Xu. 2021. "Predicting Multiple Properties of Pervious Concrete Through the Gaussian Process Regression." *Advances in Civil Engineering Materials* 10 (1): 56–73. ASTM International.

Zou, Hui, and Trevor Hastie. 2005. "Regularization and Variable Selection via the Elastic Net." *Journal of the Royal Statistical Society Series B: Statistical Methodology* 67 (2): 301–320. Oxford University Press.

8 Modeling of Product Formulations Using a Latent Variable Approach

Alexander Nguyen and Marlene Cardin

8.1 INTRODUCTION

Latent variable modeling is a powerful tool to model and analyze both large and small datasets.

> The goal of latent variable methods is to project the data onto lower-dimensional "latent variable spaces," thereby taking advantage of correlations in the data and simplifying analysis.

This proven methodology can handle correlation in the measurements, missing data, noisy data, and can be interpreted with intuitive plots (Wold et al. 2001). Owing to these important advantages, the methodology has a successful application history in manufacturing datasets, chemometrics, and is very well suited to handle product development datasets.

This chapter will focus on the latent variable regression method PLS (partial least squares or projection to latent structures).

8.2 LATENT VARIABLE METHODS FOR PRODUCT DEVELOPMENT

There are several methods to build a PLS model, such as kernel PLS (Dayal & MacGregor 1997), singular value decomposition (Abdi 2010), canonical PLS (Indahl et al. 2009), and Non-linear Iterative Partial Least Squares (NIPALS) (Wold et al. 2001).

With the rise of computational power in recent decades, NIPALS is one of the more commonly used methods to calculate PLS models and thus, will be discussed here. Furthermore, the NIPALS algorithm has the advantage of handling missing data (by substitution using model predictions (Wold et al. 2001) and can be computed sequentially and easily handled by modern computers.

DOI: 10.1201/9781003385974-11

8.2.1 Non-Linear Iterative Partial Least Squares (NIPALS)

The NIPALS algorithm is an iterative algorithm that is used to calculate the parameters in the PLS modeling framework. In PLS, one aims to compute the principal components that capture the explanation of variance in the data variance in relation to the correlation between the input and output variables. The simple NIPALS algorithm (Wold et al. 2001; Dunn 2023) is used to complete this task.

Once all of the components have been calculated, often another version of weights w is reported, since the weights calculated earlier are based on deflated matrices. It is desirable to have weights that can be used directly. These weights, named w^* (or r depending on the source), can be calculated as shown in Equation 8.1 (Wold et al. 2001).

$$w^* = w\left(p^T w\right)^{-1} \qquad \text{(Equation 8.1)}$$

w^* is then used to calculate predicted scores, and subsequently predicted outputs.

8.2.2 Number of Components

There are several methods to determine the number of components to use such as i) when the residuals of the Y variables between two consecutive components become sufficiently small (Gelaldi & Kowalski 1986); ii) using the broken-stick rule (Nomikos & MacGregor 1995); iii) using cross-validation with the prediction residual sum of squares (PRESS) (Wold 1978); or iv) using a "total" cross-validation method (Wold et al. 2001). Detailed explanations for the aforementioned methods can be found in (Wold et al. 2001) and (Dunn 2023). Some details of the third and fourth methods are briefly discussed here.

In the third method, the PRESS can be calculated for G number of cross-validation groups (where the data is split into G number of groups). In this method, 1 group is set aside as a testing set (denoted as group g), and the remainder of the data is used as the training set. A PLS model is then generated on the training set, and the testing data is projected onto the trained model to obtain the PRESS for testing group g for all Y variables m as shown in Equation 8.2.

$$PRESS_g = \sum_i^{n_g} \sum_m^M \left(y_{i,m} - \hat{y}_{i,m}\right)^2 \qquad \text{(Equation 8.2)}$$

where n_g is the number of observations in the testing group g, M being the number of quality variables, and y and $\hat{y}$ are the observed and predicted quality variables, respectively. Repeat this calculation for all groups G until there exists a group of PRESS values. These values are then added together to give a final PRESS value and then is divided by the sum of squares of the original data as per Equation 8.3.

$$SS_y = \sum_i^n \sum_m^M \left(y_{im} - \underline{y}_{im}\right)^2 \qquad \text{(Equation 8.3)}$$

where $\underline{y}$ is the mean of the data. Finally, the Q^2 metric, which is used as a measure of model predictive ability, can be calculated according to Equation 8.4.

$$Q^2 = 1 - \frac{PRESS}{SS_y} \qquad \text{(Equation 8.4)}$$

Generally, Q^2 increases with each successive component but will eventually begin to decrease. Once a decline begins, it may be an indication that the new component is not able to explain the cross-validation data and is not significant, and that is where many practitioners will terminate the number of components.

> It should be noted that cross-validation may not be a feasible method of determining the number of components in a product development dataset, since the number of observations may be low.

Suggestions for other model validation procedures are discussed in Chapter 9.

The "total" cross-validation method is similar to the second method, but a predesignated number of components is selected. The PRESS is calculated for each successive number of components, and then the value $PRESS/(N-A-1)$ is calculated. The model with the lowest value is selected and gives similar results to the previous method.

8.3 DATA PREPARATION FOR PLS MODELING

For modeling product development problems using PLS, it is important to use data that represents the development space as well as possible. Much of this data was presented in Chapter 4 and, thus, only a brief description of some of the possible data types will be discussed here. Once the data has been selected, it is crucial to pre-process the data prior to modeling using PLS in order to scale and normalize the data, then modify the importance of select variables, if necessary.

8.3.1 Mixture Properties Revisited

A complete discussion on mixture properties can be found in Chapter 4. For brevity, the simplest case of mixture property calculations will be discussed here.

As shown in Figure 8.1, the mixture properties can be calculated by multiplying the ratio block matrix by the properties of the ingredients (for the linear case).

The R matrix lists the formulations down the rows and each individual ingredient across the columns, and each element is the fraction of the ingredient in the formulation (e.g., in $R_{1,1}$, the fractional value here would be the amount of ingredient 1 in formulation 1; in $R_{1,2}$ the value is the amount of ingredient 2 in formulation 1); the X matrix lists the ingredients down the rows and the properties across the rows; and X_{mix} is the resulting mixture properties block with each element representing the mixture property for each formulation. If the formulations are partitioned into classes, then a separate X_{mix} block must be calculated for each class and will result in multiple mixture property blocks. Additionally, global mixture properties can be calculated if all classes share a common attribute. Once the mixture properties are calculated, they can be used for modeling in PLS.

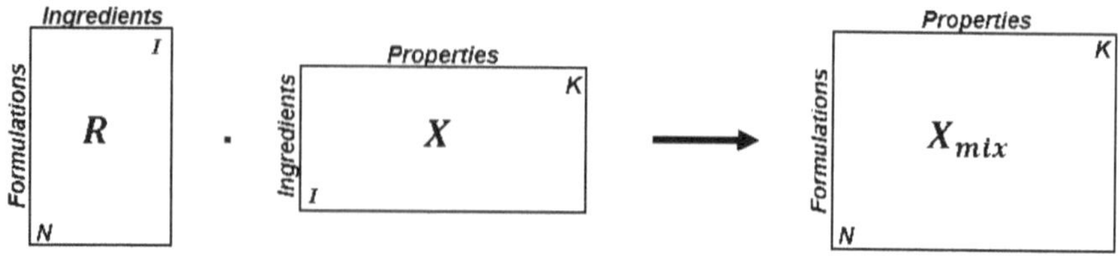

FIGURE 8.1 Calculation of mixture properties (X_{mix}) from formulation ratios (R) and ingredient properties (X).

TABLE 8.1
Example of Converting a Descriptor (String) Variable into a Set of Categorical (Binary) Variables

DescVar	DescVar_A	DescVar_B	DescVar_C
B	0	1	0
A	1	0	0
C	0	0	1

8.3.2 Process Conditions, Numerical Variables, Descriptors (Strings), and Categorical Variables

Variables that contain numerical information, such as processing conditions (temperatures, pressures, flow rates, etc.), discrete data points from curves, or calculated variables, can be included in the PLS model.

Descriptors, or string variables, can also be included in the PLS model but first must be converted into categorical variables where new variables are generated based on the number of unique strings and zeros and ones are assigned to the new categorical variables. An example is illustrated in Table 8.1.

This same conversion can be made for ingredient properties and the mixture properties can be calculated in the usual fashion (the resulting mixture property will be nonzero for the matching descriptor and zero for the remaining descriptors).

8.3.3 Grouping and Assembling the Data

Once the data has been selected, it is standard to organize the information in blocks in order to help improve interpretability of the models when the number of variables is large and the variables can be siloed and aggregated into meaningful groups (Westerhuis et al. 1998). The organization of the blocks varies depending on the dataset and the variables contained within the dataset. In the case of product development, it is helpful to group variables by type. There would typically be a block for mixture properties of each ingredient class and global properties, a block of

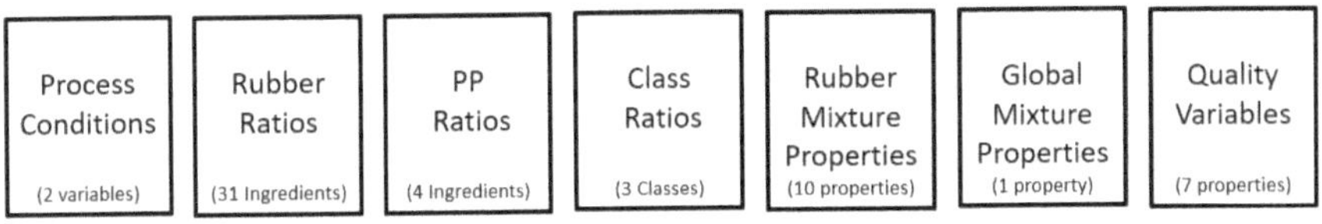

FIGURE 8.2 Data separated into blocks for a functional polymers example.

ingredient ratios for each class (if used in model), and a block of variables associated with each process unit.

An example of structuring a product development dataset for the development of functional polymers is shown in Figure 8.2 (Wallace & Cardin 2023). Encompassing ~100 historical data points where three classes of ingredients (rubbers, oil, and polypropylenes) are combined in a relatively simple process to create functional polymers for which seven key quality attributes are measured. Ten ingredient properties are measured for each rubber, while only density is measured for the oil and polypropylene class. The dataset would then be split into blocks and the modeler could select which blocks to include based on the data and model objectives.

8.3.4 Pre-Processing, Scaling, and Block-Scaling

Once the data is organized into their respective blocks of data, the data must be transformed and scaled since the results of the PLS method depend on the scaling. The first step is to apply any transformations, such as logarithm transforms, as they are helpful to linearize a non-linear system. For example, variables that range more than a magnitude of ten are often logarithmically transformed.

Next, it is standard multivariate practice to mean-center and scale the data; this is done by subtracting each data point within a variable by its respective average and then dividing by its respective standard deviation. This gives all variables the same relative weight (centered around 0 and scaled by standard deviation) and allows one to increase the importance of specific variables by multiplying additional weights to these variables after the fact.

Finally, block-scaling can be performed via several methods such as hierarchical PLS and multiblock PLS. The multiblock PLS method can be applied as a single-block model, where instead of calculating the block super scores, each variable is scaled by the square root of the number of the variables in its respective block (Westerhuis et al. 1998), as shown in Equation 8.5.

$$X = \left[\frac{X_1}{\sqrt{m_{X_1}}}, \frac{X_2}{\sqrt{m_{X_2}}}, \ldots, \frac{X_B}{\sqrt{m_{X_B}}}\right] \qquad \text{(Equation 8.5)}$$

where X_B are the blocks of input variables and m_{xB} are the number of variables in a block. X_B By scaling each block by the number of variables in the block, the scores obtained in the standard PLS method are the same as the superscores obtained in the multiblock methods. The derivation and proof are described in (Westerhuis et al. 1998).

8.4 BUILDING THE PLS MODEL USING PRODUCT DEVELOPMENT DATA

8.4.1 Block Structure for PLS

Once the data is fully assembled, transformed, scaled, and block-scaled, the blocks can be then combined as one matrix for processing through the NIPALS algorithm. With the combined data blocks, the NIPALS algorithm can now be applied and the model parameters (scores, loadings, weights) can be calculated with a determined number of components.

8.4.2 Useful Model Statistics

While prediction is generally the main goal when projecting a new observation onto the PLS model, it is helpful to understand the characteristics of the model itself to draw conclusions from the model and to know how well new data fits into the existing model. Several statistics are used to describe the model, and most of these are used when a new observation is projected onto the model.

Squared Prediction Error (SPE) and Hotelling's T^2 (HT2)

In order to judge how well a new observation follows a PLS model, squared prediction error (SPE) and Hotelling's T^2 (HT2) are two important metrics to consider. The SPE is the sum of squares between the observation's real data and the observation's projection onto the model plane (i.e., model prediction), otherwise known as the residuals, and is calculated for each observation as follows, in matrix notation (assuming X is a $1 \times K^{\circ}$ vector) as per Equation 8.6.

$$SPE = \left(X - \hat{X}\right)\left(X - \hat{X}\right)^T \qquad \text{(Equation 8.6)}$$

Or, alternatively, in summation notation as per Equation 8.7.

$$SPE = \sqrt{\sum_k^K \left(x_k - \hat{x}_k\right)^2} \qquad \text{(Equation 8.7)}$$

The SPE indicates how well the observation follows the correlation structure, or hyperplane, of the PLS model. Thus, a lower SPE indicates that the observation follows the model relatively well and vice versa. It is often measured against confidence intervals based on Chi-squared distributions. Alternatively, limits can be determined empirically using the distribution of SPE values of the training set (MacGregor & Kourti 1995).

In addition to SPE, Hotelling's T^2 (HT2) indicates how an observation compares to the average of the data (in a mean-centered and scaled PLS model, this is the origin of the hyperplane) and is calculated according to Equation 8.8.

$$\text{HT2} = \sum_a^A \frac{t_a^2}{s_a^2} \qquad \text{(Equation 8.8)}$$

where A is the final component, t_a is the score for the observation for component a, and s_a is the variance across all t_a in the training set. Larger Hotelling's T^2 values indicate that the observation is increasingly different from the average of the data.

Similar to SPE, Hotelling's T^2 is often measured against upper control limits using F-distributions. Also similar to SPE, limits can be determined empirically using the distribution of HT2 values of the training set (MacGregor & Kourti 1995).

Analyzing both the SPE and HT2 can indicate outliers in the dataset, where larger SPE values will indicate abnormal points based on correlation structure, and HT2 will indicate abnormal points based on the average.

In the context of a material formulation, an unusual correlation structure indicated by a large SPE value signifies an unusual combination of ingredients or ingredient properties – somewhat different from what an expert formulator would do. New predicted formulations with high SPE values could represent an intriguing step outside a formulator's "comfort zone," but they could equally as likely be something nonsensical, such as a drug formulation with no API added. High HT2 values represent formulations that are extreme compared to a "typical" material. Both types of residual can be used within the training data to identify data errors, and Chapter 12 will discuss how these values can also be used to prevent excessive extrapolation when using optimization to design new materials.

When using SPE and HT2 to prepare the data, caution should be taken when removing outliers – the limits of each statistic are based on the data itself, and removing too many observations from the training set may result in biased limits (i.e., new observations being projected onto the model may show up as abnormal, even if they are in-line with the training data). In some cases, it is desirable to keep minor outliers in the training set in order to recognize outliers with new observations. The same equations can be applied to new observations being projected onto the training model in order to judge the prediction of the new data.

Correlation Coefficient (R2)

Along with analyzing the characteristics of each observation and new observations, R^2 exists for PLS models and is analogous to R^2 for a linear regression.

R^2 compares how well the predicted values of the model compare to the observed data and can be interpreted as how well the PLS hyperplane predicts values from given inputs.

As shown in Equation 8.9, the R^2 can be calculated by dividing the sum of squares of the residuals by the sum of squares of Y corrected for the mean:

$$R^2_{y,m} = 1 - \frac{\sum_i^N \left(y_{i,m} - \hat{y}_{i,m}\right)^2}{\sum_i^N \left(y_{i,m} - \underline{y_m}\right)^2}, m \in M \qquad \text{(Equation 8.9)}$$

where $y_{i,m}$ is the observed data for observation i for Y variable m, $\hat{y}_{i,m}$ is the predicted value, and $\underline{y_m}$ is the mean of Y variable (Alexander et al. 2015). While this

R^2 considers each individual Y variable, a "total" R^2 value can be calculated by summing across all variables, which considers all Y variables and observations together. Additionally, there exists an equivalent version of R^2s in the X-space, calculated in a similar fashion.

Root Mean Squared Error of Estimation (RMSEE)

In order to measure the fit and performance of the trained PLS model, the root mean squared error of estimation (RMSEE) is calculated. The RMSEE is an estimate of the standard deviation of the fitted residuals from the model, corrected by the number of observations and degrees of freedom, which in the case of PLS models is the number of components. Thus, the RMSEE for Y variable m° is calculated using Equation 8.10.

$$RMSEE_m = \sqrt{\frac{\sum_i^N \left(y_{i,m} - \hat{y}_{i,m}\right)^2}{N - A - 1}}, m \in M \qquad \text{(Equation 8.10)}$$

In Equation 8.10, N is the number of observations, A is the number of components, and M is the list of all quality variables.

As the RMSEE is in units of the quality variable and is an estimate of the standard deviation of the variable, it can be used as an approximate confidence interval for predicting future observations, where $\hat{y} \pm 2RMSEE$ for 95% confidence (Alexander et al. 2015).

8.4.3 Available Tools

While PLS can be implemented from scratch using the aforementioned methods, there are packages available for the implementation of PLS, much in the same way that other machine learning methods such as neural networks (e.g., PyTorch®, TensorFlow®) are available. Some of the following packages are free to use, while others require a subscription or purchase to obtain access:

- R: The **pls** package, freely available through the CRAN website and licensed under the Gnu General Public License (GPL), computes PCR (principal component regression) and PLS for given data. The PLS algorithm employed in this package is canonical PLS using a singular value decomposition approach.
  - Kristian Hovde Liland [aut, cre], Bjørn-Helge Mevik [aut], Ron Wehrens [aut], Paul Hiemstra [ctb]. *"pls: Partial Least Squares and Principal Component Regression." V2.8–3. R. 2023.*
- Python: The **scikit-learn cross_decomposition** package offers a **PLS-Regression** module that can build a PLS regression model with given

data. The package offers several versions of PLS, including canonical PLS, singular value decomposition, and NIPALS. The **PyPhi** package (developed by Salvador García Muñoz, Imperial College of London) offers both PCA and PLS and some variations of PLS such as Locally Weighted PLS and contains features such as the ability to perform batch analysis and plotting.

  - Scikit-learn: Machine Learning in Python, Pedregosa *et al.*, JMLR 12, pp. 2825–2830, 2011.
  - PyPhi: Salvador García Muñoz. "PyPhi." V3.0. Imperial College of London. 2023.
- MATLAB: The **plsregress** module is available through the Statistics and Machine Learning Toolbox and calculates the PLS model using the SIMPLS algorithm
  - de Jong, Sijmen. "SIMPLS: An Alternative Approach to Partial Least Squares Regression." Chemometrics and Intelligent Laboratory Systems 18, no. 3 (March 1993): 251–63. https://doi.org/10.1016/0169-7439(93)85002-X).
- AspenTech ProMV: ProMV is a no-code software application that employs PCA and PLS on imported data using a variation of the NIPALS algorithm.
  - AspenTech. *"ProMV."* V14.0. AspenTech. 2022.
- Sartorious SIMCA®: SIMCA® is a no-code software application that employs PCA and PLS on imported data using the NIPALS algorithm.
  - Sartorius. *"Umetrics SIMCA 18."* Sartorius. 2023.
- ProSensus FormuSense: FormuSense is a no-code software application that cleans, organizes, and sets up product development-specific datasets and employs PLS to build predictive models. Digital simulation and back-calculation to find new formulations using mathematical optimization are available.
  - ProSensus. *"FormuSense."* V4.0. ProSensus. 2023.

There are pros and cons associated with using an existing tool versus implementing a solution from scratch. For example, using an existing no-code tool can reduce project resource requirements while expediting the model-building process, so long as the data is in a format that can be readily ingested to the tool. However, an existing tool may not include the flexibility to handle unique characteristics of the dataset, which could be handled more readily in a custom implementation. While a commercial tool may help unify and streamline R&D workflows toward a proven framework, a from-scratch approach may eliminate steps that are irrelevant to the specific project at hand or include additional steps that may be important. All this goes to say that careful consideration is needed when deciding on a project path. Fortunately, some commercial tool providers offer free trial licenses (including FormuSense) so that potential users can evaluate the solution. This is a valuable way to quickly assess the suitability of an existing tool and to complete some exploratory data analysis (as discussed in Chapter 3) prior to proceeding with a full license purchase or development of a custom solution.

8.5 INTERPRETATION OF PLS MODELS

Given that PLS is not a black-box method, there are many plots that can be useful to interpret and fine-tune the model.

In this section, the plots have been broken down into model fit, model parameters, model validity, and model coefficients. A 5-component model based on the dataset introduced in Section 8.3.3, as well as ProSensus FormuSense software (Wallace & Cardin 2023), were used to generate these plots. For this model, rubber ratios were excluded (as shown in Figure 8.3), since the rubber properties defined the characteristics well. For this model structure, new rubbers could be considered when applying the model, so long as rubber properties are available for the new rubbers.

8.5.1 Model Fit

There are a few metrics to evaluate the model fit as described in Section 8.4.2. The correlation coefficient, or R^2, can be calculated for each individual Y variable, or on a per principal component basis across all the X or Y variables.

In Figure 8.4, the R2 is shown for each quality variable for a training set (first bar), a validation set (second bar), where 20% of the data has been set aside for

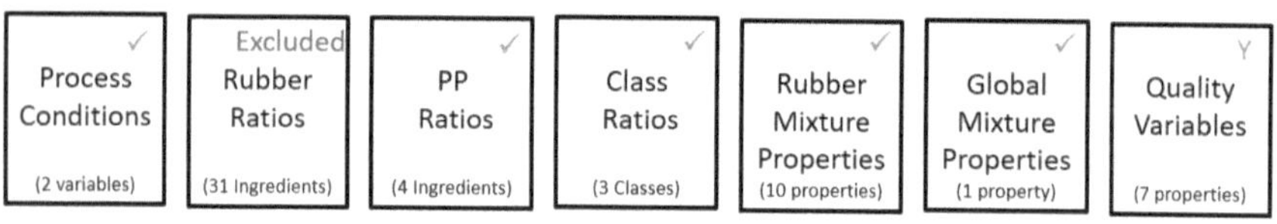

FIGURE 8.3 Data blocks included in example PLS model.

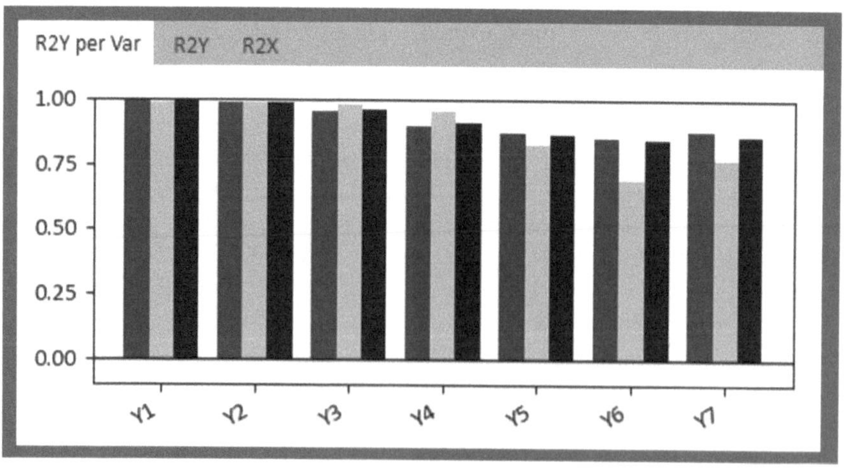

FIGURE 8.4 R^2 per quality variable for a training set (first bar), a validation set (second bar), and the final model (includes the training and validation set, and depicted as the third bar).

testing purposes, and for the final model (includes the training and validation set, and depicted as the third bar). Generally, the fit on the test set will be slightly lower than the fit on the training set.

It is important to test the model on unseen data to select the appropriate number of components and gain confidence in the model.

The observed versus predicted plot of the output variables is valuable in analyzing the predicted outputs. In this plot, shown in Figure 8.5, each observation's observed (real) output is plotted against the predicted output, with the predicted value along the horizontal axis and the observed value along the vertical axis. A line of unity (a linear line with a slope of 1 and origin of 0) is plotted on the graph as well. Observations along the line indicate a good prediction of the output variable relative to the real data and vice versa. Additionally, more observations that are further from the unity line will result in a higher RMSEE.

In Figures 8.6 and 8.7, the R^2X and R^2Y is shown for each component training set (first bar), a validation set (second bar), where 20% of the data has been set aside for testing purposes, and for the final model (includes the training and validation set, and depicted as the third bar).

The appropriate number of components to use for the model is usually determined by both the amount of variation a new component fits and the performance on the test set. Modelers must also keep in mind the number of degrees of freedom in their dataset and the measurement noise in the dataset when selecting the number of components.

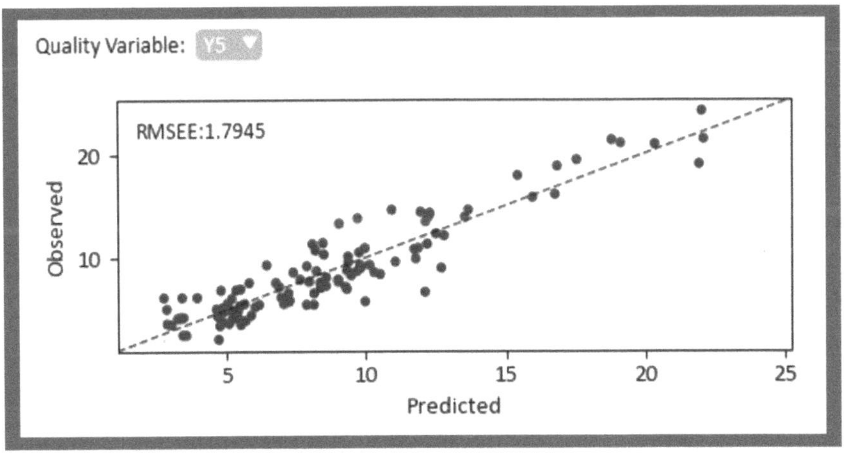

FIGURE 8.5 Observed vs. predicted plot for quality variable Y5.

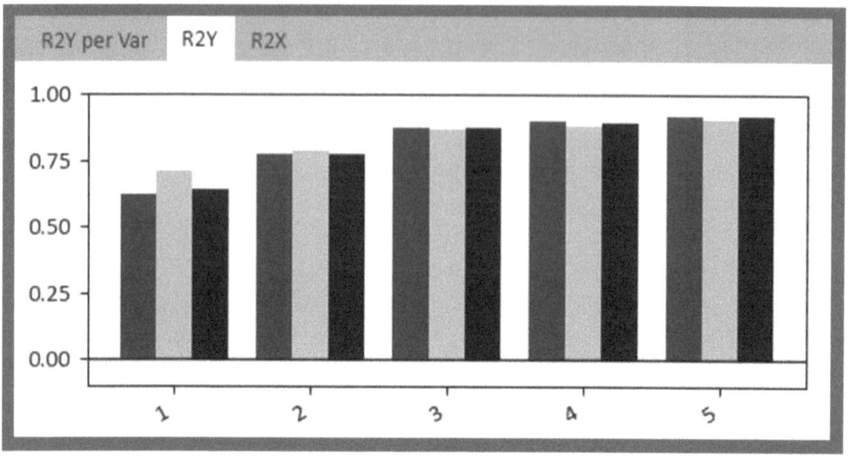

FIGURE 8.6 R^2 for the output space (Y) per component for a training set (first bar), a validation set (second bar), and the final model (includes the training and validation set, and depicted as the third bar).

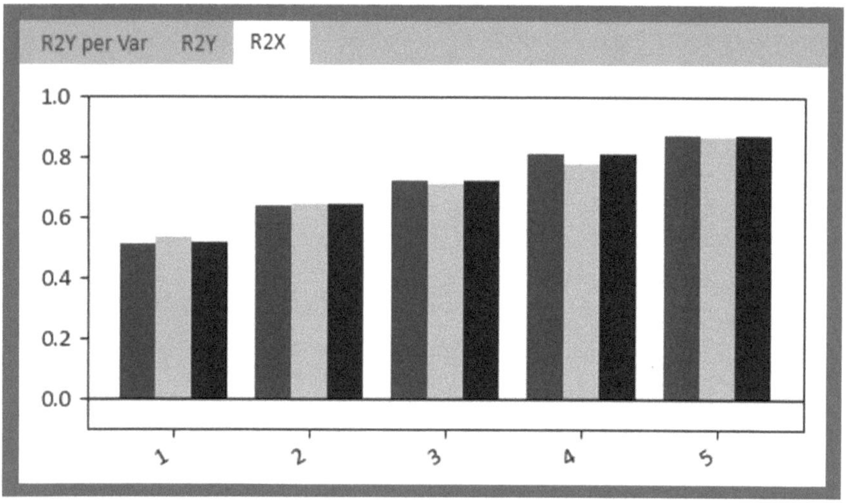

FIGURE 8.7 R^2 for the input space (X) per component for a training set (first bar), a validation set (second bar), and the final model (includes the training and validation set, and depicted as the third bar).

8.5.2 Model Parameters

Since the PLS model is a new representation of the original variables (X and Y) into the latent variable space (*t* and *u*), it is helpful to consider these new latent variables as linear combinations of the original variables and then to use these latent variables for prediction. The model parameters that are calculated are as follows, for the X-space and Y-space, respectively:

- Scores, *t* (X-space) and *u* (Y-space)
- Weights, *w* (X-space) and *c* (Y-space)
- Loadings, *p* (X-space)

Score scatterplots, as shown in Figure 8.8, are a helpful first look at the data to observe how the data is distributed, observe distinguished groups or clusters if they exist (clusters of data indicate observations with similar characteristics described by the designated components), and to find outliers.

Score plots are often shown in two dimensions with the two axes designated as the components and the 95% and 99% confidence interval ellipses drawn on the chart. The T1–T2 plot (first and second component) often describes the most variation explained by the model and is the most often shown. However, it is still useful to plot different components against each other, as each component may explain a different part of the variation in the data.

Points that lie near the origin are closer to the average of the model, and points that lie outside of the confidence interval ellipses can indicate outliers.

In tandem with score plots, the loading/weight scatterplots can be used to map out the correlation direction of the components. Variable weights that are grouped together indicate variables that correlate together. For example, weights that have a positive value will contribute to a *positive* score when the data for that variable is *positive* and vice versa. On the other hand, weights that have a negative value will contribute to a *negative* score when the data for that variable is *positive*.

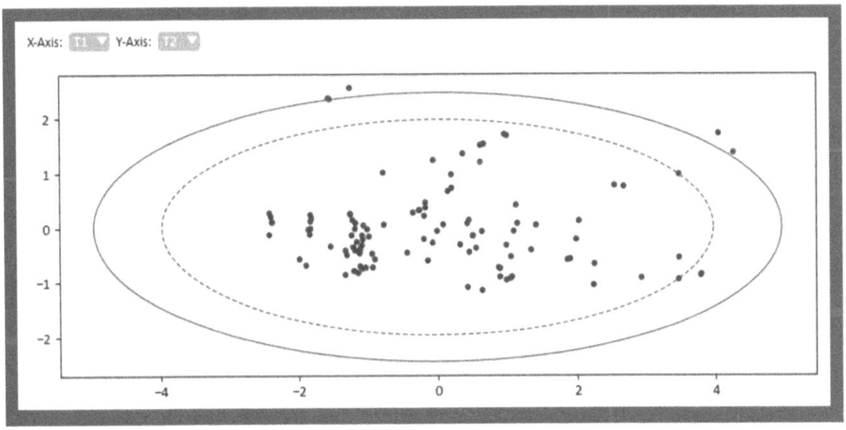

FIGURE 8.8 Score scatter plot for component 1 (x-axis) and component 2 (y-axis).

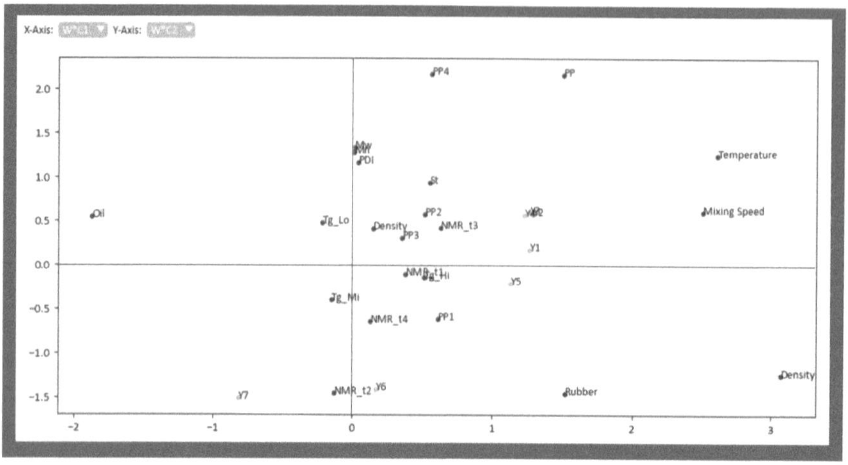

FIGURE 8.9 Loading/weight scatter plot for component 1 (x-axis) and component 2 (y-axis) with input and output variables.

Figure 8.9 shows the x and y variable weights for the first two components. Given the close proximity for Y1–Y5 on the plot, it can be seen that these five quality metrics move together and also in opposite direction of Y7. The two process variables (temperature and mixing speed) also move together and are positively correlated with Y1-Y5.

8.5.3 Model Validity Metrics

To further identify model patterns, it is important to look at HT2 and SPE plots in order to find outliers and if the observations fall within limits. Both of these plots (Figures 8.10 and 8.11) are plotted with the observation along the horizontal axis and the value (SPE/HT2) along the vertical axis, with the 95% and 99% confidence intervals plotted as horizontal lines. HT2 plots are useful as they are a cumulative summary of the scores for all components, as opposed to viewing score plots that only show two components. A higher value of HT2 indicates a distance further away from the average of the PLS hyperplane for all dimensions.

SPE plots can be used in a similar fashion as HT2 plots, with observations along the horizontal axis, SPE along the vertical axis, and horizontal lines for the confidence intervals.

It should be noted that if observations exceed the HT2 limits but not the SPE limits, the observations may still be acceptable, as it indicates that they follow the correlation structure of the model but are different than the average. In other words, the observations are extrapolating from the model. On the other hand, if SPE is high and HT2 is not, this indicates that the PLS model does not model the behavior of those observations and cannot be sure if the predictions are valid. In both cases, the modeler should take caution and carefully consider leaving the observations in the model or removing them.

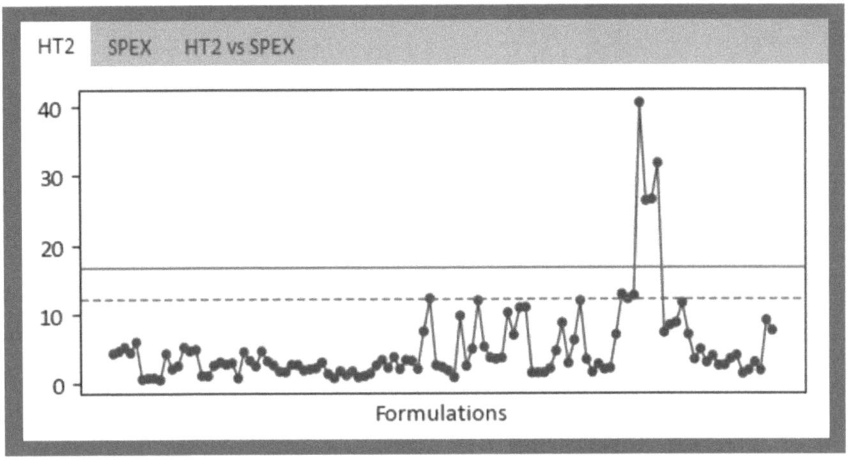

FIGURE 8.10 Hotelling's T^2 plot showing the value for each observation.

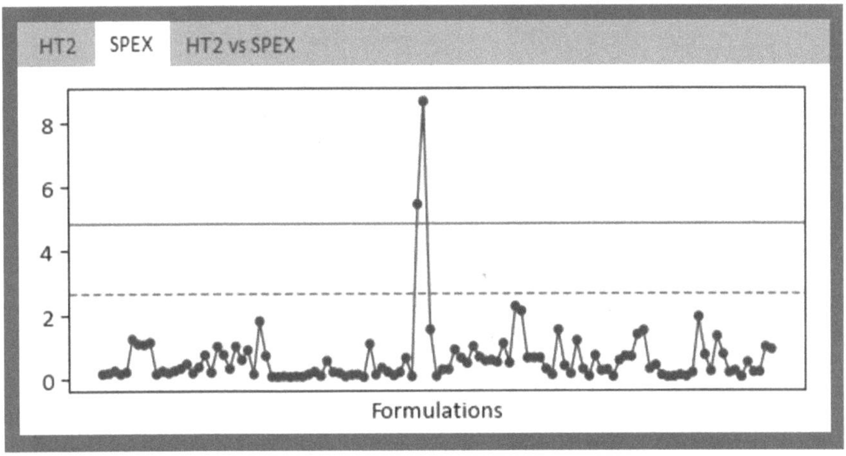

FIGURE 8.11 SPE plot showing the value for each observation.

Contribution bar plots for SPE and HT2 show the effect of each variable on the SPE and HT2, where each contribution value is plotted with the variables as the indices. Contributions to SPE can be calculated using Equation 8.11:

$$c_{SPE,i,k} = \left(x_{i,k} - \hat{x}_{i,k}\right)^2, k \in K, i \in N \qquad \text{(Equation 8.11)}$$

where $c_{SPE,i,j}$ is the SPE contribution of observation i for variable k, and $x_{i,k}$ and $\hat{x}_{i,k}$ are the data points. The contributions to HT2 can be calculated with Equation 8.12:

$$c_{HT^2,i,k} = w_{a,k}\left(x_{i,k} - \underline{x_k}\right), k \in K, i \in N, a \in A \qquad \text{(Equation 8.12)}$$

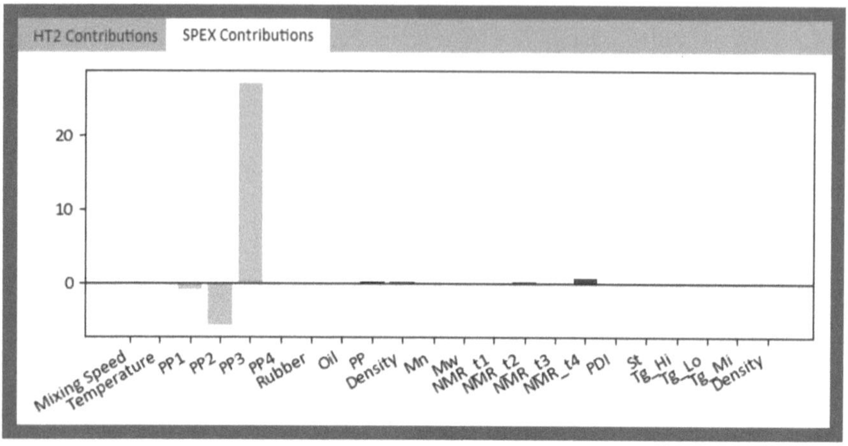

FIGURE 8.12 SPE contribution for the observation with the highest SPE, colored by data block.

where $c_{HT^2,i,k}$ is the HT2 contribution of observation i for variable k, $w_{a,k}$ are the weights for component a and variable k (Kourti 2006).

An example of the contribution to SPE is shown in Figure 8.12. This is for the observation with the largest SPE value and shows that a polypropylene, PP3, was used in larger-than-expected amounts in this formulation. The modeler may find incorrect data using such a plot, may decide to remove this observation, or may adjust the number of principal components to tune the model to better fit this behavior.

8.5.4 Coefficients

Similar to contribution plots, coefficient plots can be plotted on a single plot to show the relative correlation strength of each input variable to the output variable. PLS-regression coefficients can be calculated and used to visualize the relationship between X variables and Y variables. Starting with Equation 8.13 for the calculation of predicted Y:

$$\hat{y} = tc^T \quad \text{(Equation 8.13)}$$

where t and c^T are indexed from the component 1 through to the last component A, scores t can be substituted using:

$$t = xw^* \quad \text{(Equation 8.14)}$$

where x is the input data and w^* is the un-deflated weight matrix. Thus, the resulting relationship in Equation 8.15:

$$\hat{y} = \left(xw^*\right)c^T = x\beta \quad \text{(Equation 8.15)}$$

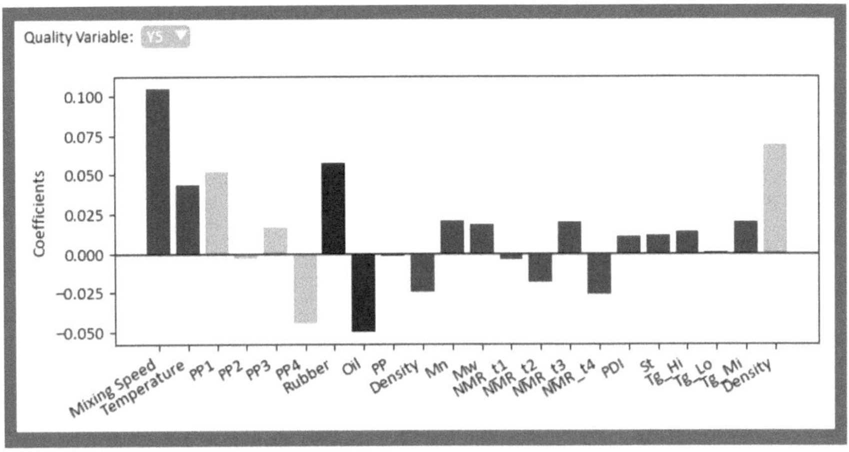

FIGURE 8.13 Coefficient plot for quality variable Y5, colored by data block.

where β is the vector of coefficients that relates x to $\hat{y}$. It is not advised to use these coefficients to directly predict the outputs with inputs alone, as it is important to check if the inputs follow the model structure (via SPE and HT2) (Dunn 2023).

In Figure 8.13, we see the simultaneous impact of all the input variables on quality variable Y5. Higher-than-average values of variables with positive bars combined with lower-than-average values of variables with negative bars will result in maximum Y5. It is important to remember that the model does not imply that each variable can be moved on their own, rather takes into account any expected correlations of the dataset.

8.6 USING PLS MODELS FOR PREDICTION

Once one obtains a training model, the model can then be used to predict quality variables using input variables from new observations and applying the model parameters, as well as evaluating if the new observation x_{new} is within the limits of the model (Wold et al. 2001). First, the new observation is projected onto the model to obtain the scores using Equation 8.16:

$$t_{new} = x_{new} w^{*}_{model} \quad \text{(Equation 8.16)}$$

Next, the SPE and HT2 should be calculated using Equations 8.17 and 8.18 to ensure that the new data is within the limits of the model:

$$SPE_{new} = \sqrt{\left(x_{new} - \hat{x}_{new}\right)\left(x_{new} - \hat{x}_{new}\right)^{T}} \quad \text{(Equation 8.17)}$$

$$HT^{2}_{new} = \sum_{a=1}^{A} \frac{t^{2}_{new,a}}{s^{2}_{model,a}} \quad \text{(Equation 8.18)}$$

where $s_{model,a}$ is the variance of the scores from the training model for component a, and $\hat{x}_{new}$ (predicted x) is calculated using Equation 8.19:

$$\hat{x}_{new} = t_{new} p_{model} \quad \text{(Equation 8.19)}$$

If the calculated SPE and HT2 are within the acceptable limits from the model (generally 95% and 99% confidence intervals are used for comparison), then one can proceed to calculate the predicted output using Equation 8.20:

$$\hat{y}_{new} = t_{new} c_{model}^{T} \quad \text{(Equation 8.20)}$$

Finally, the root mean squared error of prediction (RMSEP) for the entire set of new observations can be calculated using Equation 8.21 to quantify the prediction error from the model in units of the output variables (Zhang & Garcia-Munoz 2009):

$$RMSEP_m = \sqrt{\frac{\sum_{i}^{N_{new}} \left(y_{i,m} - \hat{y}_{i,m}\right)^2}{N_{new}}}, m \in M \quad \text{(Equation 8.21)}$$

where N_{new} is the number of new observations, and m is the output variable in the set of all M output variables. The RMSEP can then be compared to the RMSEE.

8.7 CONCLUSION

Latent variable modeling has many advantages for product development datasets. The ability to handle missing data, collinear variables, and applications to large and small datasets are key advantages. Additionally, latent variable models are unique and can be inverted in a constrained optimization framework. As with any modeling method, careful preparation of datasets is required in order to develop good models for product formulation. This may include the calculation of mixture properties, transformation of inputs/outputs to linearize the system (log, exponential, etc.), or cross-terms, and categorical variables. The next chapter will discuss how to gain confidence in a model and ensure that it meets the objectives of the formulation problem.

REFERENCES

Abdi, H. 2010. "Partial least squares regression and projection to latent structure regression (PLS regression)." *WIRE s Computational Statistics*. https://doi.org/10.1002/wics.51

Alexander, D., A. Tropsha, and D.A. Winkler. 2015. "Beware of R2: simple, unambiguous assessment of the prediction accuracy of QSAR and QSPR models." *Journal of Chemical Information and Mo deling* 55 (7), 1316–1322. https://doi.org/10.1021/acs.jcim.5b00206

Dayal, B.S., and J.F. MacGregor. 1997. "Improved PLS algorithms." *Journal of Chemometrics* 11, 73–85. https://doi.org/10.1002/(SICI)1099-128X(199701)11:1%3C73::AID-CEM435%3E3.0.CO;2-%23

Dunn, K. 2023. *Proc ess Improvement Using Data*. https://learnche.org/pid/contents

Gelaldi, P., and B.R. Kowalski. 1986. "Partial least-squares regression: a tutorial." *Analyti ca Chimica Acta* 185, 1–17. https://doi.org/10.1016/0003-2670(86)80028-9

Indahl, U.G., K.H. Liland, and T. Naes. 2009. "Canonical partial least squares – a unified PLS approach to classification and regression problems." *Journa l of Chemometrics* 495–504. https://doi.org/10.1002/cem.1243

Kourti, T. 2006. "Process analytical technology beyond real-time analyzers; the role of multivariate analysis." *Critical Reviews in Analytical C hemistry* 36 (3), 257–278. https://doi.org/10.1080/10408340600969957

MacGregor, J.F., and T. Kourti. 1995. "Statistical process control of multivariate processes." *Control Engineering Practice* 3 (3), 403–414. https://doi.org/10.1016/0967-0661(95)00014-L

Nomikos, P., and J.F. MacGregor. 1995. "Multivariate SPC charts for monitoring batch processes. " *Technometrics* 37, 41–57. https://doi.org/10.2307/1269152

Wallace, K., and M. Cardin. 2023. *Pro Sensus FormuSense Software*. https://prosensus.com/new-rapid-product-development/formusense/

Westerhuis, J.A., T. Kourti, and J.F. MacGregor. 1998. "Analysis of multiblock and hierarchical PCA and PLS models." *Journal of Chemometrics* 12, 301–321. https://doi.org/10.1002/(SICI)1099-128X(199809/10)12:5%3C301::AID-CEM515%3E3.0.CO;2-S

Wold, S. 1978. "Cross-validatory estimation of the number of components in factor and principal components models." *Technom etrics* 20 (4), 397–405. https://doi.org/10.2307/1267639

Wold, S., M. Sjostrom, and L. Eriksson. 2001. "PLS-regression: a basic tool of chemometrics." *Chemometrics and Intelligent Laboratory S ystems* 58 (2), 109–130. https://doi.org/10.1016/S0169-7439(01)00155-1

Zhang, L., and S. Garcia-Munoz. 2009. "A comparison of different methods to estimate prediction uncertainty using Partial Least Squares (PLS): a practitioner's perspective." *Chemometrics and Intelligent Laboratory Systems* 97, 152–158. https://doi.org/10.1016/j.chemolab.2009.03.007

9 Gaining Trust in Your Model

Marlene Cardin

9.1 INTRODUCTION

Now that a model has been built, it is important to take a step back and ensure that it can be used for its intended application. It has been said quite well by George Box that

"all models are bad, but some are useful."

Ensuring that a developed model can be trusted to make accurate predictions is a critical step in any AI-driven product development project. This chapter will cover how to gain trust in a model to ensure that it can be useful for the intended application(s).

9.1.1 Realistic Expectations for the Model

Before beginning the task of validating a model, it is important to have an understanding of realistic expectations for the model's usefulness and ability to predict quality metrics.

This is always related to the range covered by the dataset used to build the model, the structure of the model, and the amount of noise in the dataset. As discussed in Chapter 3, it is important to understand the characteristics of your data and appropriately match them to the prediction problem and associated business outcomes of your project.

Examples of unrealistic expectations:

- A model being able to predict the outcome of an experiment that uses a vastly different ingredient with unmeasured properties
- A quality metric that has a measurement error of +/– 20% cannot be predicted within +/– 5%

Instead, a useful model should be able to predict the results of experiments that use similar ingredients, combinations of those ingredients, and process conditions to those

DOI: 10.1201/9781003385974-12

used to build the model. In addition, the accuracy that the model can be expected to predict each quality attribute should be determined based on a well-designed repeatability study that quantifies the noise in the system. This will provide both realistic expectations for model fit (R^2) and the model's predictive ability on new data.

Of course, the goal of the model may also be to extrapolate to new regions not covered by the dataset. The degree that a model can extrapolate will again depend on the model's structure and training data.

Examples of possible extrapolation:

- A model built using only ingredient ratios in the input space will not be able to significantly extrapolate from historical data.
- A model incorporating both mixture properties that properly define the mixture and first principles knowledge will be able to extrapolate by predicting outcomes of experiments with previously unused ingredients.

9.1.2 Three-Step Validation Procedure

Once realistic expectations based on the dataset and model structure are understood, the task of validating the model can begin. In this chapter, a three-step process to validate a model will be detailed:

1. Using a validation set during model building to establish model structure and evaluate performance on unseen data
2. Verifying models with subject matter experts
3. Validating the model with a few targeted experiments in the desired application areas

9.2 STEP 1: VERIFYING MODELS WITH TEST SETS

Gaining trust in a model requires the testing of the model predictions on data not used during the model-building phase. This is accomplished by using a validation or test set and evaluating how well the model can predict the output variables of this previously unused dataset. The ideal result is that the error in predicting the performance properties of the validation set is similar to the error of prediction on the model training set.

Large differences in accuracy between seen and unseen data indicate a model that is either underfit or overfit and thus may not be useful for predictive new formulations.

Many modeling software tools have automated ways to generate a test set. This may include leaving out every nth observation or allowing the user to specify a test set. A rule of thumb is to leave out every 5th observation (or 20% of the data), but it is up to the user to decide if the resulting validation set is reasonable. This is especially important in the case of small datasets (less than 30 observations) or datasets that contain repeats or clusters. Both the input and output space of the training and test sets should be verified to ensure they have similar distributions of the data.

9.2.1 Input Space Test Set Verification

An ideal test set would span the range of input variable values and would cover the multivariate input space.

Figures 9.1 shows three examples where this is not the case. If possible, modifications should be made to these test sets to avoid having the test set extrapolate or cluster in a specific combination of input variables.

9.2.2 Output Space Test Set Verification

Examples of quality variables of test sets are shown on observed versus predicted plots in both Figures 9.2 and 9.3. In Figure 9.2, a poor test set is shown on the left where only the lower region of the quality variable range is covered.

A good test set would include data over the entire output variable range, as shown on the right of Figure 9.2.

In the model on the right, the distance from the error of the test set is similar to that of the training set, which is the desired outcome.

In the case where a dataset contains repeats (i.e., replicated samples), it is important to ensure that the validation set does not contain data points that are also included in the training dataset. In such a case as shown in Figure 9.3, it is best to choose entire repeat sets as the validation observations, rather than automated selection (such as every 5th data point).

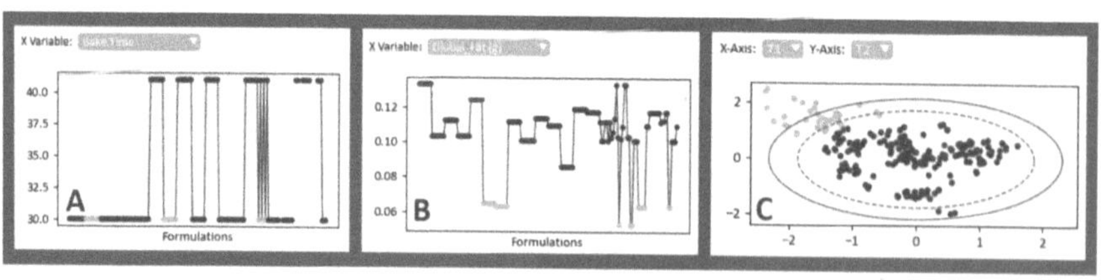

FIGURE 9.1 Examples of test sets (in light grey) with challenges.

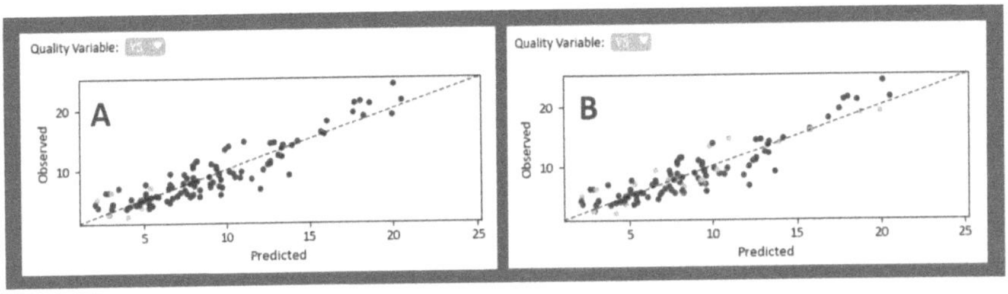

FIGURE 9.2 Test set (in light grey) and training set (in dark grey) examples showing a poor selection of test set in A and appropriate selection in B.

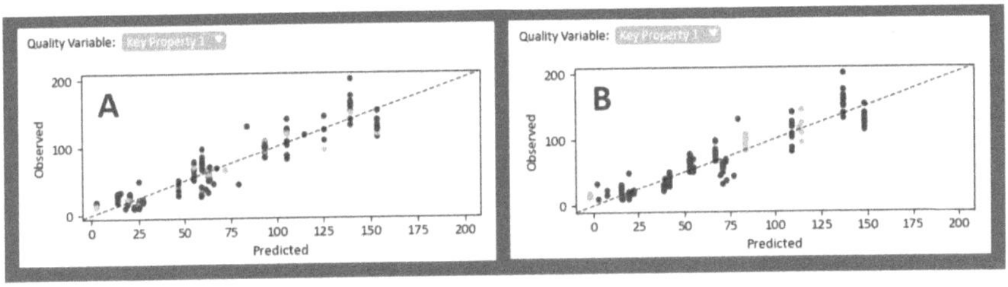

FIGURE 9.3 Test set (in light grey) and training set (in dark grey) examples showing a poor selection of test set in A, and appropriate selection in B for a dataset containing repeats.

9.2.3 Verifying Test Set Extrapolation

Certain modeling methods have metrics to ensure that data points do not extrapolate too far from the model's historical values. This should be verified in both absolute values and correlation structure. For example, if three temperatures are included that always increase or decrease together, a new data point where one temperature is at the higher end, and two are at the lower end, would be unexpected. Multivariate modeling metrics such as SPE and HT2 (as described in Chapter 8) could be used the verify that the validation set is not an outlier (above statistical limits) in either of these metrics. This could be performed as a quick check, even if latent variable modeling is not the main modeling method that has been selected for the project, for example, via principal components analysis (PCA).

9.2.4 Cross-Validation

Cross-validation is another tool that can be considered but is often not ideal for product development applications that are small, contain clusters, or repeat. In such cases, the validation set does require careful selection and does not typically lend itself well to either cross-validation or automated selection.

9.2.5 Metrics to Quantify Model Fit

An observed versus predicted plot can be used to visually confirm that the spread around the unity line is similar for the training set and validation set. This can be quantified using the root mean squared error of estimation for the prediction set (Chapter 8).

9.3 STEP 2: VERIFYING MODELS WITH SUBJECT MATTER EXPERTS

Subject matter experts will have some expectations of the correlations between the model's input and output space based on their years of experience and mechanistic understanding of the system. As such, it is important to verify that the model matches these expectations in order to gain their trust in the model.

9.3.1 Tools to Verify Model Relationships

There are a variety of tools that can be used to understand the underlying relationships in the data that a model has fit. More tools are being developed as the importance of "explainable/understandable AI" has recently become a popular topic. While this is not an exhaustive list, some tools to understand the relationships identified by the model include the following:

Coefficient Plot

For non-black box models, such as a PLS model (see Chapter 8), a coefficient plot can be used to understand the simultaneous impact of all of the input variables on an output variable. When the data has been scaled before model training, the coefficients can be used to interpret the direction and relative magnitude of the impact of each variable. In Figure 9.4, a PLS model coefficient plot shows that

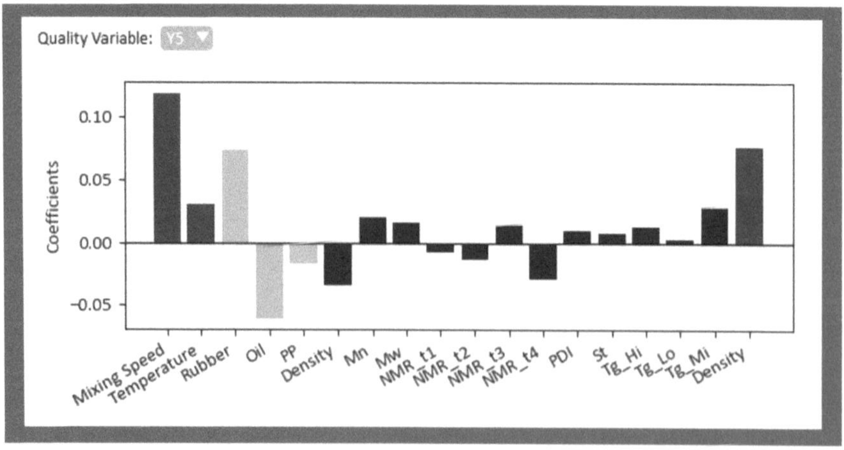

FIGURE 9.4 PLS coefficient plot.

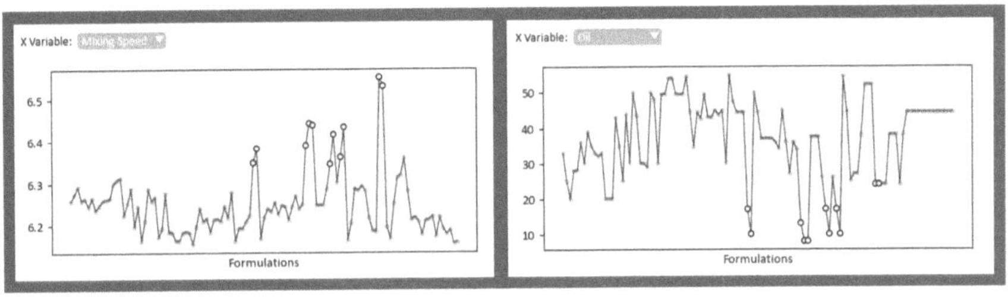

FIGURE 9.5 Raw data plots of mixing speed and oil.

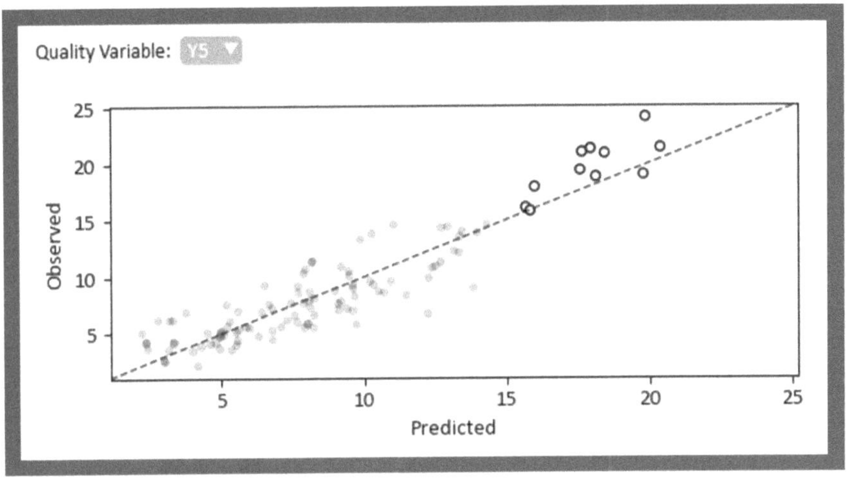

FIGURE 9.6 Observed vs. predicted plot for Y5, with high values outlined in dark grey and filled in white.

higher-than-average mixing speed, temperature, PP1, Rubber, and Global Density, in combination with lower-than-average values of PP4 and Oil will result in the highest predicted values of Y5. Relationships should be confirmed using raw data plots as shown in Figure 9.5 and observed versus predicted plots in Figure 9.6 (Wallace and Cardin 2023).

Shapley Chart

Shapley value is a method to interpret machine learning models that may otherwise be black-box in nature (Lundberg and Lee 2017). These are also known as SHAP (Shapley Additive exPlanations) and are calculated using the principles of cooperative game theory. Various permutations are generated for the input variables, predictions are calculated, and the resulting SHAP values are obtained for each input variable. These can then be used in charts to interpret the model similar to a coefficient plot (Molnar 2023).

Small Perturbations to Input Space

In the case of a black-box model, small perturbations can be made to key input variables to determine how the model's predictions will respond. When selecting the perturbations, it is important to keep in mind that not all variables can move independently. For example, if ingredient ratios are part of the input space, ensure that the formulation sums to 100% when making perturbations in an ingredient ratio.

It should be noted that causality is not implied in the aforementioned methods, and the best results will be obtained when interpreting these results with subject matter experts, keeping correlations in mind, and verifying results with the raw data used to build the model.

9.3.2 Model Mismatch with SME Expectations

In the ideal case, the model matches the domain experts' expectations. This would provide an important confidence boost in the model. However, in the case where the model doesn't fit the expectations, there can be a number of reasons including the following:

- *Incorrect data used in model building.* When a model doesn't fit expectations, the first step is to review data accuracy and ensure there were no mistakes in data assembly. As discussed in Chapter 4, typical problems may include data recorded in various units, missing ingredients in a formulation, zeros in place of missing data, etc.
- *Not enough variation in an input variable to capture the expected relationship.* This is a common occurrence in modeling and often occurs when datasets are built on subsets of larger datasets. It may be wise to consider removing some input or output variables that did not span the anticipated range. Such a model will have a narrower application but may still be useful despite these limitations.
- *Unexpected correlations are fit due to happen-stance correlation in the dataset.* For example, all experiments with high viscosity (an output variable) were run during the summer months with high ambient temperature (an input variable). Of course, the model will fit this relationship, but this does not imply causality. If such correlations are found, it may be necessary to run a few additional experiments to decouple the relationship or remove some variables from the model.
- *Over-fitting or under-fitting results in incorrect relationships being fit by the model.* Ensuring that the correct model structure and the number of model parameters has been selected is best evaluated through a validation set.
- *Missing information resulting in poor model fit.* Often when an expert looks at a first model, they will have recommendations for additional variables that were omitted in the first model. They may also have input on calculations (such as a cross-product between two variables) or transformations

(log, exponential, etc.) to better handle any non-linearities that may exist in the dataset. A new model should be built incorporating this valuable input.
- *The model may actually be correct!* There may be previously unknown combinations of input variables identified that can lead to a quality outcome. In this case, it is best to validate the findings through physical experimentation.

It is quite common to require a few model iterations as dataset challenges or model structure improvements are identified and addressed. As discussed in Chapter 3, having SME team members who are bought into the *process* of model building will help prevent frustration during these iterations.

It is important to set your team's expectations at the beginning of the project that the model will probably not be perfect on the first try but that as you combine insights learned from the data and their expertise, the model will improve.

Step 3: Verifying Models by Testing Intended Applications of the Model

Once a model has been verified with a validation set and domain knowledge, the next step is to test the intended use of the model. This step is especially important if the goal of the model is to extrapolate from historical data. The following approaches could be taken:

- Use digital simulation to evaluate the predictions on a few example formulations that test the model in its intended use. Subsequently performing a few physical experiments is recommended for many cases.
- Use constrained, numerical optimization to create products with specific targets. This is the backward direction, where the targeted output variables are set and new formulations and process conditions are calculated. Again, following optimization by performing a few physical experiments may be required for many models.

It is important to ensure significant extrapolation is avoided when applying a model. This can be done by limiting the range that input variables can have, ensuring various constraints are met (e.g., sum to 100% for the formulation ratios), and ensuring that correlated variables are not varied independently. The ease with which extrapolation can be quantified will of course depend on the model complexity and modeling method used.

9.4 CONCLUSION

Gaining trust in a model is an important step in AI-driven product development. It will require a team that includes the modeler and the subject matter expert, and the intended use of the model must be clear. In this chapter, a three-step validation process has been described that aims to provide some guidelines on this important task.

REFERENCES

Lundberg, Scott M., and Sun-In Lee. 2017. "A unified approach to interpreting model predictions." *NIPS'17: Proceedings of the 31st International Conference on Neural Information Processing Systems*, 4768–4777.

Molnar, C. 2023. *Interpretable Machine Learnin g*. Independently Published. https://christophm.github.io/interpretable-ml-book/.

Wallace, K., and M. Cardin. 2023. *Pro Sensus FormuSense Software*. https://prosensus.com/new-rapid-product-development/formusense/

Section 4

Optimization and Inverse Design

Alix Schmidt and Kristin Wallace

Using the methods in Section 3, you have hopefully arrived at a predictive model for your formulated system that the team is happy with. Depending on your objective or business case, you may be done and already creating value! The domain experts in your area can now test hypotheses rapidly and explore their ideas at minimal cost using experimentation with your model rather than in the lab. Once a user has settled on a new product formulation, a validation experiment in lab can provide the real-world basis for further commercialization or publishing *and* that new data can be fed back into the model training dataset to continually grow its design scope and improve accuracy.

However, we have often found that once formulators and materials scientists grow accustomed to using the model to test new formulations, shortly thereafter comes a new request: "can the computer just automatically suggest and test new formulations for me?" Users envision a future where they can tell the program what they want to do, let it go, and when they come back from a coffee break the perfect answer is right on their screen. There is a name for that utopian vision: inverse materials design.

Much like most data science, inverse materials design is an old concept with a new name. Building a model between raw materials and outcomes in order to target preferred outcomes has long been the goal of the fields of quantitative structure-property relationship (QSPR)/quantitative structure-activity relationship (QSAR) modeling.

DOI: 10.1201/9781003385974-13

But in the formulation context, inverse materials design refers not to molecular structure discovery but rather to the combinatorial challenge of formulated products. In the formulated products space – and similarly in inorganic materials, semiconductors, and alloys – the design problem is one of selecting one to many of a finite but large list of possible raw materials, determining their proportions, and defining process conditions to combine them to form a new functional material. Due to the combinatorial nature of the problem, the design space of possible formulations is impractically enormous; a reasonable case of choosing 15 ingredients from an available pool of 500 gives on the order of 10^{28} possible combinations, *before* contemplating their formulation ratios. And so, it makes sense that the domain experts arrive at the conclusion that we should use computational methods to search this space for us. Plus, we do not just want a possible material, we want *the best* material. So, what key unlocks that door? Optimization.

For the purposes of this text, inverse design refers to the application of optimization methods on a predictive model of formulation performance to generate candidate materials meeting user criteria. True "model inversion" refers specifically to a certain set of optimization techniques and model forms that accommodate them, but here we will discuss a wide range of possible approaches to the design problem under the more general term. Section 4 covers basic to advanced concepts of optimization for the digital transformation of material formulation.

First, Chapter 10 introduces optimization concepts such as the objective function, decision variables, constraints, and inversion or search strategies. Examples of how these concepts apply to formulation inverse design are provided in the case of a partial least squares (PLS) model. Chapter 10 represents the base case of new materials discovery exploiting sufficient historical data. Chapter 11 presents a complementary approach: Bayesian Optimization. As previously discussed, in some cases you may not have enough high-quality data in the region you are targeting to make a sufficiently accurate predictive model. In these cases, if you can make a model that captures a quantitative measurement of the uncertainty of its prediction, you can use adaptive experimental design methods (also known as sequential learning) to most efficiently explore the formulation design space. This allows your team to make data-driven decisions about which materials to formulate next to both improve the model and drive toward your performance objectives. Lastly, Chapter 12 considers how you might use optimization and a performance model, which includes processing conditions, to jointly define the raw material formulation and processing design space to guarantee product quality during manufacturing. While this chapter uses the same mathematical programming approach to PLS inversion as Chapter 10, the focus is less on ingredient selection for material discovery and more on exploiting the causal relationships in the PLS latent space for process control in industries such as pharmaceutical formulation. In this way, we present methods to further utilize your model for scale-up of new materials. In summary, Section 4 provides your team with guidance on how to make the best decisions using your model, whether that is in the early stages of new product exploration or scaling up to production.

10 Introduction to Formulation Optimization

Alix Schmidt, Luis Briceno-Mena, Sreekanth Rajagopalan, Kaiwen Ma, Benjamin Reiner and Birgit Braun

10.1 INTRODUCTION TO OPTIMIZATION: DECISION-MAKING AND SEARCH

Development of a new formulation starts with a rather plain but heavy assumption: the formulation we are looking for exists, we just need to *find* it. The simplest approach to find the right formulation would be to just try them all until we stumble upon the one that works for us. We know, however, that leaving this to chance is not practical and we then must find some systematic way of going about it. This systematization is what we call *search*, and although the term resembles the everyday word, its formal – mathematical – definition is useful to understand the idea of optimization.

10.1.1 Toy Example – Finding the Median of a Set

Formally, search is the task of finding the element or elements in a set, also called *search space*, that fits some criteria. Consider, for example, a search space of the set of integer numbers from 1 to 3, $S = \{1,2,3\}$, and the criterion of finding a value in which an equal number of the remaining integers in the set are greater than the value as those which are less than the value (i.e., the median). To search for this, we could start at the beginning of the set and compare each element to the next one, writing down these relationships. The result of this exercise would be $R = [1 < 2; 1 < 3; 2 > 1; 2 < 3; 3 > 1; 3 > 2]$ and can be summarized as shown in Table 10.1. The element for which the condition holds that number of elements greater than and less than are equal is the one we are searching for: two.

From this seemingly trivial example, we could identify potential issues in scaling the technique to harder problems. For example, the idea of evaluating every and all pairs of elements can become very computationally expensive as the number of elements in the set becomes larger. This is what is called an *exhaustive search*, and it is generally avoided. A possible improvement to our search strategy is to stop evaluating after the second element, since no new information will be obtained; the relationship between 1 and 3, and 2 and 3 has already been evaluated and repeating it adds unnecessary computation time. Issues like this give rise to the need of *search algorithms*, which are systematic ways of conducting search efficiently.

DOI: 10.1201/9781003385974-14

TABLE 10.1
Results for Search Example

Element	Number of Elements Greater Than	Number of Elements Less Than	Equal?
1	2	0	No
2	1	1	Yes
3	0	2	No

Search algorithms exploit any information known *a priori* about the search space to reduce the number of evaluations needed to find the right answer.

A helpful aspect of the previous example is that we have a clear way of applying the criterion to each element in the search space, which may not always be the case. Our criterion can be expressed as an evaluation function defined over two elements of the set S, s_i and s_j as shown in Equation 10.1.[1] Since the outcome of the evaluation function is calculated from s_i and s_j, they are typically referred to as *decision variables*.

$$\sum_{j\in S\setminus i} f\left(s_i, s_j\right) = 0 \quad \forall i \in S; \quad f = \begin{cases} -1 & if \quad s_i < s_j \\ 1 & if \quad s_i > s_j \end{cases} \tag{10.1}$$

On top of evaluation functions, we may want to define some additional criteria for our search. In formulations, for example, we may want to be restrictive with the ingredients or combinations of ingredients. These restrictions, also known as *constraints*, are necessary but not sufficient for finding a solution, meaning that in addition to satisfying them, we still want to satisfy the evaluation functions.

When the criterion for the search involves finding the extreme values, maximum or minimum of the search space, the task is called *optimization*. Given a set of evaluation functions $F = \{f_1, f_2, \ldots, f_n\}$ and a set of constraints $G = \{g_1, g_2, \ldots, g_m\}$ defined over a set of decision variables $x = \{x_1, x_2, \ldots, x_k\}$, an optimization problem can be defined to search for the values of x that minimize (or maximize) the values returned by the functions in F, subject to the constraints in G (Nocedal and Wright 2006). The earlier definition can be written as an equation as shown in Equation 10.2.

$$\begin{cases} \min\limits_{x} & f_i\left(x\right), i = 1, \ldots, n \\ subject\ to & g_j\left(x\right) \le 0, j = 1, \ldots, m \end{cases} \tag{10.2}$$

As stated earlier, evaluation functions for the behavior of the decision variables are not always readily available. In such cases, one must first find a way of *predicting* the results of the evaluation function from a given set of decision variables, without having to go to the laboratory or the plant to measure it directly. To do this, predictive models built from experimental observations of the search space, that is,

1 The reader is encouraged to verify that Equation 10.1 returns the values in Table 10.1 for all cases.

experimental data from previous formulations, can be used. The predictive model then becomes the evaluation function. As will be shown later in this chapter, the nature of the evaluation function (the predictive model) deeply affects which search approach works best for a given situation.

10.1.2 Components of a Formulation Optimization Problem

The median of the set example given earlier has walked us intuitively through the elements of data-driven decision-making. We have an objective to find a particular integer (the median), we have a defined search space (the set), we know our decision variables (pairs of elements to compare), we have an evaluation function to assess a possible solution relative to our objective, we have checked for constraints (none), and we executed an exhaustive search approach. These will be the same fundamental parts we also see for a formulation optimization problem.

> The skill of optimization is in artfully translating your real-world problems into such a decision framework and then translating those to a mathematical framework.

Let's briefly define the components of the optimization framework within the context of inverse formulation design; subsequent sections will then dive deeper into the mathematical translations of these components and options for making formulation decisions.

The first step is clearly and comprehensively defining *what you are trying to do.* Construction of the optimization problem is where the rubber hits the road in properly aligning the business goals of the project to data science. The components of the optimization are derived from the statement of your goal. Take this goal for example:

> *Elastomer production is at capacity but we have not hit our growth goal for the year. We need to increase the average margin on the mixer asset, but there can't be any disruption to our customers. And we can't spend any new capital.*

How do we formulate this mathematically as an inverse design problem? We may start with the objective, of which there are several. To increase margin, we may increase price or decrease cost. Since there are to be no customer-facing changes, this means our goal is to decrease the average cost of products made on the mixer, either by reducing the materials cost or the processing cost (or both). Another objective that is incorporated into avoiding disruption of our customers is that the product quality remains the same. Lastly, as always, we need to keep the mixer operators happy (more on this later). These objectives are summarized in Table 10.2.

What variables do we have at our disposal to accomplish these goals? Since we have posed this as an inverse formulation design problem (as opposed to an equally valid process engineering problem), we have the product formulations as decision

TABLE 10.2
Objectives for Elastomer Reformulation Example

Objectives
1. Decrease product total cost
2. Quality attributes unchanged
3. No undue burden to mixer operators

variables. But which products are we manipulating? To make the problem simpler, instead of reformulating every product on the mixer, the team does a cursory economic evaluation and finds that if we decrease the material cost of the highest volume product by 20% and keep processing costs the same, we will meet our stakeholders' financial target. We verify with our experts that changing the composition of the product – which ingredients are used to make it – is okay as long as we keep all the critical quality attributes unchanged. Therefore, our formulation decision variables are the formulation ratios for all ingredients available to the mixer: which ingredients to use and how much. Additionally, any process variables that may need to be adjusted for the new formulation to hit the target quality are decision variables. These decision variables are added to the objectives in Table 10.3.

TABLE 10.3
Objectives and Decision Variables for Elastomer Reformulation Example

Objectives	Decision Variables
1. Decrease product total cost	1. Which product(s)
2. Quality attributes unchanged	2. Formulation Ratios
3. No undue burden to mixer operators	3. Process Conditions

What constraints do we need to be aware of? First, the team discusses the implications of no new capital spending. The team determines this means that processing conditions cannot deviate beyond current ranges and that formulation ratios cannot exceed 20 wt% total solids loaded into the mixer (which would require additional ventilation equipment). Another consideration is not creating an undue burden on the operators. The team agrees that bringing in any new ingredients to the plant would introduce unnecessary complexity to the change, so the search is limited to just the existing ingredients available. Table 10.4 summarizes the objectives, decision variables, and constraints for this example.

Lastly, we must consider how we will evaluate whether the set of decision variables chosen meets our objectives – the evaluation functions. Foremost, in order to know whether or not we have changed the material quality, we must have a model to predict quality attributes for a given candidate formulation and process. This is the predictive model the team has trained using the machine learning techniques of previous chapters, and it has been validated to be accurate within the normal

TABLE 10.4
Objectives, Decision Variables, and Constraints for Elastomer Reformulation Example

Objectives	Decision Variables	Constraints
1. Decrease product total cost 2. Quality attributes unchanged 3. No undue burden to mixer operators	1. Which product(s) 2. Formulation ratios 3. Process conditions	1. Process conditions limited to current range 2. No more than 20% solids 3. No new ingredients

batch-to-batch product variation. Next, in order to know if we have achieved the 20% cost reduction goal, we must have analytical expressions for the total cost of materials and processing. Lastly, to keep the effort level of the mixer operators the same or below current, we want to minimize the number of ingredients to add in each formulation; an evaluation function counting the ingredients with a formulation ratio greater than zero will do the job. Table 10.5 now contains the qualitative framework for our formulation optimization.

TABLE 10.5
All Optimization Criteria for Elastomer Reformulation Example

Objectives	Decision Variables	Constraints	Evaluation Functions
1. Decrease product total cost 2. Quality attributes unchanged 3. No undue burden to mixer operators	1. Which product(s) 2. Formulation ratios 3. Process conditions	1. Process conditions limited to current range 2. No more than 20% solids 3. No new ingredients	1. Predictive model for elastomer quality 2. Analytic equation for materials cost 3. Analytic equation for process cost 4. Analytic equation for count of ingredients

This example illustrates the translation process of business goals into the decision framework; the next two sections will describe further translation of these components into mathematical expressions within the context of data-driven materials design. Then in Section 10.4, we will discuss the final component of solving the problem: the search algorithm.

10.2 FORMULATION OBJECTIVES

Thus far we have defined our decision as a search – finding formulations that meet the criteria. But most often, we strive to find *the best* formulations that meet our criteria. As previously noted, a search becomes an optimization when you are searching for the maximum or minimum of the evaluation functions. Thus, it is useful to pose our objectives as evaluation functions that we seek to minimize – an *objective function.*

An objective function has an eponymous name denoting precisely "what" a researcher is looking to optimize. Like any mathematical function, an objective function can be simplistic, carrying only one term, or sophisticated, containing dozens of terms with numerous transformations.

The complexity of the objective function reflects the complexity of the system.

Objective functions are commonly represented as linear functions such as:

$$Z = a * X + b * Y$$

where a and b are constraints, X and Y are variables, and Z is the optimization target. However, other mathematical operations such as modulos, logarithms, or trigonometric functions can also be used as appropriate. Objective functions do not need to be continuous or differentiable. Piecemeal functions may be appropriate estimates of complex systems that traverse several physical regimes. Objective functions should be set up as minimization problems where possible; resolving items closest to zero is an easier mathematical problem than finding the value that is arbitrarily highest, especially in high-dimensional spaces. Keen researchers should strive for parsimony when constructing objective functions in order to mitigate poorly behaved models.

Design of objective functions is simpler than the name implies. An imaginary business start-up, *TechTown*, can serve well as a toy exercise. *TechTown* produces ultra-low heat processors for high-performance computers. The research team first wants to optimize the heat profile of their processors, which leads to a simple objective function: $Z = 3 * heat\ output$. Here, the best processor will have the highest Z value (note in a real system we might invert this function to produce a minimization problem where the best processor has a value closest to zero). As *TechTown* grows in size and revenue, the objective function will evolve. The marketing team may want to add an additional consideration, cost of materials. A second-generation objective function might then be: $Z = 3 * heat\ output - 5 * cost$. Here, processors that perform excellently but are expensive may be a worse choice than processors with intermediate behavior but a low price point. The scalar values – 3 and 5, often called *weights* – can be changed to reflect the relative importance of the two variables. Scaling functions such as logarithms may be necessary to normalize the variables if heat output and cost are measured at different orders of magnitude.

An important application of an objective function is in inverse design where a set of properties are used as input and a material of interest is the expected output. Industrially relevant materials such as pharmaceutical formulations, adhesives, elastomers, barrier products (such as coatings or potting compounds), cosmetics, and even foods are often formulated through combinatorial design and can have enormous search domains. Appropriately tuned objective functions ensure that optimization solutions suggesting the next best experiments in a campaign produce a reasonable library of candidates. These functions should contain holistic information about new candidates. Our toy example from the previous example remains a convenient prop for contextualizing what "holistic" can mean. The heroic data science team at *TechTown* has built an inverse model that allows researchers to choose processors based on a variety of relevant metrics such as heat output, conductivity, footprint, price point, and logistics. A typical objective function for an inverse design in this system might contain additional terms for processor behavior, fabrication time, raw materials cost, price point, and transportation cost. This holistic lens (built from decision variables, constraints, and evaluation functions) allows a mathematical programming model to search the domain of possible processors and select candidates that are physically the best and most financially viable on an industrial scale. The definition of candidate success can be non-trivial to describe, but objective functions help translate the complexity of business value to a mathematically rigorous optimizer.

This is one area where extensive discussion and iteration with subject-matter experts on the team is especially helpful; no matter how good the model and search algorithm, if the team is operating under an incorrect definition of success the project could fail.

This is an excellent time to refer back to the project definition and actionable path to value, which was discussed in Chapter 3 – the objective function could be considered the mathematical representation of that value case.

10.2.1 Common Objectives for Inverse Materials Design

The *TechTown* processor and elastomer reformulation examples give us a glimpse into typical objectives for a materials design problem. There is nearly always a material performance objective, or multiple depending on the availability of predictive models.

Multi-objective optimization is very common due to the prevalence of two or more competing quality attributes in formulation science; the objective function provides a mechanism to quantitatively balance the trade-off between them.

Table 10.6 lists a variety of common objectives for materials performance and their mathematical formulation in the objective function. For example, if you want to minimize the tackiness of a paint, you can add the predicted tackiness value, Y_{pred}, to the objective function, weighted by some value w_i relative to other terms. To hit a specific target value, you can minimize the difference between the target and the predicted value.

There may be additional objectives pertaining to the design of the material beyond its performance properties. To write these terms, we must first define notations for the decision variables. Using the notation defined in Chapter 8, our manipulated variables are the ingredient ratios, r_i, for the N ingredients available. Each ingredient

TABLE 10.6
Objective Function Terms for Quality Attributes

Objective	Objective Function Representation (To Be Minimized over the Decision Variables)
Minimize or maximize a performance attribute	$wY_{1,\,Predicted}$ (to minimize value) or $-wY_{1,\,Predicted}$ (to maximize value)
Hit a target performance value	$w\sqrt{\left(Y_{target} - Y_{Predicted}\right)^2}$
Balance competing performance attributes, with user-input relative weights in matrix W	$w_i Y_{1,\,Predicted} + w_j Y_{2,\,predicted} + \ldots + w_n Y_{n,\,Predicted}$ (with signs set appropriately)

has K properties, stored in matrix X, where the cost per kilogram of ingredient i is particularly noted by c_i. We may use the binary variable δ_i to denote whether or not ingredient i is used in the formulation (i.e., $r_i > 0$).

$$\delta_i = \begin{cases} 1, \ where\, r_{new,i} > 0 \\ 0, \ where\, r_{new,i} = 0 \end{cases}$$

Now, with these mathematical definitions, we can easily minimize the total number of ingredients (or ingredients in a particular class), minimize the total materials cost, and add other custom analytical expressions for processing cost or other business factors.

Further preferences from the domain experts can readily be handled with additional terms to place hard or soft constraints on r or δ. We may even incorporate descriptors or mixture properties, discussed next.

10.2.2 Mixture Properties and Features

One of the challenging, but often helpful, tasks in materials data science is the efficient and responsible translation of complex physical system characteristics to objects friendly to mathematical models. One such way to incorporate fundamental properties or behaviors into a predictive model is the use of descriptors or features. As thoroughly discussed in Chapter 5, features are properties of the system that can help endow models with additional information. Feature engineering, or the specific method for curating and calculating features, is an on-going field of research.

When modeling formulations, the initial approach typically attempts to relate ingredient amounts to performance.

While intuitive, this composition-based model creates some challenges for the modeling and optimization, namely sparsity and non-relatability of ingredient characteristics to product properties.

The sparsity aspect results from the fact that the datasets resulting from large formulating projects contain a small number of non-zero values, for example, when an ingredient is only used a handful of times. In an example, paints are comprised of many different ingredient classes, such as binders, pigments, additives, and solvents; each of these ingredient classes contains multiple individual compounds that can be present one at a time or in combination, at various ratios. If the paint binder is a blend of two to three chemicals out of a set of more than 20, the columns containing the ingredient amount for the chemistries not used, which contain a zero value, represent the majority of the training data matrix. The significant uncertainty associated with each ingredient variable in a machine-learned model based on such sparse data can cause worrisome instability in the optimization.

However, addition of features can help reduce model uncertainty and add in interpretable domain information. For example, pigments can be described by a number of physically relevant features such as particle size, molecular charge, and opacity. Some of these features such as opacity can be directly measured and others such as molecular charge may require calculation. It is sensible to consider features that are actually controllable in the production process or have the potential to be altered when found to be impactful and show variability in the available dataset.

Within the optimization routine, inclusion of features represents an intermediate computational step. Descriptors are rarely decision variables; instead, they are properties calculated using decision variables and additional static data. For example, under the mixture property paradigm, the X matrix contains the ingredient property values for all ingredients. To calculate a descriptor for the formulation, as used within the predictive model, we must have an analytical expression that relates the ingredient ratios and the particular ingredient property values to the overall mixture property. The starting point is often using a simple weighted average; for example, we may calculate the density of a new candidate paint formulation by taking the weighted average density of all ingredients in that formulation. Features may also be calculated by custom, domain-based physics equations, as discussed in Chapter 5.

Datasets with well-designed features typically result in higher accuracy predictive models. However, optimization may become challenging with highly featurized datasets.

As solutions that meet objectives while respecting constraints are identified, the translation of the predictive model input space with its features to the combination of the actual ingredients may become challenging. For example, a solution may recommend a specific average pigment particle size; however, that average particle size could be achieved using many different possible combinations of ingredients. In short, the routine has identified an optimum instance of the model prediction, but it is still a "one-to-many" translation to the actual decision variable. Consequently, the optimization might involve two steps, namely first the optimization followed by the ingredient definition.

In addition, the featurized dataset may result in non-linearities that were not observed in the composition-based data, possibly requiring different and computationally more expensive optimization methods. One example is the use of viscosity as a mixture property. It is well-known that the viscosity of a polymer blend is not the weighted average of the raw viscosity values but of the logarithm of those values, which results in a log(mixture viscosity). This non-linear component of the optimization requires special handling by the solver, adding complexity and potential cost to the software.

Sometimes there are terms added to the objective function (or constraints), which are properties calculated from the decision variables but not features in the predictive model for formulation performance. These terms are not material descriptors, but they are descriptive of the system. One example is the model residuals in a PLS model. Recall from Chapter 8 that a partial least squares regression model creates a latent variable representation of the system, and when new points are predicted, we get additional diagnostic information about the prediction from the Hotelling's T^2 and Squared Prediction Error (SPE) residuals. These values can be helpful for controlling the extrapolation of solutions from the training data as well as preferring solutions with less model uncertainty. Both residuals are calculated from squared terms, meaning their inclusion may change the optimization problem from a linear program to a quadratic (non-linear) one. More discussion on the implications of this can be found in Section 10.4.3 on mathematical programming.

10.2.3 Putting It All Together

After a thorough discussion with your team on what you are trying to accomplish and translation of those success criteria to evaluation functions, you are ready to assemble your objective function. To summarize, the set of computations that occurs during a search of the formulation space is given as follows:

1. Begin with a given set of decision variables, R and x, which represent a candidate formulation
2. Use decision variables to calculate constraints
3. Use decision variables to calculate formulation mixture properties and any other model features
4. Use the feature set and the predictive model to calculate a predicted quality outcome for the candidate material
5. Calculate objective function terms from the decision variables, model outputs, and any additional user inputs
6. Scale objective function terms as appropriate, then apply weights w_i
7. Calculate the objective function value by summing all weighted terms

Conceptually, the search algorithm will use this value of the objective function to assess the preferability of the candidate solution, consider the constraints, and then select the next to test.

10.3 FORMULATION CONSTRAINTS

In the realm of formulation optimization, the process is intricately tied to the application of forward prediction models within well-defined domains, shaped by a set of constraints. These constraints draw from a wealth of sources, combining first-principles knowledge gleaned from research and subject matter expert (SME) experiences, along with valuable insights derived from historical data analytics. The role of constraints in this context is twofold: first, to guide the generation of formulations that align with first-principles knowledge, enhancing the likelihood of feasibility, and, second, to ensure the utilization of the forward-prediction model within domains characterized by high-confidence intervals.

> This synthesis of domain-specific expertise and data-driven insights forms a critical nexus in the optimization journey, enabling the creation of formulations that not only adhere to established principles but also leverage predictive models effectively within their designated confidence bounds.

Common constraints encompass various dimensions, ensuring a holistic and nuanced approach to formulation design. The key constraints typically include the following:

1. **Ingredient Quantity Bounds**: These constraints delineate the permissible range for the total number of ingredients and, further, the allocation of ingredients by class. This not only establishes the overall composition but also categorizes the formulation components.
2. **Ingredient Selection Constraints**: A crucial aspect involves modeling relationships between ingredients. This includes exclusion constraints, where certain ingredients cannot coexist in a formulation, and dependency constraints, dictating that the presence of one ingredient necessitates the inclusion of another.
3. **Formulation Ratio Bounds**: Constraints are set on the formulation ratios, ensuring that each ingredient or class total adheres to predefined limits. This fine-tunes the formulation to meet specific ratio requirements for optimal performance.
4. **Descriptor Space Bounds**: Formulation descriptors are subjected to bounds, delineating the acceptable ranges in descriptor space. This adds precision to the formulation process, aligning it with specified parameters understood by domain experts.
5. **Correlation in Ingredient Usage**: Constraints governing the relationships between ingredients extend to correlations in usage. An illustrative example is the proportional dosage of additive ingredients relative to primary components, establishing an intentional balance in formulation.
6. **Model Residual or Extrapolation Constraints**: If your model enables an expression of residual error or predicted uncertainty, then limits can be set on these values to prevent excessive extrapolation from the well-known region. The value of the constraint may be calibrated from the residuals or uncertainty within the training data.

Conditional constraints further enhance the complexity of the optimization process. These conditional rules, such as the utilization of ingredient A being contingent upon the presence of ingredient B up to a certain amount, introduce a layer of nuanced decision-making based on specific contextual conditions.

The derivation of constraints in the formulation optimization process is an iterative journey, blending insights from research, SME experiences, and the analytical examination of historical data.

The collaborative outset with SMEs initiates the identification of potential correlations among ingredients, rooted in first-principle knowledge. This qualitative foundation serves as a crucial starting point, capturing the expertise-driven nuances that are often intrinsic to formulation design. As the iterative process unfolds, the analytical study of experimental data becomes paramount. This quantitative exploration yields valuable insights, providing a numerical dimension to the relationships discerned during initial discussions. One example is simple but powerful: providing the team with graphs showing the count of ingredients within each class across all the historical formulations. In the paint example, this could look like Figure 10.1, showing that most formulations contain two binder ingredients, a similar amount contain only one, and the count of formulations tails off above five binders. Such information facilitates conversations among the team about what a preferred formulation looks like, ensuring a harmonious integration of empirical evidence and expert knowledge.

There is also a growing body of work to manage the challenge of constraints that are not easily stated or formulated into a mathematical expression. *Constraint learning* is the approach of using machine learning to construct constraints according to the data itself (Fajemisin et al. 2024).

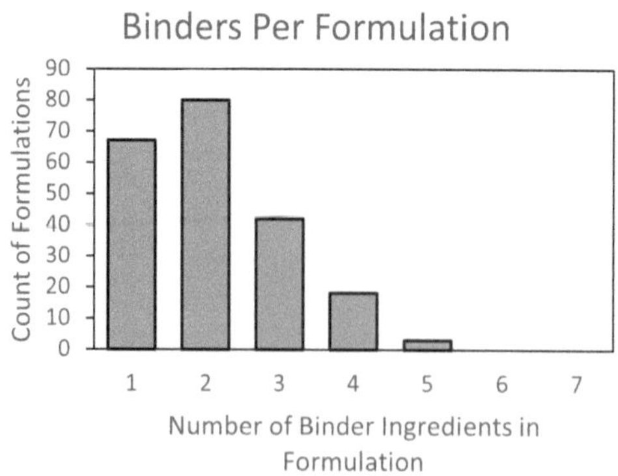

FIGURE 10.1 Example visualization of the number of binder ingredients per paint formulation in a hypothetical training data set.

For both constraint learning and manual definition of constraints using subject-matter expertise, it is imperative to acknowledge the unique characteristics of historical formulation data, which are amassed from past research endeavors and commercial necessities. The data often exhibit skewness, being densely concentrated in specific regions of interest while presenting sparse coverage in many other spaces. This inherent bias necessitates a careful understanding when extracting insights from the data. Consequently, the formulation of constraints becomes a strategic exercise aimed at leveraging the strengths of available data while mitigating the impact of its inherent limitations. Furthermore, the degree to which you want to specify your inverse design via constraints is a subjective matter loosely tied to your business objectives. The more constrained you make the system based on the research of the past, the more likely a candidate solution will succeed, but at the expense of discovering potentially unknown formulation spaces. A highly constrained model will be less risky in the short term but potentially less valuable in the long term. It is recommended to allow researchers to test a range of constraints and decide which solutions they prefer.

10.4 APPROACHES TO SEARCH

There are three general approaches for finding solutions that meet the objective while observing the defined constraints, ranging from straightforward and generally capable to complex, highly custom global optimizers. First, the predictive model can be used as an offline simulator to construct a formulation database that fills up the formulation domain with candidate materials. During the search, one finds the formulation in the database that meets the objective based on its predicted properties. Second, there are methods to adaptively explore the available space using the predictive model as the performance evaluator and gradually improve the objective based on previously evaluated information; methods following this principle are referred to as black-box optimization such as evolutionary algorithms and swarm intelligence. Third, the predictive model itself in its mathematical form can be used to identify solutions to the optimization problem using well-defined strategies, which is the deterministic form of optimization and includes powerful methods like (non) linear mathematical programming approaches.

10.4.1 Search-Based Approach

The forward prediction model plays a crucial role in formulation optimization. In cases where embedding these models into a mathematical programming framework proves impractical – often due to the sheer size of the model exceeding solver capabilities, even with attempts at mathematical reformulation and dimensionality reduction – leveraging these models as simulators becomes a valuable alternative. This simulation-based approach to finding an optimal solution is called search-based or offline simulation-based optimization.

In the search-based approach, the offline stage involves leveraging forward prediction models and constraints to build a comprehensive database.

This process entails defining the formulation domain, selecting a design of experiment method (e.g., grid search), and generating simulated predictions for formulation performance using the predictive model. Additionally, efforts need to be made to establish confidence intervals for the predictions when feasible, ideally accounting for prediction uncertainty and extrapolation from the training data space. During the online stage, user-defined formulation criteria are translated into an objective function, enabling the identification of the optimal formulation from the database, which includes both experimental and simulated data. In some cases, the best solution may not meet all criteria but is the closest (per the objective function) within the set.

10.4.2 Black-Box Optimization

Black-box optimization techniques are an alternative approach to optimization using the predictive model as a simulation tool to iteratively explore the solution space to find the optimal or near-optimal solution. The methods entail constructing an objective function based on specific user requests, as outlined in the previous section. The optimization process begins by identifying the historical formulation with the best objective function as the initial starting solution. Subsequently, the black-box optimization algorithms adaptively simulate various formulations based on previous objective function evaluations until a termination condition is met. This termination condition can be predefined, based on factors such as a simulation budget or the attainment of a satisfactory solution. The adaptive nature of this approach allows for dynamic exploration of the formulation space, offering flexibility in response to changing optimization requirements. Black-box optimization includes Bayesian-optimization (BO) and Derivative-free optimization (DFO). These optimization techniques provide efficient strategies for navigating the design space. BO leverages probabilistic models to intelligently balance exploration and exploitation and will be discussed in detail in Chapter 11. DFO employs a combination of local and global search strategies to avoid being trapped in local optima.

A DFO approach relies only on the functional value of the objective function, $f(x)$. A DFO algorithm proceeds iteratively by querying the value of the function at a specific candidate point or a set of candidate points x and relying only on the outcome of the query. DFO algorithms offer a flexible interface to solve the specific optimization problem where only an expression that returns the numerical value of the objective function to the algorithm is needed, and, as such, can be plug-and-play. However, this benefit comes at a cost: DFO algorithms can neither guarantee optimality of the solution obtained (i.e., a certificate that the solution obtained is the best possible or is close to the best within a certain bound) nor guarantee that the solution is even locally optimally (i.e., there may exist a small neighborhood around the solution where a better solution exists). However, for practical applications, a well-defined objective function subject to system-relevant constraints ensures that the identified solution meets the expectation targets; while global optimality is ideal it does not have practical impact when the objectives can be met with local optima or even solutions that do not represent a true local optimum. Lastly, one additional drawback to DFO is that algorithms do not scale well in terms of computational performance beyond a few tens to a few hundred decision variables.

Any DFO algorithm can be used to handle the constraints $g(x)$ by extending the objective function to return a very large negative (positive) value for maximization (minimization) problem when any constraint is infeasible. Some DFO algorithms can handle the constraints explicitly where the specific optimization evaluation expression can be setup to return a set of Boolean values representing the corresponding feasibility of constraints $g(x)$. Some DFO solvers allow direct specification of simple constraints such as box constraints that represent bounds on the decision variables or a set of linear constraints in decision variables.

DFO algorithms typically vary in how they balance *diversification* and *exploitation* of search. Diversification is when the algorithm searches over large regions of the search space to find potentially better solutions and is typically a randomized step that does not guarantee a better candidate solution. Exploitation is when the algorithm exploits a known region of the search space to improve the existing solution but only marginally. The diversification step is where stochastic approaches – such as evolutionary algorithms including Genetic Algorithm, Particle Swarm Optimization, and Ant Colony Optimization – excel. Algorithms such as Simulated Annealing and Tabu Search prioritize exploitation over diversification. Some DFO algorithms called *model-based* DFOs build an internal surrogate/proxy model of the function $f(x)$ and use the derivative information or the functional form to accelerate exploitation and balance it with diversification, as depicted generally in Figure 10.2. Bayesian Optimization falls under this class of DFO. See Rios and Sahinidis (2013) and Ploskas and Sahinidis (2022) for a more thorough review of DFO algorithms including computational comparisons.

To leverage Derivative-Free Optimization (DFO) packages effectively, the optimization problem must first be translated into a simulator. This simulator is designed to take formulations as input and provide the evaluation of objective functions and constraints as output. In turn, DFO packages utilize these evaluation results as input, suggesting the next batch of formulations for assessment. Typically, a DFO algorithm necessitates the provision of an initial starting point and the configuration of

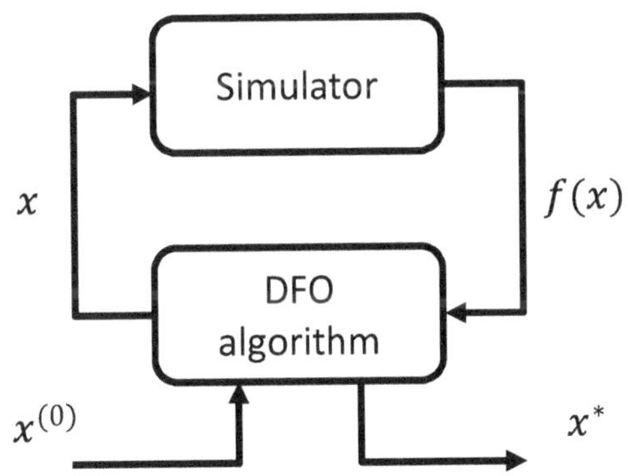

FIGURE 10.2 Schematic of the operation of a DFO package iterating between the search algorithm and the simulator.

hyperparameters, including the evaluation budget or other termination conditions. However, Rios and Sahinidis (2013) emphasize that due to variations in problem characteristics and algorithmic essence, there is no universal DFO package that caters to all problems.

Among the best-performing DFO algorithms, SNOBFIT (Huyer and Neumaier 2008) and BAM (Ma et al. 2023) stand out as deterministic global-search solvers relying on search space partition and model building. DYCORS (Regis and Shoemaker 2013) utilize Radial Basis Function (RBF) interpolation models to explore the search space. For stochastic global-search methods, CMA-ES (Hansen and Auger 2011) and NSGA-II (Deb et al. 2002) have demonstrated superior performance. NOMAD (Audet and Dennis 2006), recognized for its effective empirical performance, combines local search capabilities with clever restart strategies. BOBYQA (Powell 2009) is a model-based local search that constructs a sequence of quadratic interpolation models to guide the search. Many DFO algorithms are implemented in FORTRAN, C++, MATLAB, or Python. There are also interfaces available for accessing several DFO packages. The following list includes a set of such software interfaces.

C++:

- DAKOTA (Adams et al. 2020): Design Analysis Kit for Optimization and Terascale Application (DAKOTA) is an open-source package supported by Sandia National Laboratories.

MATLAB:

- TOMLAB (Holmström et al. 2010): TOMLAB is a commercial MATLAB environment.

Python:

- BoTorch (Balandat et al. 2020): An open-source Python framework for Bayesian Optimization.
- Scikit-Quant (Lavrijsen et al. 2020): A Python interface for several DFO solvers.
- Pymoo (Blank and Deb 2020): A collection of DFO packages in Python to address one or multi-objective problems.

10.4.3 Mathematical Programming

In contrast to black-box optimization, mathematical programming uses the functional forms of $f(x)$ and $g(x)$ to find provably optimal solutions, that is, provide a certificate of optimality. For certain functional forms, for example, when $f(x)$ and $g(x)$ are linear or convex, problem instances with hundreds of thousands or even millions of decision variables and constraints may be solved to optimality. However, these require deeper expertise in mathematical optimization to develop models and maintain. In certain instances, such as the use of PLS, random forests, gradient-boosted trees, or feed-forward neural networks predictive models, a generic mathematical model and software may be developed that is near plug-and-play. Here, we provide some classes of mathematical programming models that are applicable to formulation design based on the structure of the evaluation functions and constraints you have previously defined.

In all classes of mathematical programming outlined subsequently, the objective function is assumed to be a scalar. When we have several objective functions or a vector of objective functions, we resort to multi-objective optimization, which can be addressed in several ways (Hwang et al. 1980; Rosenthal 1985; Chinchuluun and Pardalos 2007; Deb et al. 2016). The following classes of single-objective optimization still serve as building blocks for multi-objective optimization.

Linear Programming: When $f(x)$ and $g(x)$ are both linear and all decision variables are continuous, the mathematical programming model is a linear program (LP). There are highly efficient and robust algorithms to solve these optimization problems. These problems are limited to simple formulation designs where the predictive model is a linear regression, SVM, or PLS model and when all decision variables do not need any transformation other than a linear transformation to be represented in the model. A less accurate PLS model-based linear programming formulation design optimization may serve as a starting point to seed a more accurate predictive model-based design optimization problem. See Bertsimas and Tsitsiklis (1997) and Vanderbei (2020) for details.

Nonlinear Programming: When either $f(x)$ or $g(x)$ is nonlinear and all decision variables are continuous, the optimization instance is a nonlinear program (NLP). These problems arise when the predictive model used in the design problem is a nonlinear model such as symbolic regression models or feed-forward neural networks with smooth nonlinear activation functions. When certain nonlinear transformation such as log, exponential, or normalization are required to represent the decision variable in a predictive model that is linear, the resulting optimization problem is a nonlinear programming model with some constraints in $g(x)$ representing these transformations. Often, a formulation design problem to meet a specific target property vector may be modeled with $f(x)$ being a loss function with appropriate norm. When this is two-norm and $g(x)$ is linear, we have a quadratic programming model. A quadratic programming model is a special case of nonlinear programs where both $f(x)$ and $g(x)$ are convex functions. These classes of convex optimization problems can be solved efficiently. A general nonlinear program (i.e., at least one of $g(x)$ is nonconvex) may have multiple locally optimal solutions. Rigorous optimization algorithms that seek globally optimal solutions belong to a class of mathematical programs called *global optimization*. See Nocedal and Wright (2006), Boyd and Vandenberghe (2004), and Biegler (2010) for details.

Mixed-Integer Linear Programming: When any of the decision variables is discrete (i.e., not continuous) and $f(x)$ and $g(x)$ are linear, the optimization problem is called mixed-integer linear programming (MILP). This is typical when an input or an output variable in the design formulation model is categorical. Some of the design constraints such as the number of ingredients to use also require additional discrete variables to write the optimization model formulation. Also, predictive models such as neural networks with ReLU (rectified linear unit) activation function and gradient-boosted trees require an MILP formulation. It is typically a lot harder to solve a

MILP, but the algorithms to solve MILPs are highly mature and problems of sizes up to a few million variables can be solved in practice. See Chen et al. (2011), Conforti et al. (2014), Jünger et al. (2009), and Wolsey (2020) for details.

Mixed-Integer Nonlinear Programming: When either $f(x)$ or $g(x)$ is nonlinear and any of the decision variables are discrete, the resulting optimization problem is a mixed-integer nonlinear program (MINLP). This is the most general of mathematical problems that cover a wide variety of engineering problems. Further, the convexity of the objective function and the constraint right-hand side functions determine how efficiently these problems can be solved to optimality. Global optimization methods address MINLP with nonconvex functions. When we include constraints such as a limit on the number of ingredients to use in design problems that include predictive models for NLP mentioned earlier, the resulting optimization problems are MINLPs. See Lee and Leyffer (2011), Belotti et al. (2013), Horst and Tuy (2013), and Tawarmalani and Sahinidis (2013) for details.

In Section 10.6, we provide specific software and tools to solve the various classes of optimization problems.

10.4.4 Overcoming Optimization Challenges

One of the best ways to embark on a modeling enterprise is to make sure we are asking the right questions. One of the most common curses of any data science campaign is trying to tackle a problem that is disproportionately complex. A simple example is differentiating animals and numbers. In this toy exercise a data scientist is tasked with developing a program to positively identify whether the submitted data is a giraffe or a number. A very ambitious researcher may develop a sophisticated machine vision model. A keen research might ask what is the structure of the raw data and learn that the data is split into two groups: jpg's of only giraffes and csv's of numbers. The determination of the best program is left to the reader. Industry and academia enjoy much more complex problems than differentiating giraffes and numbers, but the same intellectual path of first finding the right question through judicious questions is still relevant.

The complexity of modeling problems can ultimately be reduced through leveraging expert knowledge.

The precise machinery to do so may vary by context, but domain knowledge is one of the best tools the data scientist carries.

Optimization algorithms (in particular DFO) can start to struggle to find solutions as problems increase in dimensionality (Rios and Sahinidis 2013). However, there are numerous research efforts that are looking to improve the resiliency of algorithms to high-dimensional problems. Mathematically, utilization of expert knowledge translates to dimension reduction. This can be accomplished in several ways including 1) dimension reduction techniques such as PCA, UMAP, or t-SNE; 2) factor reduction through pre-experiments; and 3) mathematical reformulation.

Dimension reduction, factor reduction, and mathematical reformulation reduce the complexity of an optimization problem by removing low-impact dimensions or recasting the problem into a numerically friendlier format. *Dimension reduction* techniques accomplish this by striving to cast data into a new lower dimensional space that looks to maximize the amount of explained variance (or higher-order moments of a probability distribution). A three-dimensional projection of a ten-dimensional space may capture 80% of a system's variance and dramatically reduce the complexity of the optimization problem. This happens inherently in approaches such as partial least squares (PLS) inversion but can be applied to other models through an additional step. Occasionally application of tools like PCA will reveal that high-dimensional spaces can be represented well (>90%) by just one or two dimensions dominated by one or two features or predictors. This exercise can help identify that a training set may be over-featurized or contain predictors that are heavily correlated.

Factor reduction can simplify systems by helping to identify which factors of an experimental design have a low impact. This leverages expert knowledge by conducting pre-experiments that are used to cull irrelevant physical phenomena. For example, in a formulated system with multiple classes of ingredients, dimensionality of the system may be high due to the availability of tens or even hundreds of candidate ingredients per class. However, if experimentation or hypothesis testing on training data shows that the properties of a particular class that vary between candidate ingredients are not significant to the model prediction, then instead of including every ingredient as a decision variable you may instead include just the formulation ratio of the overall class.

Finally, *mathematical reformulation* is an excellent way to simplify system complexity. The set of constraints, parameters, variables, expressions, and optima that comprise a mathematical programming model are an alphabet that can be re-written in ways that are more or less friendly to various optimization algorithms. Reformulating a problem to be amenable to a specific algorithm can often dramatically decrease the optimization computational time. Most often, this occurs by converting elements of the objective function (sometime referred to as "soft" constraints) into hard constraints that can significantly decrease the size of the searchable design space. Examples can be found in Hou et al. (2020) and Ashby (2000). An additional way to reformulate the problem is to convert the structure of the predictive model itself into another form that is suited to a particular search algorithm, such as representing a nonlinear neural network as binary variables for use in a mixed-integer optimization with linear constraints (Maragno et al. 2023; Wang et al. 2024).

Lastly, it is important to discuss how to balance practical versus fundamental strengths of competing optimization approaches. Naturally, this balance should be based on what you are using optimization to accomplish. As previously discussed, mathematical programming approaches come with a guarantee of global optimality – you can be certain you are obtaining the best solution available. In contrast, black-box optimization approaches come with no such guarantee and many better solutions (e.g., lower cost, higher performing) could be possible but not identified.

When choosing which path to follow, it is helpful to understand the risk to your project that is associated with a sub-optimal solution.

For example, consider a project where the program is to be used only within your team to accelerate development of a new, higher-performing material, and because the goal is to be first to the market with the better product, your total budget is measured in time from project start to new product commercialization, rather than in the number of experiments. In this case, the cost of a sub-optimal prediction is low (i.e., the lab process is inexpensive and a failed experiment poses no strategic risk), and the increased speed of configuring an adaptive experimental design campaign may more than account for potentially increased numbers of experiments required to find a sufficiently optimum solution. Contrast this to a project attempting to secure new customers by providing a tool online that recommends the best formulation for a customer to incorporate your product into their application. If there are many competing products, the customer may only tolerate one or two experiments with your product before they move on. In this case, your team might prioritize finding globally optimum solutions and staying within the known product space to avoid unexpected inaccuracies in the model prediction. That problem is well-suited to the mathematical programming approach. An additional example is when the cost of an uncertain prediction is very high, such as when defining operating condition boundaries associated with a given formulation and its raw material variability. In this case, the result is not a failed lab experiment but potentially an entire production batch of out-of-specification material. Mathematical programming for this case is the subject of Chapter 12.

A related subject is the notion of causality. The strength of the traditional design of experiment approaches is the ability to isolate sources of variation to prove causality in the system. In most machine learning applications, the historical data is undesigned and does not contain the necessary variable independence to accomplish this, hence the perpetual caveat of "correlation not causation." However, when a predictive model is used in *active* applications such as optimization (vs. passive use in descriptive or predictive analytics), we are making an assumption that the variable values identified in the optimum solution *will cause* the expected result. Strictly speaking, most machine learning models cannot provide this guarantee, as argued by MacGregor et al. (2015) and MacGregor (2018). MacGregor points to PLS models as a mechanism for causal optimization, because the variable manipulation occurs in the latent space, where latent variables are in fact orthogonal to each other. This is further discussed in Chapter 12.

Thus, in considering the modeling and optimization approach for a project, it is a worthwhile exercise for the team to discuss the risk of misuse of non-causal models and the likelihood of the model/training data missing the controlling variables of the system.

10.5 EXAMPLE APPLICATION: PLS MODEL INVERSION

PLS models are a mature technology for formulation inverse design and thus make a good case study for the reader. Review of the literature can complement this section with further details of implementation (Muteki et al. 2006; Muteki and MacGregor 2007; Nichols 2011; García-Muñoz and Polizzi 2012; Tomba et al. 2014; Palací-López et al. 2020).

Let us revisit the elastomer reformulation example and assume that the team used a PLS model for prediction of formulation performance. The model uses the formulation ratio data table, R, process variables Maximum Temperature and Batch Time, and a mixture properties block X_{mix} calculated with linear mixing rules, to predict $L = 3$ quality variables. The PLS model used the NIPALS algorithm for training, and cross-validation was used to select $A = 7$ components. Figure 10.3 provides a schematic of the data structure.

Matrix R contains the historical formulation data for N materials made from I ingredients, and X_{Prop} provides the ingredient property data. Linear mixing by the dot product results in X_{mix}. This $N \times K$ matrix is joined to R and the two processing variables to form the overall X data for modeling. PLS relates this X matrix to the Y matrix containing the three performance variables for each formulation.

From Equation 8.13 we have our prediction expression

$$\hat{y} = tc^T = \left(x_{new} w^*\right) c^T = x_{new} \beta,$$

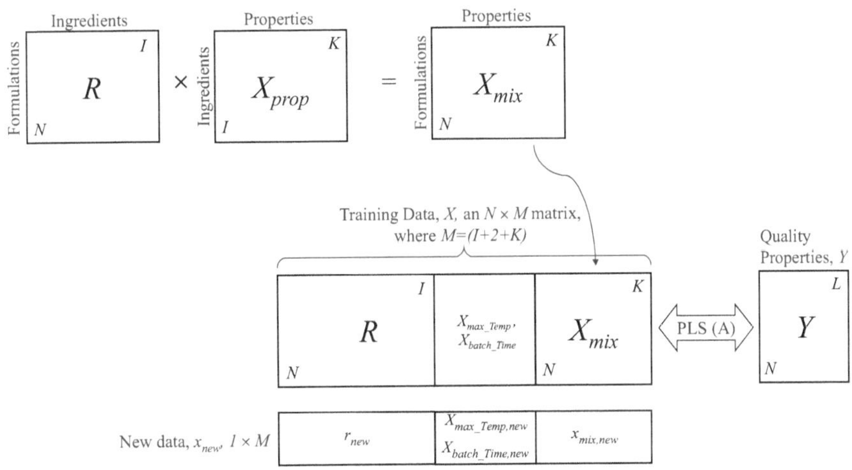

FIGURE 10.3 Schematic of the data structure for the PLS inversion example.

where w^* and c are the x- and y-space weights learned by NIPALS, respectively, t are x-space scores, and x_{new} is a vector of new input values with the structure shown in Figure 10.3: $[r_{new}, X_{max\_Temp,new}, X_{Batch\_Time,new}, X_{mix,new}]$.

We can now construct the optimization framework for this system. First, the decision variables. As presented earlier, the decision variables are the formulation ratio amounts for each available ingredient, i, plus the two process conditions.

$$0 \leq r_{new,i} \leq 1$$

The final elements of x_{new}, $x_{mix,new}$, are not decision variables themselves but are calculated by the linear mixing rule:

$$x_{mix,new} = r_{new} \cdot x_{Prop}$$

We also calculated the ingredient selection binary variable δ_i, which denotes whether or not ingredient i is used in the formulation.

$$\delta_i = \begin{cases} 1, \ where\ r_{new,i} > 0 \\ 0, \ where\ r_{new,i} = 0 \end{cases}$$

Now, we may consult Table 10.5 to review our constraints: within current processing conditions, no more than 20% solid ingredients, no new ingredients. The latter is already accounted for by the ingredient pool being limited to the I ingredients from R. These expressions provide constraints on the processing conditions:

$$\min\left(x_{\max temp}\right) \leq x_{\max temp,new} \leq \max\left(x_{\max temp}\right)$$

$$\min\left(x_{batch\ time}\right) \leq x_{batch\ time,new} \leq \max\left(x_{batch\ time}\right)$$

And, lastly, for the solids constraint we assume there are class labels within R such that we may sum the formulation ratios for all ingredients within the solids class:

$$\sum_{j}^{J,solids} r_{j,solids} < 0.20,$$

The inverse design problem also comes with some constraints of its own, namely the mixture constraint that all formulation ratios sum to 1.

$$\sum_{i}^{I} r_{new,i} = 1,$$

For the final constraints, we provide hard limits on the PLS squared prediction error (SPE) and Hotelling's T^2 values.

$$SPE = \sum_{m}^{M} (x_m - \hat{x}_m)^2 \leq \varepsilon$$

$$HT^2 = \sum_{a}^{A} \frac{t_a^2}{s_a^2} \leq T_{max}^2$$

where ε and T^2_{max} are manually specified or derived from the training data. Now to assemble the evaluation functions for our objectives. Recall the goal is for the changes not to affect the quality of the product, so the predicted Y values should match the current average production values, $Y_{current}$, meaning we want to minimize $(y_{current} - \hat{y}_{new})$. For a model with a single Y, this would often be the root squared error, but because the PLS model has three Ys and its parameters are matrices, we use this form:

$$\left(y_{current} - \beta^T x_{new}\right)^T \cdot W_1 \cdot \left(y_{current} - \beta^T x_{new}\right)$$

where W_1 is a 3 × 3 matrix with the relative weights of the accuracy objective for each of the three Ys along the diagonal (i.e., we need not prioritize all quality properties the same). Next, we have our objectives for minimizing materials cost and processing cost, as previously described in Table 10.7.

$$w_2 \sum_{i=1}^{N} r_{new,i} \times c_i$$

$$w_3 \times \$0.05\left(x_{\max temp,new} \times x_{batch\,time,new}\right)$$

Lastly, per our objective to minimize the number of ingredients in the new formulation:

$$w_4 \sum_{i=1}^{N} \delta_i$$

We now have the optimization framework specified, as summarized in Table 10.8.

To implement the optimization, all the decision variables, calculated expressions, objective function, and constraints must be programmed into a modeling system and solved with a solver – both topics discussed in the following section.

TABLE 10.7
Objective Function Terms to Control Formulation Composition

Objective	Objective Function Representation (To Be Minimized over the Decision Variables)
Minimize materials cost	$w \sum_{i=1}^{N} r_{new,i} \times c_i$
Minimize number of ingredients	$w \sum_{i=1}^{N} \delta_i$
Minimize processing cost	Highly custom expressions relating one or more process variables to cost, e.g.: $w \times \$0.05\left(X_{Max\_Temperature} \times X_{Batch\,Time}\right)$

TABLE 10.8
Optimization Framework for Elastomer Reformulation Using PLS Inversion

Objective function, $f(r_{new})$	$\min_{r_{new}} \left(y_{current} - \beta^T x_{new}\right)^T \cdot W_1 \cdot \left(y_{current} - \beta^T x_{new}\right) + w_2 \sum_{i=1}^{N} r_{new,i} \times c_i + w_3 \times \$0.05\left(x_{\max temp,new} \times x_{batch\ time,new}\right) + w_4 \sum_{i=1}^{N} \delta_i$
Constraints, $g(r_{new})$	subject to $\{0 \le r_{new,i} \le 1,\ \sum_{i}^{I} r_{new,i} = 1,\ \min\left(x_{\max temp}\right) \le x_{\max temp,new} \le \max\left(x_{\max temp}\right),\ \min\left(x_{batch\ time}\right) \le x_{batch\ time,new} \le \max\left(x_{batch\ time}\right),\ \sum_{j}^{J,solids} r_{j,solids} < 0.20,\ SPE = \sum_{m}^{M}(x_m - \hat{x}_m)^2 \le \varepsilon, T^2 = \sum_{a}^{A} \frac{t_a^2}{s_a^2} \le T_{max}^2\}$

Fine tuning of the optimization to achieve the best results and runtimes can be done by adjusting the weights and investigating if any further scaling of the objective function terms is required.

This process is often iterative with the subject-matter experts who will be using the program, as inspection of solutions may inspire new constraints or objectives to be added.

10.6 IMPLEMENTATION

We previously introduced some important classes of mathematical programming models. To solve these optimization problems in practice effectively, there are two different software tools that are required.

- *Solver*: A solver is a software that finds an optimal solution for a specific canonical form of the mathematical program where all variables and constraints attributes are passed to the solver interface in a compact vector and matrix form.
- *Modeling System*: A modeling system is a software that allows the modeler to define different sets of variables and constraints in a logical way as how the optimization model is formulated on paper. Commercial modeling systems also have other development capabilities such as debuggers and visual inspectors of model that enhance the model development.

There are several open-source and commercial solvers and modeling systems options. Typically, a modeling system supports a wide variety of solvers including open-source solvers.

Table 10.9 provides a list of solvers and the types of models that can be solved.

TABLE 10.9
Math Program Solvers and Their Capabilities

Solver	License	LP	NLP	MILP	MINLP	Global
BARON	Commercial	X	X	X	X	X
Bonmin	EPL	X	X	X	X	
CBC	EPL	X		X		
CPLEX	Commercial	X		X		
CONOPT	Commercial	X	X			
COPT	Commercial	X		X		
DICOPT	Commercial	X	X	X	X	
GLPK	GPL	X		X		
Gurobi	Commercial	X		X		
HiGHS	MIT	X		X		
Ipopt	EPL	X	X			
Knitro	Commercial	X	X			
MINOS	Commercial	X	X			
SCIP	Apache	X	X	X	X	X
SNOPT	Commercial	X	X			
Xpress	Commercial	X		X		

The following is a list of major optimization modeling systems and the year they were introduced. Most of them support the solvers listed in Table 10.9. See the respective documentation for a complete list of solvers supported.

- Open Source
  - JuMP (Lubin et al. 2023): Julia-based math programming modeling system that was developed at Massachusetts Institute of Technology (2013)
  - Pyomo (Bynum et al. 2021): Python-based math programming modeling system that is developed and maintained by Sandia National Labs. Several programs and initiatives by national labs use Pyomo (2008)
- Commercial
  - AIMMS by AIMMS B.V., formerly, Paragon Decision Technology B.V. (1993)
  - AMPL by AMPL Optimization, Inc. (1985)
  - GAMS by GAMS Development Corporation (1976)
  - Xpress Mosel by FICO (1983)

Almost all commercial solvers and modeling systems are available free for academics. There are several other solvers, both open-source and commercial, and solvers that leverage the aforementioned solvers within specific modeling systems; for example, MindtPy uses an MILP and an NLP solver in Pyomo and AOA uses an MILP and an NLP solver in AIMMS.

For an open-source setup, SCIP with Pyomo would support several model types. SCIP uses SoPlex for LP solver (included) and can leverage Ipopt for NLP.

In addition to modeling systems, there are special packages that help automatically translate a predictive model into a set of constraints in the optimization model. There are also specific solvers designed for (global) optimization of problems based on specific predictive models. The following list includes a set of such specialized software packages.

- IDAES-PSE (Lee et al. 2021): A Pyomo-based Python package that includes a variety of tools for surrogate modeling and optimization of systems engineering applications.
- OMLT (Ceccon et al. 2022): A Pyomo-based Python package that includes easy embedding of neural networks and tree-based models in math programs.
- MeLOn (Schweidtmann et al. 2021): A C++-based global optimization solver for optimization of Gaussian process regression and neural network models.

Next, Chapter 11 will further discuss black-box optimization in the context of adaptive experimental design.

REFERENCES

Adams, Brian M., William J. Bohnhoff, Keith R. Dalbey, Mohamed S. Ebeida, John P. Eddy, Michael S. Eldred, Russell W. Hooper, et al. 2020. "Dakota: a multilevel parallel object-oriented framework for design optimization, parameter estimation, uncertainty quantification, and sensitivity analysis: version 6.13 user's manual." No. SAND2020-12495. *Sandia National Lab.* (SNL-NM), Albuquerque, NM (United States).

Ashby, M.F. 2000. "Multi-objective optimization in material design and selection." *Acta Materia lia* 48 (1): 359–369. https://doi.org/10.1016/S1359-6454(99)00304-3.

Audet, Charles, and John E. Dennis Jr. 2006. "Mesh adaptive direct search algorithms for constrained optimization." *SIAM Journal on Optimizati on* 17 (1): 188–217. https://doi.org/10.1137/040603371.

Balandat, Maximilian, Brian Karrer, Daniel Jiang, Samuel Daulton, Ben Letham, Andrew G. Wilson, and Eytan Bakshy. 2020. "BoTorch: a framework for efficient Monte-Carlo Bayesian optimization." *Advances in Neural Information Processing Systems* 33: 21524–21538. https://doi.org/10.5555/3495724.3497531.

Belotti, Pietro, Christian Kirches, Sven Leyffer, Jeff Linderoth, James Luedtke, and Ashutosh Mahajan. 2013. "Mixed-integer nonlinear optimization." *Acta Numerica* 22: 1–131. https://doi.org/10.1017/S0962492913000032.

Bertsimas, Dimitris, and John N. Tsitsiklis. 1997. *Introduction to Linear Optimization*, Vol. 6. Belmont, MA: Athena Scientific.

Biegler, Lorenz T. 2010. *Nonlinear Programming: Concepts, Algorithms, and Applications to Chemical Processes*. Philadelphia, PA: SIAM. ISBN: 978-0-89871-702-0.

Blank, Julian, and Kalyanmoy Deb. 2020. "Pymoo: Multi-objective optimization in python." *IEEE Access* 8: 89497–89509. https://doi.org/10.1109/ACCESS.2020.2990567.

Boyd, Stephen P., and Lieven Vandenberghe. 2004. *Convex Optimization*. Cambridge, UK: Cambridge University Press. ISBN: 9780521833783.

Bynum, Michael L., Gabriel A. Hackebeil, William E. Hart, Carl D. Laird, Bethany L. Nicholson, John D. Siirola, Jean-Paul Watson, et al. 2021. *Pyomo-Optimization Modeling in Python*, Vol. 67. Springer. https://doi.org/10.1007/978-3-030-68928-5.

Ceccon, F., J. Jalving, J. Haddad, A. Thebelt, C. Tsay, C.D. Laird, and R. Misener. 2022. "OMLT: optimization & machine learning toolkit." *Journal of Machine Learning Research* 23 (1): 15829–15836. https://doi.org/10.5555/3586589.3586938.

Chen, Der-San, Robert G. Batson, and Yu Dang. 2011. *Applied Integer Programming: Modeling and Solution*. John Wiley & Sons. ISBN: 9780470373064, https://doi.org/10.1002/9781118166000.

Chinchuluun, Altannar, and Panos M. Pardalos. 2007. "A survey of recent developments in multiobjective optimization." *Annals of Operations Research* 154 (1): 29–50. https://doi.org/10.1007/s10479-007-0186-0.

Conforti, Michele, Gérard Cornuéjols, and Giacomo Zambelli. 2014. "Integer programming models." In *Integer Programming*. Cham: Springer Internationa l Publishing, 45–84. https://doi.org/10.1007/978-3-319-11008-0_2.

Deb, Kalyanmoy, Amrit Pratap, Sameer Agarwal, and T.A.M.T. Meyarivan. 2002. "A fast and elitist multiobjective genetic algorithm: NSGA-II." *IEEE Transactions on Evolutionary Computat ion* 6 (2): 182–197. https://doi.org/10.1109/4235.996017.

Deb, Kalyanmoy, Karthik Sindhya, and Jussi Hakanen. 2016. "Multi-objective optimization." In *DecisionSciences*. Boca Raton, FL, USA: CRC Press, 161–200. ISBN: 9781466564305.

Fajemisin, Adejuyigbe O., Donato Maragno, and Dick den Hertog. 2024. "Optimization with constraint learning: a framework and survey". *European Journal of Operational Rese arch* 314 (1): 1–14. https://doi.org/10.1016/j.ejor.2023.04.041.

García-Muñoz, Salvador, and Mark Polizzi. 2012. "WSPLS – a new approach towards mixture modeling and accelerated product development." *Chemometrics and Intelligent Laboratory Systems* 114: 116–121. https://doi.org/10.1016/j.chemolab.2012.03.009.

Hansen, Nikolaus, and Anne Auger. 2011. "CMA-ES: evolution strategies and covariance matrix adaptation." *Proceedings of the 13th Annual Conference Companion on Genetic and Evolutionary Computation*, 991–1010. https://doi.org/10.1145/2598394.2605347.

Holmström, Kenneth, Anders O. Göran, and Marcus M. Edvall. 2010. *User's Guide for TOMLAB 7*. Technical Report. Tomlab Optimization Inc.

Horst, Reiner, and Hoang Tuy. 2013. *Global Optimization: Deterministic Approaches*. Springer Science & Business Media. ISBN : 978-3-540-61038-0, https://doi.org/10.1007/978-3-662-03199-5.

Hou, Zhanglu, Cheng He, and Ran Cheng. 2020. "Reformulating preferences into constraints for evolutionary multi- and many-objective optimization." *Information Sciences* 541: 1–15. https://doi.org/10.1016/j.ins.2020.05.103.

Huyer, Waltraud, and Arnold Neumaier. 2008. "SNOBFIT – stable noisy optimization by branch and fit." *ACM Transactions on Mathematical Software (TOMS)* 35 (2): 1–25. https://doi.org/10.1145/1377612.1377613.

Hwang, Ching-Lai, Sudhakar R. Paidy, Kwangsun Yoon, and Abu Syed Md Masud. 1980. "Mathematical programming with multiple objectives: a tutorial." *Computers & Operations Research* 7 (1–2): 5–31. https://doi.org/10.1016/0305-0548(80)90011-8.

Jünger, Michael, Thomas M. Liebling, Denis Naddef, George L. Nemhauser, William R. Pulleyblank, Gerhard Reinelt, Giovanni Rinaldi, and Laurence A. Wolsey. 2009. *50 Years of Integer Programming 1958–2008: From the Early Years to the State-of-the-Art*. Springer Science & Business Media. ISBN : 978-3-540-68274-5, https://doi.org/10.1007/978-3-540-68279-0.

Lavrijsen, Wim, Ana Tudor, Juliane Müller, Costin Iancu, and Wibe De Jong. 2020. "Classical optimizers for noisy intermediate-scale quantum devices." *2020 IEEE International Conference on Quantum Computing and Enginee ring (QCE)*, 267–277. https://doi.org/10.1109/QCE49297.2020.00041.

Lee, Andrew, Jaffer H. Ghouse, John C. Eslick, Carl D. Laird, John D. Siirola, Miguel A. Zamarripa, Dan Gunter, et al. 2021. "The IDAES process modeling framework and model library – flexibility for process simulation and optimization." *Journal of Advanced Manufacturing and Processing* 3 (3): e10095. https://doi.org/10.1002/amp2.10095.

Lee, Jon, and Sven Leyffer. 2011. *Mixed Integer Nonlinear Programming*, Vol. 154. Springer Science & Business Media. https://doi.org/10.1007/978-1-4614-1927-3.

Lubin, Miles, Oscar Dowson, Joaquim Dias Garcia, Joey Huchette, Benoit Legat, and Juan Pablo Vielma. 2023. "JuMP 1.0: recent improvements to a modeling language for mathematical optimization." *Mathematical Programming Computation* 15: 581–589. https://doi.org/10.1007/s12532-023-00239-3.

Ma, Kaiwen, Luis Miguel Rios, Atharv Bhosekar, Nikolaos V. Sahinidis, and Sreekanth Rajagopalan. 2023. "Branch-and-model: a derivative-free global optimization algorithm." *Computational Optimization and Applicati ons* 85 (2): 337–367. https://doi.org/10.1007/s10589-023-00466-3.

MacGregor, John F. 2018. "Empirical models for analyzing BIG data. What's the difference?" Abstract submitted to the *2018 AIChE Spring Meeting*, Orlando, FL. https://aiche.confex.com/aiche/s18/meetingapp.cgi/Paper/511688 (accessed January 27, 2024).

MacGregor, John F., M.J. Bruwer, I. Miletic, M. Cardin, and Z. Liu. 2015. "Latent variable models and big data in the process industries." *IFAC-Paper sOnLine* 48 (8): 520–524. https://doi.org/10.1016/j.ifacol.2015.09.020.

Maragno, Donato, Holly Wiberg, Dimitris Bertsimas, S. Ilker Birbil, Dick den Hertog, and Adejuyigbe Fajemisin. 2023. "Mixed-integer optimization with constraint learning." *Opera tions Research*. https://doi.org/10.1287/opre.2021.0707.

Muteki, Koji, and John F. MacGregor. 2007. "Multi-block PLS modeling for L-shape data structures with applications to mixture modeling." *Chemometrics and Intelligent Laboratory Systems* 85 (2): 186–194. https://doi.org/10.1016/j.chemolab.2006.06.018.

Muteki, Koji, John F. MacGregor, and Toshihiro Ueda. 2006. "Rapid development of new polymer blends: the optimal selection of materials and blend ratios." *Industrial and Engineering Chemistry Research* 45 (13): 4653–4660. https://doi.org/10.1021/ie050953b.

Nichols, Emily. 2011. "Latent variable methods: case studies in the food industry." PhD Di ss., *McMaster University*. https://macsphere.mcmaster.ca/handle/11375/11317.

Nocedal, Jorge, and Stephen J. Wright. 2006. *Numerical Optimization*. Springer Series in Operations Research and Financial Engineering. https://doi.org/10.1007/978-0-387-40065-5.

Palací-López, D., P. Villalba, P. Facco, M. Barolo, and A. Ferrer. 2020. "Improved formulation of the latent variable model inversion–based optimization problem for quality by design applications." *Journal of Chemometrics* 34: e3230. https://doi.org/10.1002/cem.3230.

Ploskas, Nikolaos, and Nikolaos V. Sahinidis. 2022. "Review and comparison of algorithms and software for mixed-integer derivative-free optimization." *Journal of Global Optimization* 82: 1–30.

Powell, Michael J.D. 2009. *The BOBYQA Algorithm for Bound Constrained Optimization without Derivatives*. Cambridge NA Report NA 2009/06. Cambridge: University of Cambridge, 26.

Regis, Rommel G., and Christine A. Shoemaker. 2013. "Combining radial basis function surrogates and dynamic coordinate search in high-dimensional expensive black-box optimization." *Engineering Optimization* 45 (5): 529–555. https://doi.org/10.1080/0305215X.2012.687731.

Rios, Luis Miguel, and Nikolaos V. Sahinidis. 2013. "Derivative-free optimization: a review of algorithms and comparison of software implementations." *Journal of Global Optimization* 56: 1247–1293.

Rosenthal, Richard E. 1985. "Concepts, theory, and techniques principles of multiobjective optimization". *Decision Scienc es* 16 (2): 133–152. https://doi.org/10.1111/j.1540-5915.1985.tb01479.x.

Schweidtmann, Artur M., Dominik Bongartz, Daniel Grothe, Tim Kerkenhoff, Xiaopeng Lin, Jaromił Najman, and Alexander Mitsos. 2021. "Deterministic global optimization with Gaussian processes embedded." *Mathematical Programming Computati on* 13 (3): 553–581. https://doi.org/10.1007/s12532-021-00204-y.

Tawarmalani, Mohit, and Nikolaos V. Sahinidis. 2013. *Convexification and Global Optimization in Continuous and Mixed-Integer Nonlinear Programming: Theory, Algorithms, Software, and Applications*, Vol. 65. Springer Science & Business Media. https://doi.org/10.1007/978-1-4757-3532-1.

Tomba, Emanuele, Massimiliano Barolo, and Salvador García-Muñoz. 2014. "In-silico product formulation design through latent variable model inversion." *Chemical Engineering Research an d Design* 92 (3): 534–544. https://doi.org/10.1016/j.cherd.2013.08.027.

Vanderbei, Robert J. 2020. *Linear Programming: Foundation and Extensions*. New York, NY: Springer. https://doi.org/10.1007/978-1-4614-7630-6.

Wang, Q., X. Chen, Q. An, M. Chen, H. Guo, and Y. He. 2024. "Multi-objective optimization strategy for continuous drilling parameters of superalloys". *International Journal of Precision Engineering and Manufa cturing-Green Technology*. https://doi.org/10.1007/s40684-023-00591-8.

Wolsey, Laurence A. 2020. *Integer Programming*. John Wiley & S ons. ISBN: 9781119606536, https://doi.org/10.1002/9781119606475.

11 Adaptive Experimental Design

Joel C.J. Strickland, Phillip F.D. Woolston and Thomas M. Whitehead

11.1 INTRODUCTION TO ADAPTIVE EXPERIMENTAL DESIGN

11.1.1 Overview of the Challenges in Optimizing Products

In the realm of product optimization, particularly materials, chemicals, and drugs, professionals are faced with a multitude of challenges. These include technical hurdles requiring cutting-edge technology and skilled personnel, the imperative for regulatory compliance, the essential safeguarding of user and environmental safety, and the daunting task of cost management. Additionally, with shifting consumer preferences, market acceptance must be earned, and sustainability goals must be met through eco-friendly practices. The pressure of rapid innovation adds the need for expedited time-to-market, while scaling up and intellectual property protection present their own unique complexities. "Adaptive Experimental Design" is the bridge that connects these multifaceted challenges to solutions by optimizing the route to a final product design (Wang and Dowling 2022; Zhang et al. 2020). This chapter delves into how adaptive experimental design, as an agile and dynamic approach, can streamline the development process, facilitate innovation, and foster efficiency in addressing these challenges.

11.1.2 Motivation for Adaptive Experimental Design (AED) in Product Development

In traditional design, the entire experimental plan is locked down at inception, whereas Adaptive Experimental Design (AED) offers the flexibility to make iterative adjustments based on emerging data. This adaptability empowers teams to focus on the most promising areas, be it formulations, processing parameters, or product designs, expediting the optimization process. Through AED, scientists can steer clear of unfruitful trajectories by utilizing preliminary data, mitigating risks while conserving time and resources.

The dynamic nature of AED permits teams to react in real time to experimental findings, propelling the product development process forward.

DOI: 10.1201/9781003385974-15

Additionally, AED operates as a learning-driven approach. It effectively handles uncertainty by making smart adjustments as new data comes in. Continuous adaptation is not just about flexibility; it also enhances decision-making by providing valuable insights into how different parameters are performing. By intelligently adjusting to new information, AED ensures that efforts are not only adaptable but also focused and informed (Lei et al. 2021).

11.1.3 Brief Overview of the Process and Key Concepts

In AED, the goal is to identify trends, correlations, or insights that contribute to a deeper understanding of the system under study and use this knowledge to effectively guide subsequent rounds of experimentation.

The journey begins by defining the objective, which entails crystal-clear identification of the end goals, such as cost reduction, product enhancement, or ingredient minimization. Subsequently, the initial design is planned, which could mean starting from scratch, building on existing literature, or utilizing historical data.

Before implementing AED, the execution of several preparatory steps is considered essential. We begin by identifying key variables, including both inputs and outputs, to be manipulated or monitored. The selection of pertinent metrics or indicators is essential for the measurement of the effects prompted by the variables during the experimentation process. At this stage, the leveraging of domain knowledge is often crucial, as it guides the choice of relevant metrics. Subsequently, the determination of initial exploratory ranges for the variables is necessitated, with domain knowledge deployed to concentrate the experimentation on areas predicted to generate meaningful insights. At this juncture, distinguishing between exploratory range and boundary conditions is seen as significant. The exploratory range delineates the parameters that can change for testing purposes, while the boundary conditions establish firm restrictions that, for reasons of safety, practicality, or feasibility, must not be exceeded. Additionally, the establishment of optimization constraints is to be remembered, which is necessitated for the assurance that the solutions generated through experimentation are not only feasible but also practical within a real-world context. Once these steps are in place, the experiment can proceed with data collection and analysis.

11.1.4 The Dynamics of Learning and Adaptability in Adaptive Experimental Design (AED)

Evolving from the preparation phase, AED employs sequential learning, using knowledge garnered from each experiment to guide subsequent ones. This progression entails striking a balance between *exploration* – a comprehensive survey of potential scenarios – and *exploitation*, where attention increasingly sharpens to the most promising avenues as the experimental design unfolds. Integral to AED is the principle of adaptation, which denotes the ability to modify the experimental design predicated on past results, thereby distinguishing it markedly from traditional Design of Experiments (DoE) that set a predetermined and fixed experimental plan (The nuances of this comparison will be elaborated in Section 11.5.3). In different fields, the application of a growing corpus of knowledge to guide progress, as utilized during AED, is also referred to as "active learning," especially in machine learning contexts. This term underscores the proactive decision-making integral to AED, in stark contrast to the passive adherence to a predetermined DoE experimental plan.

Bayesian statistics play a pivotal role in the underpinnings of AED, facilitating the updating of probability evaluations of various outcomes based on accumulated data. This process is intrinsic to the inherently iterative nature of AED, where continuously learning from each experimental cycle and adapting the methodology as necessitated is vital (Shahriari et al. 2016).

Capitalizing on this virtuous cycle of design, experimentation, and learning, the dynamic nature of AED significantly hastens the process in comparison to traditional static DoE.

11.1.5 Illustrations of Successful Applications of AED

Adaptive Experimental Design (AED) has proven to be a versatile tool, demonstrating its adaptability and utility across various fields. In the pharmaceutical industry, for instance, AED plays a pivotal role in early-stage clinical trials for biologic drugs – medications derived from living organisms. Instead of the traditional focus on maximum tolerance, it allows for a paradigm shift toward a more balanced relationship between efficacy and toxicity, signifying a notable stride in patient care (Takahashi and Suzuki 2021).

Moving onto the manufacturing and process industries, AED becomes an essential tool to discern the optimal combination of raw materials, processing conditions, and equipment settings, thereby leading to high-quality or efficient production (Zhang et al. 2021).

When considering the sphere of materials science, the value of AED is magnified. With the experimental process being costly and time-consuming, AED becomes indispensable for identifying the best blend of materials and processing techniques. Liang's review (Liang et al. 2021) showcases this point, highlighting the effectiveness of AED across a variety of experimental materials science datasets.

In the aerospace industry, AED proves to be an invaluable resource for designing robust composite laminates – layered materials made from two or more constituents. AED allows for the inclusion of typical aerospace design constraints, such as in-plane stiffness, ply balance, and laminate symmetry about the midplane. This ensures the experimental model's integrity is maintained while consistently delivering feasible solutions (Chuaqui et al. 2021).

Beyond the aforementioned industries, AED's adaptability extends to a multitude of sectors. For instance, in the automotive industry, it fine-tunes parameters in vehicle design and production, leading to enhanced performance and efficiency (Zhu et al. 2022). Likewise, within agriculture and the food industry, it facilitates an increase in crop yields and the refinement of food processing techniques (Drury et al. 2017).

Moreover, in both renewable and conventional energy sectors, AED can maximize efficiency and optimize operations (Tillmann et al. 2019). It has revolutionized trading strategies and risk management protocols in quantitative finance (Jacquier and Polson 2012) and has been beneficial in optimizing remediation strategies in environmental science (Hellan et al. 2023). Not to mention, within the realms of artificial intelligence and machine learning, AED has proven to be a key player for model training and algorithm optimization (Nguyen 2019).

In summary, the inherent adaptability and iterative nature of AED make it a universally applicable tool for optimization across diverse domains.

AED's capacity to manage uncertainties and adjust based on new data becomes a crucial feature in any scenario where decisions need to be made in dynamic environments or with minimal experimental data.

11.1.6 Case Study

This chapter will feature a detailed exploration of how *Domino Printing Sciences* successfully implemented Adaptive Experimental Design (AED) in their development R&D process for ink development. We will examine their particular approach, shaped by the complex challenges inherent in formulation chemistry, and dissect the AED workflow into three essential components: the acquisition function, optimizer, and surrogate model. By understanding how these elements synergize, we can understand the creation of effective and powerful workflows in AED.

11.2 ACQUISITION FUNCTIONS

11.2.1 Understanding the Role of Acquisition Functions

In Adaptive Experimental Design (AED), the role of an *acquisition function* – a mathematical construct – serves a dual purpose. First, it is used to estimate how well the experimental outcomes align with the predefined goals of a project. Second, it helps identify the optimum of a formulation landscape using as few iterations as possible, especially useful when each experimental run is costly or time-consuming.

The acquisition function operates as a compass, guiding the focus toward those areas within the input space that are probable candidates for the global optimum.

To do this effectively, it leverages a probabilistic understanding of the formulation landscape, often provided by Bayesian methodologies. In this capacity, it is essential for the acquisition function to maintain a delicate balance between *exploration* – probing new regions of the input space, and *exploitation* – focusing and refining the search around promising regions.

The Decision-Making behind Selecting the Next Experiment

Within the framework of AED, the selection of subsequent points for evaluation hinges on the acquisition function. This function plays an instrumental role in assessing the potential value of evaluating a point, drawing on the current knowledge of the landscape targeted for optimization. It uses a probabilistic approach, considering both the expected performance and the uncertainty associated with each point. It is worth noting here that the quality and calibration of this uncertainty, a topic covered in Chapter 7, play a crucial role in the effectiveness of the acquisition function.

Conventionally, the most favored strategy to select the next point for evaluation is to maximize the acquisition function throughout the search space. The aim is to identify the point that amplifies the expected utility, leveraging the present understanding of both the formulation landscape and the acquisition function. This often involves a balance of exploitation, selecting points with high expected performance, and exploration, selecting points with high uncertainty. This balance of exploration against exploitation distinguishes the acquisition function from the objective function of traditional function optimization. An objective function can often be considered as a scenario of complete exploitation, concentrating solely on high performance. A traditional objective function is more suited to situations with minimal or no uncertainty about the value of the function being optimized, as it does not account for the exploration aspect inherent in an acquisition function's design.

Ideally, the acquisition function should be evaluated for all points within the search space, and the set of points with the highest utility be selected for experimental testing. However, as discussed in Section 11.3, evaluating the acquisition function for all points is often infeasible, necessitating an alternative approach for selecting candidate points. Regardless of the method used for candidate selection, once these points are experimentally tested, the new observations are used to update the AED method's understanding of the broader landscape, such as formulation, process parameters, material properties, or other relevant factors. The next set of experimental points is chosen based on this updated knowledge. This iterative process ensures that each new observation brings immediate value to the AED approach, progressively refining the overall strategy with each cycle.

11.2.2 Popular Acquisition Functions

In this section we delve into various prominent acquisition functions employed in Adaptive Experimental Design (AED), elucidating the roles and characteristics of each. We will illustrate how different acquisition functions operate within a specific scenario, not limited to formulation landscapes but encompassing a range of application contexts. This is depicted in Figure 11.1, where the grey line represents the property of interest and the shaded area indicates its associated uncertainty. The

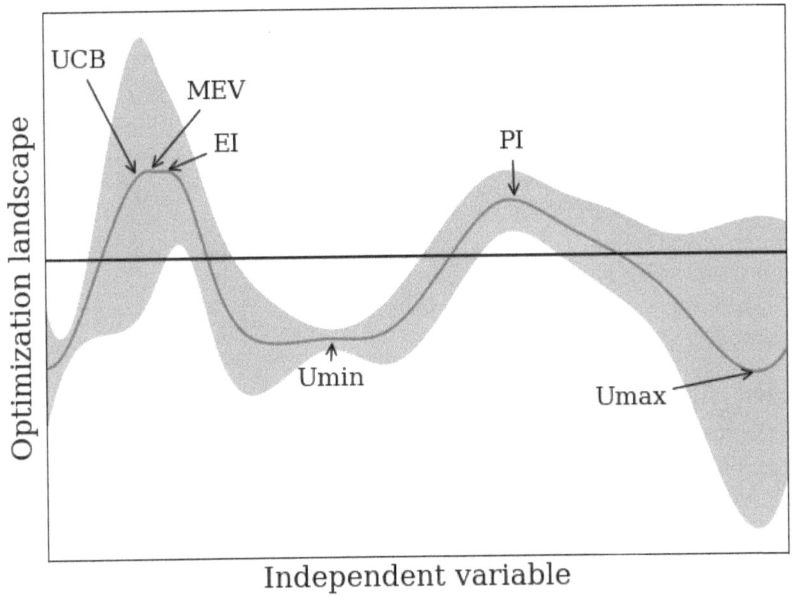

FIGURE 11.1 How different acquisition functions select different points in a formulation landscape. The acquisition functions are Upper Confidence Bound (UCB), Maximum Expected Value (MEV), Expected Improvement (EI), Uncertainty Minimization (Umin), Probability of Improvement (PI), and Uncertainty Maximization (Umax).

horizontal line in the figure represents the target value that the AED process aims to maximize above.

The Exploration-Exploitation Trade-Off

The simplest style of acquisition function is one that rewards experiments that closely match desired property values, essentially carrying out a "*greedy*" search through the formulation space. In each round of suggestions, it proposes experiments that fully leverage the existing knowledge of the optimal location. An alternative approach is to always search in new areas, in the hope of discovering superior options beyond what we have found so far so that any surrogate model (see Section 11.4) that is being used has a wider base of data to learn from. This *exploratory* approach is naturally riskier than exploiting current knowledge but has the potential to find completely novel opportunities that otherwise might be missed.

The most effective AED approaches strike a balance by combining both of these styles of acquisition function, allowing users to choose a trade-off between exploration and exploitation that best suits their needs.

Maximum Expected Value

The *Maximum Expected Value (MEV)* acquisition function aims to optimize the anticipated value of the function under optimization, completely neglecting any uncertainty in this value. It targets the point within the search space that most outperforms the current best-observed value. This acquisition function focuses purely on exploitation, simply targeting performance according to our current knowledge of the system. While this function is effective under certain conditions, it may encounter obstacles, such as local optima, when dealing with highly nonlinear functions or intricately complex search spaces. In such challenging scenarios, an acquisition function that considers uncertainty in the surrogate model can be beneficial.

Expected Improvement

Enter the *Expected Improvement (EI)* acquisition function. EI shares certain similarities with MEV but incorporates an extra layer of sophistication by considering uncertainty in the property values: instead of maximizing the expected value, EI maximizes the probability of outperforming the current best-observed value, weighted by how much improvement a proposal offers. This encourages a more comprehensive exploration of the search space, including suggestions that might not on average give such a big improvement over the current best suggestion but which we are uncertain about: if, due to this uncertainty, we are underestimating how good the suggestion will be, EI will prioritize this suggestion where MEV would neglect it.

Overall, the EI function selects the point that not only maximizes the expected degree of improvement but also has a high probability of doing so once uncertainty is taken into account.

EI includes an in-built exploration-exploitation trade-off factor but not one that is adjustable by the user.

Probability of Improvement

Another member of the acquisition function family is the *Probability of Improvement (PI)*. PI solely concentrates on the probability of any improvement over the current best-observed value, irrespective of the improvement's magnitude. This method is particularly effective in scenarios where even minor advancements are valuable. However, one limitation of PI is its potential to overlook regions that might offer substantial improvements but are more uncertain. It tends to favor more certain, albeit possibly smaller gains.

This conservative nature makes it a more exploitation-oriented approach, prioritizing safer, incremental advancements over riskier, potentially larger ones.

It is important to note that in cases where many potential points in the search space have very high uncertainty, the PI will tend toward a 50% probability for every point. This happens because the high uncertainty contributes almost equally above and below the target value, thereby inadvertently shifting the focus of PI toward exploration. In practice, PI is often used in conjunction with other acquisition functions to strike a more balanced exploration-exploitation trade-off, tailored to the specific requirements of the problem at hand.

Following this exploration of the PI function, let's now consider another subset of acquisition functions that specifically target uncertainty within the surrogate model.

Uncertainty Maximization/Minimization

Acquisition functions focused on uncertainty maximization (*Umax*) and minimization (*Umin*) are particularly well-suited for scenarios where uncertainty is of interest. These functions focus on entropy or variance of the surrogate model, selecting points in the search space that either maximize or minimize these metrics, respectively.

> Aiming for maximization of uncertainty fosters further exploration of the search space whilst minimizing uncertainty tends toward exploitation (as the search focuses on already well-understood areas).

Upper Confidence Bound

> The *Upper Confidence Bound (UCB)* acquisition function provides an explicit user-defined parameter to navigate the terrain between exploration and exploitation.

Expressing the expected value of the property being optimized as μ and its uncertainty as σ, the UCB acquisition function takes the form $\mu + b\sigma$ where b is a user-defined parameter that sets the exploration-exploitation trade-off. With $b = 0$, UCB reduces to the MEV acquisition function, while with $b \rightarrow \infty$, UCB effectively reduces to uncertainty maximization.

Other Acquisition Functions

A vast landscape of alternative acquisition functions exists for use in AED. These include *Knowledge Gradient* (Frazier et al. 2008) and *Thompson Sampling* (Thompson 1933), among others. Each of these functions represents unique trade-offs between exploration and exploitation and can be applied to solve distinct optimization problems based on the function and search space characteristics.

While our discussion has primarily focused on evaluating individual experimental proposals, many experimental processes involve running multiple samples in parallel. To harness this parallelism, the AED approach should recommend multiple candidate formulations simultaneously. The acquisition function must account for this by not merely returning multiple identical suggestions, possibly with minor variations, as this does not generate as much extra information. Instead, it is more beneficial to provide a diverse set of suggestions. Even if the "best" suggestion is focused toward exploitation, the full array of suggestions also inherently promotes exploration. Typically, this balance of exploration and exploitation is maintained by considering the dissimilarities or differences between the set of suggested points, ensuring the suggestions are not too similar to one another. This introduces an additional user-defined parameter into the search configuration. Given these various options, how does one decide on the right approach?

11.2.3 Selecting an Acquisition Function

While certain acquisition functions like MEV prioritize exploitation and tend to converge quickly, they risk getting trapped in local optima. In contrast, functions like uncertainty maximization or UCB, which emphasize exploration, typically require more function evaluations but may prove more effective in finding the approximate location of the global optimum.

> Choosing the appropriate acquisition function for a specific problem depends on several factors, including the complexity of the search space, the smoothness and nonlinearity of the function, and the available computational resources.

Therefore, a careful examination and comparison of the performance of different acquisition functions, considering the number of function evaluations required to converge to the optimal solution, often provide valuable insights.

In addition to the choice of acquisition function, other factors such as the choice of optimization algorithm (see Section 11.3) and surrogate model (Section 11.4) can influence the balance between exploration and exploitation. By carefully selecting these parameters, one can tailor the optimization algorithm to the unique characteristics of the problem, achieving an effective balance between exploration and exploitation.

11.2.4 Case Study

In their investigation to reformulate an existing ink product, Domino Printing Sciences required both exploration and exploitation during the course of the project. At the beginning of the project, it was important to explore new alternative ingredients to understand the options open for reformulation, while toward the end of the project it was necessary to focus on identifying the best possible single formulation, exploiting all the knowledge gathered over the project. To this end, Domino utilized an acquisition function within the Alchemite™ machine learning software (Conduit et al. 2017; Whitehead et al. 2022). This function combines two key components: an Uncertainty Maximization term that emphasizes exploration and a Probability of

Improvement term that facilitates exploitation. The Alchemite™ software provides users with a slider to control the weighting of these two terms, allowing them to specify the desired trade-off between exploration and exploitation for each round of experimental suggestions. Additionally, the software incorporates factors to account for correlations between multiple output variables and employs similarity measures to ensure that the batches of suggested inks (12 at a time) did not overly concentrate in the same region of the formulation space. At the conclusion of the project, ink formulations selected for comprehensive testing predominantly employed an acquisition function focused on exploitation in line with the expectation that this will give the suggested next experiments most likely to achieve the project goals.

11.3 OPTIMIZATION ALGORITHMS

11.3.1 Introduction to Optimization Algorithms

Once the objectives of an Adaptive Experimental Design (AED) campaign are defined and encapsulated within an acquisition function, the next step is to determine which new experiments to conduct in order to achieve these objectives. There are potentially infinitely many combinations of different ingredients and processing conditions that we could try, but given our finite knowledge of this space we need to select realistic candidates for the ingredients and processing conditions that can be tested in the acquisition function to most efficiently get us to the optimal selection.

Optimization algorithms serve as essential tools to explore the vast landscape of possibilities and propose the most efficient next steps.

Chapter 12 addresses the solution to this challenge when the formulation landscape can be accurately represented by a linear model. In such cases, it becomes feasible to immediately identify the global optimum within the landscape. However, in more complicated formulation landscapes, immediate discovery of the optimum is typically unattainable. Instead, we must judiciously select combinations of ingredients and processing conditions, guided by informed decisions, to expedite our journey toward the optimum outcome.

11.3.2 Static Optimization Algorithms

The easiest way to select the next set of experiments to try is to ignore those that have been carried out previously and try to get a good coverage of the full set of possibilities in the formulation space. The most naive approach to this is to try everything: enumerate all the possible combinations of ingredients and processing conditions and try all of them. If there are N input variables, each of which can take d possible values, this is known as a "full factorial" experimental design (Oehlert 2000), requiring d^N experiments. As N increases, a "full factorial" experimental design becomes practically impossible. For instance, with only 10 variables, each having just 3 possible values, we would already need to carry out 60,000 new experiments. To mitigate this substantial number of experiments, various methods aim to select only the most informative ones, rather than the complete factorial set. Examples include *Box-Behnken designs* (Box and Behnken 1960), *Plackett-Burman designs* (Plackett and Burman 1946), *Taguchi arrays* (Taguchi and Tsai 1995), and more recently *generalized optimal designs* (NIST/SEMATECH 2023), which together all sit under the banner of "Design of Experiments" (DoE). However, these methods, in general, do not take account of any existing experimental data within a system. Their primary aim is to comprehensively cover as many possibilities as possible within the formulation space, rather than directing the experimenter toward the global optimum in the most efficient manner.

Additionally, DoE also provides access to further statistical analysis of a design space, beyond the search for optimal formulations directly. The structured layout of DoE-suggested experiments, often accompanied by the use of replicates, facilitates hypothesis testing, analysis of repeatability and reproducibility, and casual investigations. These analyses enable scientists to gain a deeper understanding of the formulation domain but do not directly contribute to the primary objective of AED, which is the search for optimal formulations. Approaches more commonly used in AED for suggesting new experiments tend to emphasize this optimization search over broader knowledge capture.

11.3.3 Genetic Algorithms

One way to take advantage of existing experimental data when proposing new experiments is to be inspired by the way that nature leverages the success of earlier generations of living creatures when generating new entities. For many species, this is achieved through the amalgamation of genetic material from the most successful examples of each generation, resulting in offspring that inherit genetic features from (typically two) parents. It might be expected that the combination of two successful sets of genetic material might give rise to an even more successful offspring, which is a fundamental principle behind evolution by natural selection. In formulation development, the same approach can be used: the ingredients and processing conditions that gave rise to the most successful historical experiments are combined (typically also with some random variations, analogous to biological mutations) and the resulting formulations taken forward as the next generation of proposals. This is known as a "*genetic algorithm*."

> Genetic algorithms are typically very interpretable, in that it is clear which parent formulations informed the new selection, but can suffer from two drawbacks that are particularly egregious for high-dimensional chemical formulation problems: slow convergence and lack of extrapolation.

Nature does not have a time limit for how long it takes to find an optimal set of genetics (and indeed the environment often changes on a similar timescale to the fitness changes from the genetic updates, so the optimum itself is moving), but chemical products need to be designed in as time- and money-efficient way as possible. Chemical formulations that can contain a selection from hundreds or thousands of different ingredients are particularly badly affected by the slow convergence of genetic algorithms, which in these high-dimensional spaces can converge exponentially slowly (Chen and He 2021). Because genetic algorithms combine existing formulations, they also limit the ability to explore new areas of formulation space that lie outside that explored previously, limiting their utility in high-dimensional applications, which almost always amount to extrapolation (Balestriero et al. 2021). To overcome these limitations, we can instead turn to a tool known as "Bayes' theorem."

11.3.4 Introduction to Bayes' Theorem

Bayes' theorem formalizes the concept of utilizing pre-existing (or "prior") knowledge about a system into a solid mathematical framework. This can be summarized as our updated (or "posterior") confidence in achieving a result being the product of our prior knowledge on the probability of the result $P(result)$ and the likelihood of the result given what has come before $L(result \mid history)$:

$$P(result \mid history) \propto P(result)\, L(result \mid history)$$

If we have a way of estimating these two terms, then we can apply this theorem to find the new formulation that maximizes the probability of achieving our goal, as encapsulated in the acquisition function.

11.3.5 Tree-Structured Parzen Estimators

One approach to explicitly model $P(result)$ and $L(result \mid history)$ is to effectively count how many historical samples achieved a given level of success, and focus on areas of the formulation space where this predicts future successful results. *Tree-structured Parzen Estimators (TPE)* provide one way to do this: they set a threshold value (which sets the desired exploration/exploitation tradeoff – see also Section 11.2) and then explicitly look for the densities of historic results that outperform and underperform this threshold in formulation space. By setting this threshold to be a quantile of the observed results, we get an implicit estimate of $P(result)$ for free (Bergstra et al. 2011). Using these implicit and explicit representations of the terms in Bayes' theorem, we can then efficiently try multiple new formulations to identify which give the best performance against our acquisition function.

TPE has the advantage of scaling linearly with the number of dimensions and volume of data being examined and providing a fully Bayesian utilization of all historic data values. The flexibility of the density construction approach for $L(result \mid history)$ also enables the searching of complicated (potentially non-convex) formulation landscapes, where an evolutionary approach that combines existing results would find it difficult to stay within the feasible region (see also Section 11.3.8).

However, TPE considers each component in its search space independently, so if there are correlations between ingredients or processes in a formulation, these may be missed by the TPE optimization. TPE also does not consider any gradient information about a result – it utilizes the value of a result only, and not any information about how this changes as we make small adjustments to the inputs. However, there are other methods that can take this extra information and use it to sometimes accelerate the identification of new suggested formulations.

11.3.6 Gradient-Based Optimization

Taking the gradient of a function into account when optimizing provides us access to more information about the function than simply its value at each point. For example, it may point us in the direction of increasing performance, giving us a head-start on which type of formulations to test next. Optimizers that can take this information into account are called *gradient-based* or *derivative-based* optimizers. There are dozens of different algorithms for gradient-based optimization, ranging from very simple (gradient descent: follow the direction of the gradient at each step of the optimization) to the very involved gradient-based methods used to optimize deep neural networks (Reddi et al. 2018).

While recent academic research has extensively explored these gradient-based optimizers due to their effectiveness in optimizing deep neural networks, they are generally not well-suited for formulation development problems when working directly with experimental results.

This limitation arises from the fact that most experimental data related to formulations only yield point values for properties, without the availability of gradient information that is essential for gradient-based optimization methods.

However, when using a surrogate model (see Section 11.4) that does provide convenient access to gradient information, gradient-based optimizers come back into contention. Nevertheless, it is worth noting that these methods can also sometimes struggle when dealing with more complex constraints within the feasible formulation search space.

11.3.7 Reinforcement Learning

Although optimizers designed for improving deep neural networks are not themselves always suitable for use in formulation development, there are classes of deep neural networks and other artificial intelligence (AI) tools that are helpful in this context. One in particular is known as "*reinforcement learning*": this involves an AI acting as an "*agent*," able to take one of a set of "*actions*" in the "*state*" of the system being examined (Arulkumaran et al. 2017). Applied to formulation development, this translates to an AI proposing new formulations and using the success of these formulations as a "*reward*." This feedback loop enables the AI agent to refine its neural network, progressively improving its ability to propose more effective formulations. Particular examples of reinforcement learning approaches include *deep Q networks* (Mnih et al. 2015) that use the achieved reward to update their own estimate of the reward a new action will bring, and combinations of neural networks and random searching in the *AlphaGo* framework that learned to play Go without human oversight (Silver et al. 2016).

Reinforcement learning is a powerful way to leverage the ever-growing strengths of AI tools for experimental design, but the deep neural networks typically used for these applications require a large volume of data for effective training.

This can be often impractical for active formulation development projects, where data might be limited. Moreover, reinforcement learning might struggle with small or complex feasible regions, as hard boundaries in these spaces challenge the neural network's reliance on differentiable functions.

11.3.8 Advanced and Recent Optimizers

The field of optimizer development is still rapidly progressing, driven in part by the use of optimizers for the tuning of large neural network models. This section provides a brief overview of some notable recent advancements in optimizer technology, building on the concepts discussed earlier. A key innovation comes from Huawei with the *Heteroscedastic Evolutionary Bayesian Optimization (HEBO)* algorithm (Cowen-Rivers et al. 2022). HEBO is a sophisticated algorithm that merges the principles of genetic algorithms with empirical data transformations. This blend makes it particularly suitable for navigating complex search spaces, offering a more versatile approach to optimization.

Meanwhile, Uber's *Trust Region Bayesian Optimization (TuRBO)* (Eriksson et al. 2019) takes a different track. TuRBO attempts global optimization by employing a collection of local Bayesian models. This strategy allows for efficient allocation of resources across these models, providing a nuanced balance between exploration-exploitation.

Adding to these advancements is the *MetaBO* framework (Volpp et al. 2020). MetaBO takes the idea of utilizing accumulated knowledge within an AED framework a step further. It incorporates insights gained from specific source tasks, enabling the framework to rapidly adjust and respond to new optimization problems. This approach provides a further emphasis on the importance of learning from past experiences.

Finally, the *Mango* optimizer (Sandha et al. 2020) focuses on parallel optimization. Mango is designed to effectively incorporate information from multiple simultaneous experiments into the learning process. This methodology is particularly valuable in scenarios where conducting parallel experiments is feasible, ensuring that the gathered data is utilized to its fullest potential.

11.3.9 Incorporating Constraints

The previous subsections have covered different optimization approaches and algorithms, and a common thread through many has been the difficulty in applying constraints to the formulation search space. This section aims to provide an overview of common constraints encountered in the formulation development process and to elucidate the complexities they introduce.

A constraint that is simple to state (but can be complicated in practice) is the "*composition constraint*" that the components of a formulation specified in weight percentage or molar percentage, for example, must add up to 100% for the eventual product (otherwise the product becomes impossible to construct).

Mathematically this corresponds to the d components of the formulation composition existing in a $d-1$ dimensional simplex. Searching for compositions within a simplex is a well-understood problem, particularly in low numbers of dimensions; but when further constraints are added to the system (e.g., each different ingredient has different minimum and maximum amounts it is allowed to take within the formulation) the search space quickly becomes very complex. In general, a three-component formulation with different minimum and maximum amounts on each ingredient has a search space that is an irregular 2D hexagonal shape, which is a difficult shape to search over consistently – and this only becomes more complicated in higher dimensions (Griffin et al. 2020).

These composition constraints can also be *nested*: for example, in a total formulation adding up to 100% weight, between 25% and 30% might need to consist of one type of ingredient, between 40% and 50% might need to consist of a second type, and between 30% and 40% might need to consist of a third type. The intersection of these constraints leads to a very complex search space and typically requires a toolset designed specifically for handling this type of formulation development.

In particular, most approaches based on genetic algorithms, gradient-based search, and reinforcement learning struggle when the viable part of the search space is small compared to the overall range of values being searched over.

Other types of constraints include dependencies between ingredients (e.g., when creating a copolymer the ratio of different species of monomers may be set), categorical choices (e.g., we can use machine A or machine B only; again difficult for gradient-based methods to explore), and discrete choices (e.g., an oven can be heated to 200°C, 250°C, or 300°C but nothing in between). For production development (as opposed to academic research), it is also often important to minimize the number of ingredients used in the formulation, which effectively introduces another objective to the optimization but one that is discrete rather than continuous and so difficult to optimize using gradient-based methods.

11.3.10 Case Study

Domino Printing Sciences sought an optimizer to help them identify the best next experiments to carry out as part of an ink redevelopment campaign. Domino has historically used Design of Experiments approaches to generate fresh data when beginning a new project but wanted to investigate using their plentiful historic data to speed up the experimental design process. Inks have to satisfy several constraints in order to be successfully manufactured; for example, the photoinitiator content needs to be high enough that the ink cures successfully but not too high or otherwise the ink may not be stable over time leading to printer performance issues. To take full advantage of their historic data, Domino decided to use a Bayesian optimizer: Tree-structured Parzen Estimators provided the ability to encode the physical constraints on the ink, handle the several hundred ingredients being selected between in a reasonable time, and minimize the number of new experiments required for the optimization process. Alternatives in terms of genetic or reinforcement learning optimizers would have struggled with the constraints being imposed, and the experiments run by Domino did not provide the gradient information to make gradient-based optimizers more effective, leading to a Bayesian optimizer being the preferred method for this project.

11.4 SURROGATE MODELS

11.4.1 Introduction to Surrogate Models

The simplest way to use an optimizer and acquisition function is to work purely with experimental data: for example, using a genetic algorithm, the algorithm examines historic experimental data points, constructs a new combination of the ingredients and processes by combining the best historic examples, and proposes this for new physical experimentation. The result of the experiment is then evaluated against the acquisition function to see how highly to score this new data point for inclusion in future steps of the genetic algorithm. This approach has the natural obstacle that the number of formulations that can be tested is limited by the available experimental budget, much of which might be wasted on new formulations that have little chance of success. An alternative approach is to construct a "*surrogate model*," a term we use interchangeably with "*predictive model*" as discussed in Chapter 7. This is a learned model that takes the formulation and processing conditions as input and predicts the experimental properties. Typically, this model is trained on the existing experimental data.

When an optimizer proposes a new formulation, we can then evaluate how well it performs against the acquisition function using the surrogate model. This approach is significantly more cost-effective and faster than direct experimentation. Only formulations that show promise in the surrogate model, as determined by our acquisition function, are then advanced to the stage of physical experimentation.

This method streamlines the process, focusing resources and time on the most promising formulations.

11.4.2 Response Surface Methods

Perhaps the simplest type of surrogate model, and one that has been used for decades in experimental design, is the response surface. This usually involves fitting a polynomial function, often a quadratic function, to the existing data. This fitting procedure is very quick, and for a quadratic response surface the optimum of the response surface is easy to identify analytically. Response surfaces are often used in static Design of Experiments approaches, but can also be used in adaptive experimental design, where the quick fitting of the response surface when new data becomes available is an advantage. However, response surfaces quickly overfit data as the complexity of the polynomial increases, and for low-order polynomial functions are not powerful enough to capture complex relationships between inputs and properties.

The recent boom in machine learning approaches, which are explicitly designed to model complicated relationships based on finite datasets, provides a toolbox of more powerful methods for creating surrogate models that we can leverage for formulation development applications, such as those discussed in Chapters 7 and 8. From a purely theoretical standpoint, one of the most attractive of these machine learning approaches is called a "*Gaussian Process.*"

11.4.3 Gaussian Processes

Gaussian processes are a machine learning technique that aims to explicitly model the $P(result \mid history)$ term found in Bayes' theorem (see Section 11.3.4). Gaussian processes make the assumption that future formulations we might explore can be modeled by taking the joint distribution between historic and future formulations to have a Gaussian (or "bell-curve") structure. Gaussian processes provide a tool to sample the space of possible functions that would satisfy this structure and generate estimates of the chemical properties (and, importantly, the uncertainties in those estimates) for new formulations. This results in new predictions being effectively the sum of the properties of each historic formulation, weighted by how similar the inputs for the new formulation are to each historic example as measured by a "kernel" function (Phillips and Venkatasubramanian 2011). Denoting the matrix of kernelized similarities between each pair of historic formulations as K, the matrix of kernelized similarities between each historic formulation and each new potential formulation as K^*, and the vector of historic formulation properties as y, the Gaussian process prediction is given by

$$y^* = K^* K^{-1} y$$

which can be seen to match the functional form of traditional linear regression, where K plays the role of the independent variable's variance and K^* plays the role of the covariance between dependent and independent variables. The difference in Gaussian processes is that the matrices change depending on the new formulations being examined, resulting in an overall nonlinear model but one that still neatly extends the response surface polynomial methodology.

Gaussian processes are often considered the "gold standard" for surrogate models due to their minimal assumptions, generally reliable uncertainty estimates, and flexibility to apply to many different types of function-fitting problems.

However, Gaussian processes have a key disadvantage in terms of their scaling to larger datasets: calculating the matrix inverse K^{-1} scales cubically with the number of historic formulation examples being trained on, which rapidly becomes infeasible as the number of training examples grows. Calculating the K and K^{*} matrices involves measuring the similarity between each pair of formulations, which typically only scales linearly with the number of ingredients in a formulation, but even so, for moderate-sized chemical formulation datasets.

Gaussian process models can be impractically slow to use.

Although approaches exist to accelerating the calculation of Gaussian process model predictions (Furrer et al. 2006; Nguyen-Tuong et al. 2008), these typically come with significant compromises in terms of modeling accuracy.

11.4.4 Random Forests

Another popular machine learning approach to creating surrogate models are *random forests* (Breiman 2001), see also Chapter 7. These consist of ensembles of decision trees, each of which is trained on slightly different subsets of the available data to build a more robust understanding of the variation in the data. This provides random forests great flexibility in capturing different behaviors of properties as a function of the inputs. By measuring the variation across the ensemble of trees, random forests also provide access to a measure of the uncertainty in a prediction, vital for application with acquisition functions that account for uncertainty, particularly in exploration. Random forests are also much quicker to train than Gaussian processes, making them more attractive for larger datasets in particular. However, as highlighted in Chapter 7, random forests typically suffer in extrapolation. This is because a decision tree within the forest bases its prediction for points outside the training range on the values of the edge-most training points, resulting in the entire ensemble predicting a single value with an uncertainty that is reflective only of the variations across the trees. This approach often fails to accurately represent the behavior of real chemical systems outside the range of the training data, limiting the effectiveness of random forests in scenarios where extrapolation is crucial. There are extensions of the random forest concept to provide linear extrapolation behavior (Quinlan 1992), but this also introduces assumptions about the far-extrapolation behavior of the function that are often unjustified. Nevertheless, random forests provide one of the more popular machine learning methods for quickly and easily modeling chemical systems.

11.4.5 Neural Networks

The explosion in deep learning in recent years has been driven by the increasing performance of neural network machine learning architectures for image processing, time series analysis, and natural language processing applications. This development has also driven new tools for chemical applications – in particular, in the drug discovery field using recurrent neural networks (Brown et al. 2019) and graph convolutional neural networks (Chen et al. 2018). In the industrial chemistry and formulation fields, however, neural networks have not seen a comparable breakthrough moment: tree-based methods such as random forests are still typically the most performant methods for these types of applications (Grinsztajn et al. 2022), in terms of both accuracy and computational performance. The main advantage of neural networks in this field is that they are intrinsically smooth, which can be more realistic for many chemical properties, but they usually do not provide uncertainty estimates (although this can be addressed (Srivastava et al. 2014; Whitehead 2021)). However, it is noteworthy that the limitation of neural networks in providing uncertainty estimates can be addressed through their extension to Bayesian Neural Networks. For an in-depth discussion on this extension and how Bayesian Neural Networks provide a framework for incorporating uncertainty estimates in neural network predictions, readers are encouraged to refer to Chapter 7.

The data-hungry training procedure for neural networks means they are rarely practical for smaller-scale formulation chemistry problems, but the advancements in Bayesian approaches offer promising avenues for future research and application in this field.

11.4.6 Other Surrogate Models

There is a whole panoply of other machine learning methods that are sometimes used for surrogate model creation. These range from simply examining the average value of a property in the neighborhood of a new proposed formulation (*k-nearest neighbor algorithm* (Cover and Hart 1967) – quick to train but cannot extrapolate), through linear methods such as the *partial least squares* approach covered in Chapter 8, *support vector machines* (Drucker et al. 1997), and *matrix factorization* (Singh and Gordon 2008) (which all suffer from an inability to capture complex relationships), to other nonlinear methods utilizing the same "kernel trick" (Hofmann et al. 2008) as seen with Gaussian processes in Section 11.4.3 (which can learn complex relationships, usually at the cost of explainability).

11.4.7 Strategies for High-Dimensional Data

Formulation data typically involves selections from dozens or hundreds of ingredients being combined together, and in industrial research the number of formulations available to learn a surrogate model from is often much smaller than the datasets available in more data-rich environments such as financial or social network data. Naively applying many surrogate models to this type of data will result in an under-constrained model that is impossible to accurately fit.

Instead, we often have to utilize other, more information-rich, data to construct a reliable surrogate model. A key source of more information-rich data is expert-selected raw material information, as discussed in Chapter 5. For example, if a domain expert knows that the total water content of a formulation is a key parameter for understanding the viscosity of the formulation, instead of using only the amount of each ingredient in the formulation as input to the surrogate model, an extra variable can be introduced that captures the expected amount of water contributed by each ingredient, leading to a law-of-mixtures approximation to the total amount of water in the system. This feature engineering approach can be extended using more complex expert-driven tools as well, including other computational methods such as Hansen solubility parameters (Hansen 1967).

Another more automatic approach to simplifying high-dimensional data before using it in a surrogate model is to use a *dimensionality reduction technique* (Maaten et al. 2009): this is a class of tools that aim to preserve as much structure in a high-dimensional dataset as possible, while finding the smallest possible representation of the data. Common techniques include the linear *Principal Component Analysis* method (which can also be kernalized to capture nonlinear effects (Schölkopf et al. 1998)) and the fully nonlinear *Uniform Manifold Approximation* method (McInnes et al. 2020). These methods often help build a more accurate surrogate model, although they can reduce the interpretability of the result as the model is no longer in terms of chemically informative variables but rather the reduced variables.

11.4.8 Case Study

In their Adaptive Experimental Design (AED) campaign, Domino Printing Sciences were aiming to maximize the value obtained from every experimental run, particularly because the testing of multiple properties for each trial formulation is a manual and time-consuming step of the process. They therefore sought the use of a surrogate model. This strategic choice was driven by the desire to replace some aspects of physical experimentation with computer modeling, thereby streamlining the process and reducing manual labor. A major challenge encountered by Domino in implementing this approach was the sparse nature of their historical data. Not every historical formulation was measured against all of the properties of interest in a new project, but they did want to take advantage of what historical information was available.

Domino chose to use the Alchemite™ machine learning algorithm (Conduit et al. 2017; Whitehead et al. 2019) because it was especially designed to handle this type of sparse data. This surrogate model also provided uncertainty estimates in predicted properties, which supported the use of the Uncertainty Maximization acquisition function for extrapolation. With the optimizer, acquisition function, and surrogate model selected, Domino was able to move ahead with the experimental campaign, leveraging all of their historical data.

After three AED cycles of 12 samples each, an optimized formulation was achieved. The formulation met the desired target criteria of matching an existing ink's performance, while removing a material that had been reclassified. Several variations of the optimized formulation were then suggested by the model in order to replace the full ink set.

In contrast to previous static optimization campaigns, which necessitated approximately 500 samples, this AED campaign required only 36 formulations. This represents a remarkable 90+% reduction in the workload required for achieving the same objectives. Such efficiency highlights the significant benefits of employing AED in formulation development.

The surrogate model was also used to undertake in-silico exploration of the formulation space, such as counterbalancing cost against performance. These explorations provided a data rational basis for further development projects.

11.5 PRACTICAL CONSIDERATIONS FOR ADAPTIVE EXPERIMENTAL DESIGN IN PRODUCT DEVELOPMENT

The previous sections have covered the common tools and techniques available for adaptive experimental design. We now turn to how these tools are applied in practice, and the considerations and compromises that need to be made to get the best results in a product development project.

11.5.1 Combining Optimizers, Acquisition Functions, and Surrogate Models

Not all combinations of the optimizers, acquisition functions, and surrogate models discussed earlier are equally as powerful: some combinations complement each other well, while others do not. For example, most acquisition functions rely on the presence of accurate uncertainty estimates, and so surrogate models that do not provide such estimates (including many neural network or matrix factorization approaches) are not suitable for use with those acquisition functions. Similarly, gradient-based optimizers are most useful with smooth surrogate models, such as Gaussian processes or neural networks, rather than random forests, which do not have a well-defined gradient everywhere.

A common combination of tools for formulation development consists of using a genetic algorithm or TPE as the optimizer with a Gaussian process or random forest as the surrogate model, as these approaches tend to work better with smaller datasets than some alternatives.

In contrast, for large datasets, there may be enough information available to train neural networks for both a surrogate model and a reinforcement learning optimizer, providing a smoother model search space at the cost of limiting which acquisition functions can be utilized (MEV is the most applicable acquisition function without uncertainties being available).

11.5.2 Multi-Objective Optimization

In real product development applications, we need to optimize not just one property of the product but several simultaneously, for example, product performance, rheological properties, and cost. Almost always the optimal value of one property is not found at the same composition as for another property. This means that we need to compromise between the different properties when searching for the next experiment to carry out.

There are two main ways that these multiple objectives are optimized simultaneously: they are either combined into one function that is optimized using one of the tools discussed so far, or a specific multi-objective approach is taken.

Combining objectives into a single function is the easiest approach: often the "costs" returned by the acquisition functions for each objective are averaged, perhaps including a weighting term, and then this overall cost function is optimized.

Viewing this average as an L1 norm of a vector of costs per acquisition function, extensions are also possible to, for example, the $L\infty$ norm, known as the Chebyshev scalarization approach (Chugh 2020), which can capture more complex optima configurations. Where properties are related to each other in some way, extensions are required that capture this covariance so as not to double-count the performance against similar properties.

When combining objective functions is not desired, the multi-objective optimization target becomes to find the Pareto front of suggestions, none of which are "dominated" by any other suggestions.

Approximations are generally required to make the problem of finding the Pareto front tractable, often by generating several different scalarizations of the cost function (Das and Dennis 1998). A human expert then needs to select which experiments from this Pareto set to carry out, a task which is complicated by the difficulty of visualizing a more than two- or three-dimensional Pareto front. Although this approach in principle provides a better understanding of the full set of opportunities for new experiments to achieve all objectives, in practice the practical difficulties make the combination of objectives into a single cost function more realistic.

It is also often the case that although there are multiple objectives that need to be satisfied by a new product, not every trial design along the way needs to experimentally test each one of these properties. For example, if two properties are correlated, and a surrogate model can be used to model one of them well when the other is measured, then the experimental workload can be cut in half on each round of experiments by measuring one property and predicting the other. This stands in contrast to traditional static experimental design approaches, where every property for each formulation is typically measured, because when the design is created it is not known which properties will be limiting for each individual formulation. The final set of proposed products typically have all their properties measured, to confirm surrogate model predictions before the product moves toward manufacturing, but saving experimental workload during the development program. For a deeper understanding of handling correlated properties in multi-objective optimization, readers are referred to the "Correlated Predictions" section in Chapter 7.

11.5.3 The Adaptive Experimental Design Process in Practice

A realistic Adaptive Experimental Design campaign typically follows a fairly well-worn path, through a variety of decisions and results. We contrast this here with the "classical" experimental design approach, comparing the benefits and drawbacks of each approach, and focusing on the practical steps taken in each.

In the traditional Design of Experiments (DoE) approach, each project begins with a blank slate. A project domain expert decides which variables are most likely to be strongly impacting the performance of the product being developed, informed by their domain expertise. These few variables (typically not more than 5) are then provided to a computer program that generates a DoE design of the type discussed in Section 11.2.2, sometimes alongside the number of experiments that are intended to be run. Once all the experimental results are obtained, a polynomial response surface surrogate model (see Section 11.4.2) is often fitted to the results, and the optimum of this surface is analytically obtained. Another experiment may be run at this optimum to give the expected best formulation. If this does not achieve the project goals, a further full or partial DoE is run again.

In contrast, in an Adaptive Experimental Design (AED) approach, each project usually starts with a foundation of data available from previous, similar, projects. These data are used to build a surrogate model: often a Gaussian process model if there are not many historical data points, or a random forest or kernelized linear model if there is a large volume of data available. An acquisition function is then selected: often based on Uncertainty Maximization at the beginning of a project to focus on exploring uncertain regions of the formulation space. A small number of experiments are then suggested from this acquisition function and surrogate model using an optimizer, usually one based on Bayesian statistics. Importantly, the number of experiments suggested in each round of an AED campaign is small compared to the number of experiments carried out in the comparable DoE campaign.

> It is not expected that the first experimental suggestions will necessarily achieve the project goals – but instead, the results are fed back into the surrogate model, and further rounds of experiments are proposed, typically transitioning to the use of a more exploitation-focused acquisition function to narrow down on the final proposed formulation.

This process is illustrated in Figure 11.2.

A key difference between these two approaches is the number of experiments run during the process. In the DoE approach, the number of experiments that will be performed is fixed at the beginning of the project, while in the AED approach the number of experiments depends on how many rounds of experimentation are carried out. This can of course be set at the beginning of a project as well but is often adjusted to take into account results obtained in earlier experimental rounds. The AED approach does often reduce the total number of experiments carried out in a project (by up to 90% (Intellegens 2022)) but increases the variance in the number carried out. The AED approach lends itself to cumulative application, each project building on the data collected in the last, which typically requires a more mature data infrastructure to realize its full potential for application across large organizations.

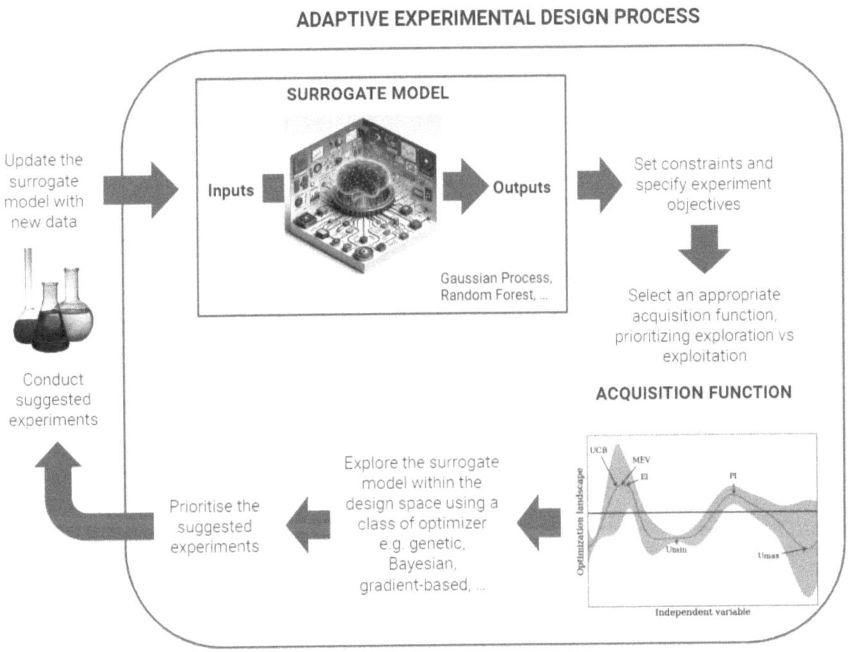

FIGURE 11.2 An illustration of the Adaptive Experimental Design process, including each of the steps covered in the previous sections.

At the beginning of this journey, organizations may find they do not have enough data to build a first surrogate model: in this case a small DoE design complements the AED approach well by building a foundation that future projects can build upon.

11.5.4 Recommended Tools

To realize the promise of Adaptive Experimental Design of more efficient formulation design and optimization, robust software tools are needed for each stage of the process. For building surrogate models, there are now many open-source statistical and machine learning packages available, for example *scikit-learn* (Pedregosa et al. 2011) for general machine learning models and *GPy* (GPy 2012) for Gaussian processes in the Python ecosystem. Many (but notably not all) of the optimization approaches covered in Section 11.3 are also open-source: Bayesian Tree-structured Parzen Estimators are implemented in the widely used Python hyperopt library (Bergstra et al. 2011), for example. There are also a variety of tools that cover the whole AED pipeline: many require users to code in Python or R (e.g., scikit-activeml (Kottke et al. 2021) in Python), while others provide an easy-to-use no-code web interface (e.g., Alchemite (Conduit et al. 2017)).

In an industrial setting, the ability for multiple users to collaborate on the same AED campaign can be key; for example, if one scientist is responsible for training and analyzing surrogate models, another is responsible for proposing new trial formulations, and a third runs the experiments, then all these users having access to the same system becomes vital.

This typically limits the utility of code-based solutions as not all team members will have the requisite skill set: instead, web-based or centrally installed software packages tend to perform better with larger teams.

11.6 CONCLUSIONS

11.6.1 Summary of Key Concepts

We have seen how the Adaptive Experimental Design (AED) approach can be broken down into three constituents: the acquisition function, which lets us define a measure of success; an optimizer, which provides efficient searching of a complex formulation landscape; and a surrogate model, which allows us to make the most of existing data at each stage of the process by sampling this understanding of the space rather than requiring every sample to be a physical experiment. The acquisition function allows the user to set their appetite for exploration of the search space against exploitation of their existing knowledge, and by tuning the selection of the acquisition function can drive the AED program toward different styles of experimental suggestions.

Different classes of optimizers provide different benefits – from nature-inspired genetic searches, through the solid foundation of Bayesian statistics, to cutting-edge reinforcement learning approaches building on the ongoing deep learning revolution, and again selecting the correct optimizer for a given problem (are constraints needed, or are gradients of the function being optimized available?) can make the approach more efficient. Finally, surrogate models can be key to efficient adaptive experimental design campaigns; for small systems, Gaussian processes remain the gold standard because of their thorough Bayesian grounding, but for larger, more realistic formulation problems, other approaches such as random forests may be more practical. Effectively combining these three pieces into a single AED campaign provides the experimentalist the best chance of maximizing the benefits from their experimental work and minimizing the time taken to achieve product optimization goals.

11.6.2 Future Directions and Developing Trends

Although the overall AED framework provides a powerful tool for experimental scientists, many of the individual pieces are still developing and improving. In particular, the requirement in the deep learning field for efficient hyperparameter optimizers for neural networks has driven the development of a new slew of optimizers, which although designed for that purpose, are easily transposed into experimental design and can prove effective in practice. AED also relies much more strongly on a solid data infrastructure to be most effective than traditional experimental design techniques: if a good surrogate model can be built using data from multiple previous projects, this gives a new project a great starting point, but all that historic data must first be gathered and kept consistent, which, as discussed in Chapter 4, can be quite challenging. Large chemical organizations are only at the beginning of the journey toward more robust data handling techniques, but the prize it will enable in terms of expedited product development means companies are progressing data projects rapidly. In a world where more efficient and more environmentally friendly product development is becoming ever more important, AED provides a clear toolbox enabling expedited formulation and product design within a flexible development approach.

REFERENCES

Arulkumaran, Kai, Marc Peter Deisenroth, Miles Brundage, and Anil Anthony Bharath. 2017. "Deep Reinforcement Learning: A Brief Survey." *IEEE Signal Processing Magazine* 34 (6): 26–38. https://doi.org/10.1109/MSP.2017.2743240.

Balestriero, Randall, Jerome Pesenti, and Yann LeCun. 2021. "Learning in High Dimension Always Amounts to Extrapolation." *ArXiv*.

Bergstra, James, Rémi Bardenet, Yoshua Bengio, and Balázs Kégl. 2011. "Algorithms for Hyper-Parameter Optimization." In *Advances in Neural Information Processing Systems*, edited by J. Shawe-Taylor, R. Zemel, P. Bartlett, F. Pereira, and K.Q. Weinberger. Vol. 24. Curran Associates, Inc.

Box, G.E.P., and D.W. Behnken. 1960, November. "Some New Three Level Designs for the Study of Quantitative Variables." *Technometrics* 2 (4).

Breiman, Leo. 2001. "Random Forests." *Machine Learning* 45: 5–32.

Brown, Nathan, Marco Fiscato, Marwin H.S. Segler, and Alain C. Vaucher. 2019. "GuacaMol: Benchmarking Models for de Novo Molecular Design." *Journal of Chemical Information and Modeling* 59 (3): 1096–1108. https://doi.org/10.1021/acs.jcim.8b00839.

Chen, Hongming, Ola Engkvist, Yinhai Wang, Marcus Olivecrona, and Thomas Blaschke. 2018. "The Rise of Deep Learning in Drug Discovery." *Drug Discovery Today* 23 (6): 1241–1250. https://doi.org/10.1016/j.drudis.2018.01.039.

Chen, Yu, and Jun He. 2021, July. "Average Convergence Rate of Evolutionary Algorithms in Continuous Optimization." *Information Sciences* 562: 200–219. https://doi.org/10.1016/J.INS.2020.12.076.

Chuaqui, T.R.C., A.T. Rhead, R. Butler, and C. Scarth. 2021. "A Data-Driven Bayesian Optimisation Framework for the Design and Stacking Sequence Selection of Increased Notched Strength Laminates." *Composites Part B: Engineering* 226: 109347. https://doi.org/10.1016/j.compositesb.2021.109347.

Chugh, Tinkle. 2020. "Scalarizing Functions in Bayesian Multiobjective Optimization." *2020 IEEE Congress on Evolutionary Computation, CEC 2020 – Conference Proceedings*. https://doi.org/10.1109/CEC48606.2020.9185706.

Conduit, B.D., N.G. Jones, H.J. Stone, and G.J. Conduit. 2017. "Design of a Nickel-Base Superalloy Using a Neural Network." *Materials and Design* 131: 358–365. https://doi.org/10.1016/j.matdes.2017.06.007.

Cover, T., and P. Hart. 1967. "Nearest Neighbor Pattern Classification." *IEEE Transactions on Information Theory* 13 (1): 21–27. https://doi.org/10.1109/TIT.1967.1053964.

Cowen-Rivers, Alexander I., Wenlong Lyu, Rasul Tutunov, Zhi Wang, Antoine Grosnit, Ryan Rhys, Alexandre Max Maravel, et al. 2022. "HEBO: Pushing the Limits of Sample-Efficient Hyperparameter Optimisation." *Journal of Artificial Intelligence Research* 74 (1): 1269–1349. https://doi.org/10.1613/JAIR.1.13643.

Das, Indraneel, and J.E. Dennis. 1998. "Normal-Boundary Intersection: A New Method for Generating the Pareto Surface in Nonlinear Multicriteria Optimization Problems." *SIAM Journal on Optimization* 8 (3): 631–657. https://doi.org/10.1137/S1052623496307510.

Drucker, Harris, Chris J.C. Surges, Linda Kaufman, Alex Smola, and Vladimir Vapnik. 1997. "Support Vector Regression Machines." *Advances in Neural Information Processing Systems* 1: 155–161.

Drury, Brett, Jorge Valverde-Rebaza, Maria-Fernanda Moura, and Alneu de Andrade Lopes. 2017. "A Survey of the Applications of Bayesian Networks in Agriculture." *Engineering Applications of Artificial Intelligence* 65: 29–42. https://doi.org/10.1016/j.engappai.2017.07.003.

Eriksson, David, Michael Pearce, Jacob R. Gardner, Ryan Turner, and Matthias Poloczek. 2019. "Scalable Global Optimization via Local Bayesian Optimization." *Advances in Neural Information Processing Systems (NeurIPS)* 32.

Frazier, Peter I., Warren B. Powell, and Savas Dayanik. 2008. "A Knowledge-Gradient Policy for Sequential Information Collection." *SIAM Journal on Optimization* 47 (5): 2410–2439.

Furrer, Reinhard, Marc G. Genton, and Douglas Nychka. 2006. "Covariance Tapering for Interpolation of Large Spatial Datasets." *Journal of Computational and Graphical Statistics* 15 (3): 502–523. https://doi.org/10.1198/106186006X132178.

GPy. 2012. "GPy: A Gaussian Process Framework in Python." http://github.com/SheffieldML/GPy.

Griffin, David, Iain Bate, and Robert I. Davis. 2020, December. "Generating Utilization Vectors for the Systematic Evaluation of Schedulability Tests." *Proceedings – Real-Time Systems Symposium*, 76–88. https://doi.org/10.1109/RTSS49844.2020.00018.

Grinsztajn, Léo, Edouard Oyallon, Gaël Varoquaux, Léo Grinsztajn, Edouard Oyallon, Gaël Varoquaux, and Edouard Oyallon. 2022. "Why Do Tree-Based Models Still Outperform Deep Learning on Tabular Data?" *36th Conference on Neural Information Processing Systems (NeurIPS 2022) Track on Datasets and Benchmarks*.

Hansen, Charles M. 1967. "The Three Dimensional Solubility Parameter and Solvent Diffusion Coefficient. Their Importance in Surface Coating Formulation." *Journal of Paint Technology* 104.

Hellan, Sigrid Passano, Christopher G. Lucas, and Nigel H. Goddard. 2023. "Bayesian Optimisation Against Climate Change: Applications and Benchmarks." *ArXiv*.

Hofmann, Thomas, Bernhard Schölkopf, and Alexander J. Smola. 2008. "Kernel Methods in Machine Learning." *Annals of Statistics* 36 (3): 1171–1220. https://doi.org/10.1214/009053607000000677.

Intellegens. 2022. "Additive Manufacturing." https://intellegens.com/applications/materials/additive-manufacturing/.

Jacquier, Eric, and Nicholas Polson. 2012, September. "Bayesian Methods in Finance." In *The Oxford Handbook of Bayesian Econometrics*. https://doi.org/10.1093/oxfordhb/9780199559084.013.0010.

Kottke, Daniel, Marek Herde, Tuan Pham Minh, Alexander Benx, Pascal Mergard, Atal Roghman, Christoph Sandrock, and Bernhard Sick. 2021. "scikit-activeml: A Library and Toolbox for Active Learning Algorithms." https://www.preprints.org/manuscript/202103.0194/v1.

Lei, Bowen, Tanner Quinn Kirk, Anirban Bhattacharya, Debdeep Pati, Xiaoning Qian, Raymundo Arroyave, and Bani K. Mallick. 2021. "Bayesian Optimization with Adaptive Surrogate Models for Automated Experimental Design." *npj Computational Materials* 7 (1): 1–12. https://doi.org/10.1038/s41524-021-00662-x.

Liang, Qiaohao, Aldair E. Gongora, Zekun Ren, Armi Tiihonen, Zhe Liu, Shijing Sun, James R. Deneault, et al. 2021. "Benchmarking the Performance of Bayesian Optimization across Multiple Experimental Materials Science Domains." *NPJ Computational Materials* 7 (1). https://doi.org/10.1038/s41524-021-00656-9.

Maaten, L.J.P. Van Der, E.O. Postma, and H.J. Van Den Herik. 2009. "Dimensionality Reduction: A Comparative Review." *Journal of Machine Learning Research* 10: 1–41. https://doi.org/10.1080/13506280444000102.

McInnes, Leland, John Healy, and James Melville. 2020. "UMAP: Uniform Manifold Approximation and Projection for Dimension Reduction." *arXiv*:1802.03426v3.

Mnih, Volodymyr, Koray Kavukcuoglu, David Silver, Andrei A. Rusu, Joel Veness, Marc G. Bellemare, Alex Graves, et al. 2015. "Human-Level Control through Deep Reinforcement Learning." *Nature* 518 (7540): 529–33. https://doi.org/10.1038/nature14236.

Nguyen, Vu. 2019. "Bayesian Optimization for Accelerating Hyper-Parameter Tuning." *2019 IEEE Second International Conference on Artificial Intelligence and Knowledge Engineering (AIKE)*, 302–305.

Nguyen-Tuong, Duy, Matthias Seeger, and Jan Peters. 2008. "Local Gaussian Process Regression for Real Time Online Model Learning." *Nips*, 1193–1200.

NIST/SEMATECH e-Handbook of Statistical Methods. 2023. "D-Optimal Designs." In *e-Handbook of Statistical Methods*. https://doi.org/10.18434/M32189, https://www.itl.nist.gov/div898/handbook/pri/section5/pri521.htm.

Oehlert, Gary W. 2000. *A First Course in Design and Analysis of Experiments*. W.H. Freeman.

Pedregosa, F., G. Varoquaux, A. Gramfort, V. Michel, B. Thirion, O. Grisel, M. Blondel, P. Prettenhofer, R. Weiss, V. Dubourg, J. Vanderplas, A. Passos, D. Cournapeau, M. Brucher, M. Perrot, and E. Duchesnay. 2011. "Scikit-Learn: Machine Learning in Python." *Journal of Machine Learning Research* 12: 2825–2830.

Phillips, Jeff M., and Suresh Venkatasubramanian. 2011. "A Gentle Introduction to the Kernel Distance." 1–9. *arXiv*:1103.1625v2.

Plackett, R.L., and J.P. Burman. 1946. "The Design of Optimum Multifactorial Experiments." *Biometrika* 33 (4): 305–325. https://doi.org/10.1093/BIOMET/33.4.305.

Quinlan, J.R. 1992. "Learning with Continuous Classes." *Australian Joint Conference on Artificial Intelligence* 92: 343–348.

Reddi, Sashank J., Satyen Kale, and Sanjiv Kumar. 2018. "On the Convergence of Adam and Beyond." *6th International Conference on Learning Representations, ICLR 2018 – Conference Track Proceedings*, 1–23.

Sandha, Sandeep Singh, Mohit Aggarwal, Igor Fedorov, and Mani Srivastava. 2020, May. "Mango: A Python Library for Parallel Hyperparameter Tuning." *ICASSP, IEEE International Conference on Acoustics, Speech and Signal Processing – Proceedings*, 3987–3991. https://doi.org/10.1109/ICASSP40776.2020.9054609.

Schölkopf, Bernhard, Alexander Smola, and Klaus-Robert Müller. 1998. "Nonlinear Component Analysis as a Kernel Eigenvalue Problem." *Neural Computation* 10 (5): 1299–1319. https://doi.org/10.1162/089976698300017467.

Shahriari, Bobak, Kevin Swersky, Ziyu Wang, Ryan P. Adams, and Nando De Freitas. 2016. "Taking the Human Out of the Loop: A Review of Bayesian Optimization." *Proceedings of the IEEE* 104 (1): 148–175. https://doi.org/10.1109/JPROC.2015.2494218.

Silver, David, Aja Huang, Chris J. Maddison, Arthur Guez, Laurent Sifre, George van den Driessche, Julian Schrittwieser, et al. 2016. "Mastering the Game of Go with Deep Neural Networks and Tree Search." *Nature* 529 (7587): 484–489. https://doi.org/10.1038/nature16961.

Singh, Ajit P., and Geoffrey J. Gordon. 2008. "Relational Learning via Collective Matrix Factorization." *Proceedings of the ACM SIGKDD International Conference on Knowledge Discovery and Data Mining*, 650–658. https://doi.org/10.1145/1401890.1401969.

Srivastava, Nitish, Geoffrey Hinton, Alex Krizhevsky, Ilya Sutskever, and Ruslan Salakhutdinov. 2014. "Dropout: A Simple Way to Prevent Neural Networks from Overfitting." *Journal of Machine Learning Research* 15: 1929–1958.

Taguchi, Genichi, and Shih Chung Tsai. 1995. "Quality Engineering (Taguchi Methods) for the Development of Electronic Circuit Technology." *IEEE Transactions on Reliability* 44 (2): 225–229. https://doi.org/10.1109/24.387375.

Takahashi, Ami, and Taiji Suzuki. 2021. "Bayesian Optimization Design for Dose-Finding Based on Toxicity and Efficacy Outcomes in Phase I/II Clinical Trials." *Pharmaceutical Statistics* 20 (3): 422–439. https://doi.org/10.1002/pst.2085.

Thompson, William R. 1933. "On the Likelihood That One Unknown Probability Exceeds Another in View of the Evidence of Two Samples." *Biometrika* 25 (3–4): 285. https://doi.org/10.2307/2332286.

Tillmann, Peter, Klaus Jäger, and Christiane Becker. 2019. "Minimising the Levelised Cost of Electricity for Bifacial Solar Panel Arrays Using Bayesian Optimisation." *Sustainable Energy and Fuels* 4 (1): 254–264. https://doi.org/10.1039/c9se00750d.

Volpp, Michael, Lukas P. Fröhlich, Kirsten Fischer, Andreas Doerr, Stefan Falkner, Frank Hutter, and Christian Daniel. 2020. "Meta-Learning Acquisition Functions for Transfer Learning in Bayesian Optimization." *arXiv*:1904.02642v5.

Wang, Ke, and Alexander W. Dowling. 2022. "Bayesian Optimization for Chemical Products and Functional Materials." *Current Opinion in Chemical Engineering* 36 (c).

Whitehead, T.M. 2021. "Hyperparameter-Free Regularization by Sampling from an Infinite Space of Neural Networks." *International Journal on Artificial Intelligence Tools* 30 (2): 2150008. https://doi.org/10.1142/s0218213021500081.

Whitehead, T.M., Flora Chen, Christopher Daly, and Gareth J. Conduit. 2022. "Accelerating the Design of Automotive Catalyst Products Using Machine Learning." *Johnson Matthey Technology Review* 66 (2): 130–136. https://doi.org/10.1595/205651322x16270488736796.

Whitehead, T.M., B.W.J. Irwin, P. Hunt, M.D. Segall, and G.J. Conduit. 2019. "Imputation of Assay Bioactivity Data Using Deep Learning." *Journal of Chemical Information and Modeling* 59 (3): 1197–1204. https://doi.org/10.1021/acs.jcim.8b00768.

Zhang, Lei, Haitao Mao, Qilei Liu, and Rafiqul Gani. 2019, November/2020. "Chemical Product Design – Recent Advances and Perspectives." *Current Opinion in Chemical Engineering* 27: 22–34. https://doi.org/10.1016/j.coche.2019.10.005.
Zhang, Mimi, Andrew Parnell, Dermot Brabazon, and Alessio Benavoli. 2021. "Bayesian Optimisation for Sequential Experimental Design with Applications in Additive Manufacturing." *ArXiv*.
Zhu, Ling, Yan Wang, Anuj Pal, and Guoming G. Zhu. 2022. "Adaptive Design of Experiments for Automotive Engine Applications Using Concurrent Bayesian Optimization." *ASME Letters in Dynamic Systems and Control* 2 (3). https://doi.org/10.1115/1.4054222.

12 Inverse Design via PLS Model Inversion

Daniel Palací-López, Joan Borràs-Ferrís, Pierantonio Facco, Massimiliano Barolo and Alberto Ferrer

12.1 INTRODUCTION

12.1.1 The Link between Inverse Design, Product Formulation, and Optimization

Inverse design can be understood as the identification of the ingredients, ingredient ratios, ingredient property specifications, and/or process conditions required to guarantee the desired specifications on the *Critical to Quality Attributes* (*CQAs*) of the resulting product. In other words, it requires exploiting previously acquired knowledge with regard to how the raw materials and process conditions behave and interact with each other and with the CQAs (i.e., the predictive model developed using the techniques discussed in the previous section), for the purpose of delivering products of the desired quality. Therefore, inverse design constitutes an integral part of product formulation.

Intuitively, inverse design could also be seen as a methodology that, rather than predicting the outputs of a process from its inputs, attempts to do the opposite. However, notwithstanding the issues associated with trying to do so in such a way, which will be addressed later on,

> the end goal of inverse design is not limited to just guaranteeing a final product whose characteristics fall within previously defined specifications. Rather, the aim is to achieve this optimally.

But what does that mean, exactly?

Within the framework of inverse design for product formulation, optimization can be carried out through two complementary approaches:

- Given a product or a process, for which specific characteristics or outputs are desired, define the subspace of all possible combinations of process conditions and ingredients ratios and their properties that guarantee the desired CQAs or process outputs. Section 12.2 will focus on the basis of this approach.

DOI: 10.1201/9781003385974-16

- Given a series of constraints on the process conditions under which an operation can be carried out, and availability of raw materials, find the combination of them all that provides the best CQAs for the end product, or the best process outputs. This will be the main topic addressed in Section 12.3.

While the first of these two approaches may not appear to be an example of optimization, it can be seen as such when it is formulated in a slightly different way. Specifically, to define the combination – out of all possible combinations of process conditions, ingredient ratios, and their properties – that minimizes the risk of obtaining unacceptable product characteristics or process outputs. This may, however, not be good enough in some cases, since no unique solution would be arrived upon, given the specific circumstances surrounding the production process at a given point in time.

The second approach presents the advantage of providing such a solution without ambiguity, while meeting any constraints, whenever possible. However, this also means that any changes in the constraints imposed or the optimality criterion considered will require solving the optimization problem again, if not reformulating it altogether. Depending on the models the optimization is based upon, and the optimization algorithm itself, this could be quite time-consuming. Therefore, this is not optimal in environments such as Industry 4.0, where conditions are expected to vary frequently and the capability to adapt to them must be high, despite the large amounts of data being continuously generated and regardless of the complexity of the models built around them.

Therefore, inverse design for product formulation will require tackling the optimization problem in an optimal way, combining the advantages of these two approaches while minimizing their shortcomings.

12.1.2 Inverse Design, Quality by Design, and the Design Space

In its simplest form, inverse design is just another way of using the available expertise and resources to plan ahead of time in order to achieve a desired goal before even starting. However, when this approach is scaled to a higher level of complexity, some standardization is required.

In line with this, inverse design is very closely tied to the *Quality-by-Design* (*QbD*) (FDA 2004) initiative, which promotes implementing science-based methodologies to increase the flexibility and robustness of production processes, that is, their capability to tolerate changes in their inputs (process conditions and raw materials) without negatively affecting the outputs (i.e., guarantying acceptable products). The design space[1] is a key concept in this context, as it refers to "the multidimensional combination and interaction of input variables (e.g., material attributes) and process parameters that have been demonstrated to provide assurance of quality" (ICH 2009).

Applying the QbD philosophy and defining the design space require a proper understanding of the corresponding process and materials involved. This is because, as opposed to an environment where a model would be used for, for example, calibration or monitoring purposes, the QbD is meant to be applied in an active environment, where the desired outputs are known, and the way to achieve them has to be defined, be it for control, optimization, or product formulation.

Consider, as an example, a polynomial model representing a hypothetical input-output causal relationship between the output (y) and five apparent inputs (x_1 to x_5), where three of the apparently independent inputs (x_3 to x_5) are actually just the squared values of the first two inputs (x_1 and x_2) and their interaction. Figure 12.1a shows the response surface plot for the real, underlying mathematical model relating the output y and the inputs x_1 and x_2, and Figure 12.1b illustrates the design space for a desired output, $y = y_{DES}$ (continuous white line), as well as for a range of acceptable values of the output, $y_{min} < y < y_{max}$ (subspace delimited by discontinuous white lines), obtained after inverting such mathematical model.

Although simple, this example should serve to illustrate some important points to address with regard to model inversion and the definition of the design space:

- For this example, the actual existing relationships between the different input variables and the output variable are known. Excluding instances where first principles models are known to be enough, this will usually not be the case. Therefore, the uncertainty in the model should generally be accounted for.
- No "noise" was considered in this example, and therefore a perfectly smooth response surface and design space limits were obtained. Again, uncertainty in the data due to, for example, measurement errors should also be taken into account in practice.
- The design space in this example, which only considers five input variables and one output variable, is not a convex space; that is, if any two points within this design space are connected by a segment, there is no guarantee that all points belonging to this segment will also belong to the design space. Furthermore, the design space cannot be properly restricted by simply defining univariate specifications on the input variables.

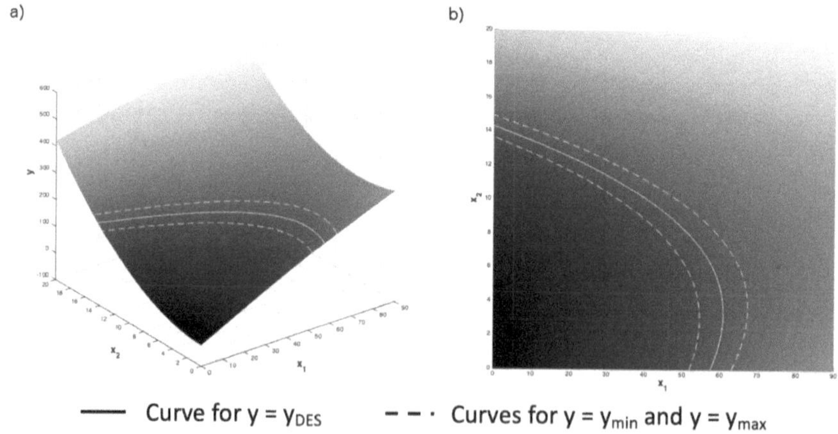

FIGURE 12.1 a) Response surface plot for a second-degree polynomial model representing a hypothetical input-output causal relationship between an output variable (y) and two input variables (x_1 and x_2), and b) design space containing all combinations of $\{x_1, x_2\}$ leading to $y = y_{DES}$ (continuous white line), and any combinations of $\{x_1, x_2\}$ leading to $y_{min} < y < y_{max}$ (subspace delimited by discontinuous white lines).

To further and more intuitively illustrate the problems associated with this last point, consider a more general example to illustrate the differences between the knowledge space, defined as the subspace comprised by the set of historical products that have already been developed in the past (MacGregor and Bruwer 2008; Jaeckle and Macgregor 1996), the design space and the so-called *Normal Operating Conditions* (*NOC*), which encompasses the set of operating conditions, raw materials, and raw material properties that respect the manufacturing process instructions. In this example, only two input variables (x_1 and x_2) and two output variables (y_1 and y_2) are to be accounted for. The graphical representation of such scenarios can be seen in Figure 12.2.

As expected, the NOC represent only a fraction of the knowledge space, which is within the univariate historical limits but cannot be defined simply by these limits as univariate constraints. Furthermore, a small portion of the product obtained while working in NOC will not be within the "acceptable product space." Lastly, the "acceptable product space" in the X space (i.e., the design space) encompasses a larger area of process conditions outside of the NOC. Consequently, properly identifying the design space would, in this scenario, increase the flexibility and robustness of the process, allowing for a wider range of operating conditions to be run while producing within specifications and preventing the manufacturing of out-of-specification product. However, this would require the definition of multivariate constraints on the input variables, as opposed to the more usual univariate constraints imposed in many production processes in practice. Further emphasis on this will be made in the practical example illustrated in Chapter 14.3 – "Setting Multivariate Raw Material Specifications."

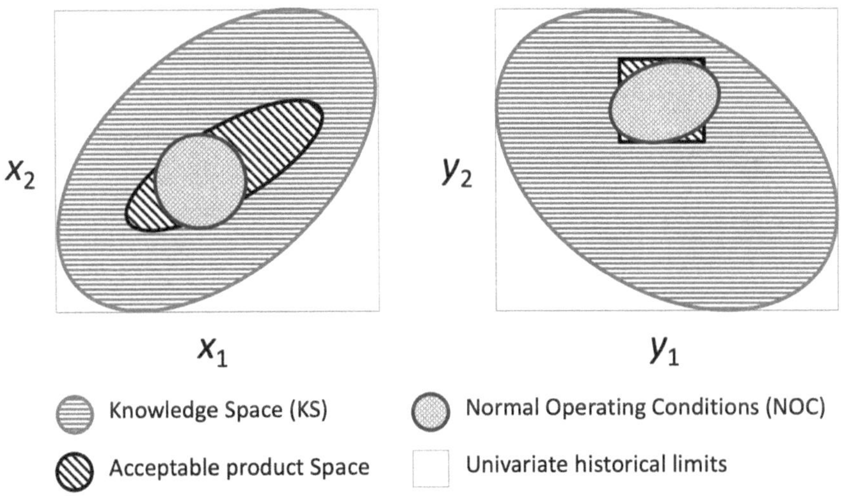

FIGURE 12.2 Knowledge space (KS), Normal Operating Conditions (NOC), "acceptable product space" (equivalent in the Y space to the design space in the X space), and univariate historical limits in the inputs (X) and outputs (Y) spaces for a hypothetical process with two input and output variables involved.

12.1.3 The Importance of Causality in Model Inversion

Despite the large volume of "happenstance" data being continuously generated in the Industry 4.0 environment, the variation in the inputs is usually not independent, that is, data are not obtained from a Design of Experiments (DOE) that would guarantee such independent variation. This is of great relevance for model inversion, since causality is required for such inversion to be meaningful. Therefore,

> when using data-driven approaches, independent variation in the input variables is required to guarantee causality (Box et al. 2005).

Given this, caution must be exercised when classical linear regression and even some machine learning methods are to be used for product formulation involving model inversion with "happenstance" data involving not just ingredient rates and their properties but also process conditions. This is because, especially for process variables, the number of independent variations is usually much smaller than that of measured variables, many of which tend to be highly correlated.

In any case, models fitted using these methods can still be resorted to when independent variation in the inputs can be guaranteed for the data used to build such models (e.g., by means of a DOE in the original space) or by explicitly accounting for the existing correlations, if and when they are known (e.g., by imposing such relationships as additional constraints in the optimization problem or during the model inversion). Nonetheless, even then, or if first principles models are available, most optimization algorithms may suffer from prohibitive computational cost and present convergence issues due to the large number of variables involved in the optimization, and the presumably also large number of constraints imposed, when working in the space of the original variables.

In order to overcome these difficulties,

> methods based on projection onto latent structures, such as *partial least squares (PLS)* regression-based techniques, allow analyzing large datasets with highly correlated data by relating inputs and outputs through a smaller number of uncorrelated latent variables.

These latent variables permit exploiting the underlying causal relationship between X and Y variables and provide models for both spaces of X and Y. And, since causal relationships can be inferred in the latent space (Liu et al. 2011), historical datasets can then be used for model inversion and optimization, significantly reducing or even eliminating the need for experimentation (MacGregor et al. 2015).

As with any data-driven approach, PLS-regression-based techniques present several advantages but also suffer from a series of limitations that must be accounted for. The aim of the following section(s) is to present how to make the best out of such advantages, without failing to address the accompanying limitations.

12.2 DEFINING THE DESIGN SPACE IN THE LATENT SPACE

12.2.1 The Utility of Partial Least Squares-Based Approaches for Model Inversion and Optimization

As many other methods based on projection onto latent structures, PLS regression-based techniques can be used to exploit large datasets with highly correlated data by relating the inputs (X) and outputs (Y) through the so-called latent variables. Because these latent variables are uncorrelated to each other, and they are fewer in number than the original ones (X and Y), PLS-based model inversion and optimization will present several advantages:

- The capability of inferring causality in the latent space allows exploiting large historical datasets with minimal requirement for experimentation.
- Because the number of variables involved (in X and Y) has been reduced to a smaller one (that of latent variables), the computational cost of any optimization problem in the latent space will be significantly reduced when compared to that of the equivalent optimization problem in the space of the original variables.
- Because the LV are uncorrelated to each other but also serve to model both the X and Y spaces, and the correlation between them, any acceptable range of values on the outputs Y can be immediately transferred to the latent space, where a design space in the latent space can be estimated and backpropagated to the inputs X, while respecting any existing correlations between the original variables.

The latest point is especially relevant, since it guarantees that, while operating within the knowledge space, any combination of process inputs (i.e., process conditions, ingredients rates, and their properties) inside such estimated design space can be executed in practice (i.e., it is a feasible solution to the optimization problem), and achieves, at least theoretically, the stated goal of the QbD initiative and, by extension, that of product formulation.

Any effort to define the design space will be pointless, however, if it does not exist to begin with.

This may happen in the following scenarios:

- At least one of the output specifications cannot be met in conjunction with the rest, independently of the inputs. This may be solved by relaxing one or more of such specifications, if possible.
- All the specifications can be met in theory but not within the knowledge space. Since the design space is assumed to be a subspace of the knowledge space, the only way to mitigate this is to expand the later by making changes in the process toward new operating conditions and/or the use of new raw materials, if one is willing and capable of doing so. More information on using the knowledge space to guide experimentation toward the preferred quality region can be found in Chapter 11.

Lastly, it is important to consider that, as with any data-driven approach, PLS regression-based techniques will be affected by the uncertainty inherent to the data and the model, which will be back-propagated during the model inversion, and shall therefore be accounted for. In line with this, the term design space will be used for the remaining of this chapter to refer to the estimation of the true design space through model inversion.

12.2.2 Direct Inversion-Dependent Definitions of the Design Space

Assuming the existence of at least one combination of inputs leading to the desired outputs, the direct inversion can be performed to obtain an estimation of such a set of input values. If more than one combination exists, some approaches based on Latent Variable Regression Model Inversion (LVRMI) have also been proposed to define the *Null Space* (*NS*), that is the projection onto the latent space of all combinations of inputs that theoretically guarantee the desired outputs (i.e., variations in the outputs when moving through the NS are, in fact, null according to the model).

One such method was initially proposed by Jaeckle and MacGregor (2000), and later expanded upon by García-Muñoz et al (2006). While the former required that the NS satisfies the specifications for all outputs simultaneously, the later identified such subspace as the "combined pseudo-NS," and instead defined the NS (associated to a given output) as the subspace that satisfies the specifications of an individual output but not necessarily the rest. By identifying the null components (i.e., directions of null variability in one or all outputs, depending on how the NS is defined), it is then possible to find different points along the NS that guarantee the desired output or outputs, according to the model. Alternatively, different optimization problems, with different weights given to each term of the objective function, can also be solved to obtain such combinations of inputs (Tomba et al. 2012).

These approaches present some drawbacks for the purpose of product formulation, though:

- They do not explicitly account for the uncertainty in the data when defining the NS nor when solving the optimization problem, although this has been addressed by Zhang and Garcia-Munoz (2009) and Pantelides et al. (2009). In response, direct inversion-based methods to bracket the design space by considering this uncertainty have also been suggested (Facco et al. 2015; Bano et al. 2017), in order to define a subspace of the knowledge space within which the design space is expected to lay.
- They provide no analytical expression[2] of the NS and therefore do not allow using it, nor the limits of the associated confidence region, as constraints in, for example, the optimization approach formulated by Tomba et al. (2012). Consequently, since no equality constraints associated with each output specification would be imposed, a combination of inputs that exactly leads to all desired outputs may exist and still not be obtained as a solution to the optimization problem, depending on the weights given to each term in the objective function.

Despite these shortcomings, such approaches provide a solid basis for overcoming their own limitations, as will be shown in the next section.

12.2.3 Analytical Definition of the Null Space and Bracketing of the Design Space

In its most restricted form, given a series of specifications on the CQAs of the outputs and constraints in the process conditions and ingredients involved in a process, the design space is a subregion of the knowledge space that satisfies such specifications without violating any of the constraints. Furthermore, when accounting for the uncertainty inherent to the use of data-based approaches, two scenarios may be of interest in the estimation of the design space:

- If the goal is to maximize the flexibility of a process or range of acceptable ingredients and their properties, defining the subspace most likely to contain the true design space provides a way to make sure that few combinations of process conditions and ingredients leading to the desired outputs would be left out, at the cost of considering additional combinations that do not belong to the true design space.
- If the goal is to minimize the proportion of unacceptable product, then the opposite is true, and defining the subspace least likely to fall outside of the true design space will be better, at the cost of potentially reducing the flexibility of the process and/or the range of acceptable ingredients and their properties.

Although the first scenario is clearly tied to product discovery, the second one is also particularly relevant in the context of product formulation, as it is the conceptual basis for defining multivariate raw material specifications, as will be illustrated in Chapter 14.3 – "Setting Multivariate Raw Material Specifications."

Nonetheless, in both scenarios, providing analytical expressions for the constraints that delimit the corresponding subspace is essential. Similarly, these expressions should be meaningful and not trivial, that is, they should be made in a way that allows comparing and combining constraints imposed on either the inputs or the outputs with each other. Therefore, for PLS model inversion, they can be defined in the latent space.

Thus, as discussed by Palací-López et al. (2019), the proposed approach for the analytical definition of the DS will require:

- Transferring constraints to the latent space
- Providing an analytical expression for the NS for the specifications on each CQA
- Providing an analytical expression for the confidence regions for such NSs

Transferring Constraints to the Latent Space

In order to formulate constraints originally imposed on the X or Y spaces, as constraints on the latent variables or the PLS regression model, the model structure must be considered. As illustrated in Chapter 8, the relationship between the X and Y variables through the latent space can be formulated as:

$$\mathbf{X} = \mathbf{T} \cdot \mathbf{P}^{\mathrm{T}}$$
$$\mathbf{Y} = \mathbf{T} \cdot \mathbf{Q}^{\mathrm{T}}$$
$$\mathbf{T} = \mathbf{X} \cdot \mathbf{W} \cdot \left(\mathbf{P}^{\mathrm{T}} \cdot \mathbf{W}\right)^{-1} = \mathbf{X} \cdot \mathbf{W}^{*}$$

where both the **X** and **Y** data matrices have already been pre-processed (e.g., centering the data by removing the average value of each variable from all observations, and then scaling all variables to unit variance).

For an individual observation (row vector $\mathbf{x}^{\mathrm{T}}$) in the X space:

$$\mathbf{x} = \mathbf{P} \cdot \boldsymbol{\tau}$$

where $\boldsymbol{\tau}$ is the vector of scores associated with $\mathbf{x}$.

Then, for the most general, linear case, consider a series of constraints imposed on different linear combinations of inputs, expressed as

$$\mathbf{A}_{\mathrm{x}} \cdot \mathbf{x} \leq \mathbf{d}_{\mathrm{x}}$$

with $\mathbf{A}_{\mathrm{x}}$ being a $[I \times M]$ matrix whose i-th row contains the coefficients of the i-th linear combination of inputs, and $\mathbf{d}_{\mathrm{x}}$ the $[I \times 1]$ column vector whose i-th element is the value that must not be surpassed by the corresponding linear combination of inputs.

Then, the constraints on the inputs (X variables) in the original space can be transferred to the latent space as:

$$\mathbf{A}_{\tau} \cdot \boldsymbol{\tau} = \mathbf{A}_{\mathrm{x}} \cdot \mathbf{P} \cdot \boldsymbol{\tau} \leq \mathbf{d}_{\tau} = \mathbf{d}_{\mathrm{x}}$$

Note that the same procedure can be applied to constraints on the output variables, and with equality constraints. Furthermore, it is also important to consider that, in this example, the input variables are assumed to have already been pretreated, that is, centered and scaled to unit variance and that this is also implicitly accounted for in the formulation of the constraints of the linear combinations of input variables as presented earlier. If not, such pretreatment should be explicitly considered, making the previous transferal of constraints onto the latent space, and its impact for optimization, slightly less trivial, as discussed by Palací-López et al. (2019, 2020).

Transferring more complex constraints from the original to the latent space, however, requires a more thorough and customized approach. Furthermore, when such a situation arises

> it is important to consider if such constraints could be linearized by either transforming the original variables into new ones prior even to fitting the PLS-regression model, or by restricting the scope of the study to a smaller subspace where an approximation of linearity becomes reasonable.

In any case, the complexity of such a scenario would exceed the scope of this chapter and will not be addressed here.

Providing an Analytical Expression for Each NS

Assuming specifications to be of, at most, the same complexity as the constraints in the previous subsections, the analytical expression of the NS associated to a specific value for a given CQA can be obtained very similarly to that of an equality constraint.

For the purpose of greater generality, consider R CQAs defined as R different linear combinations of the L output variables, and the r-th CQA to be defined as $\mathbf{d}_r = \mathbf{a}_r^{\mathrm{T}} \cdot \mathbf{y}$, with $\mathbf{a}^{\mathrm{r}}$ being the $[L \times 1]$ vector of coefficients that relate the L output variables with the r-th CQA. Then, given a desired value $d_{DES,r}$ for the r-th CQA, a vector of outputs $\mathbf{y}_{\mathrm{NS}_r}$, associated to a set of scores $\boldsymbol{\tau}_{\mathrm{NS}_r}$ on the r-th NS, can be found such that $\mathbf{a}_r^{\mathrm{T}} \cdot \mathbf{y}_{\mathrm{NS}_r} = d_{DES,r}$.

Given that, following the PLS-regression model structure:

$$\mathbf{y}_{\mathrm{NS}_r} = \mathbf{Q} \cdot \boldsymbol{\tau}_{\mathrm{NS}_r}$$

then:

$$\mathbf{a}_r^{\mathrm{T}} \cdot \mathbf{Q} \cdot \boldsymbol{\tau}_{\mathrm{NS}_r} = d_{DES,r}$$

or, when applying this to R CQAs:

$$\mathbf{v}_0 + \mathbf{V} \cdot \boldsymbol{\tau}_{\mathrm{NS}} = 0$$

$$\mathbf{v}_0 = -\mathbf{d}_{\mathrm{DES}} \quad ; \quad \mathbf{V} = \mathbf{A}_{\mathrm{r}} \cdot \mathbf{Q}$$

The formulation of these equations is different to that of the constraints in the previous subsections for two main reasons:

- When presented in this way, this equality represents a system of R equations with L unknowns, similar to the one in the methodology of García-Muñoz et al. (2006). Such a system of equations can serve the purpose of evaluating whether a solution exists or not that satisfies the desired values for all CQAs, assuming specific values are desired for all of them. If it does, then the intersection of the R NSs will be an (A-R)-dimensional subspace, where A is the number of LVs selected when fitting the model. Furthermore, if the CQAs are equal to the L output variables, this subspace is also coincident with the "combined pseudo-NS" that would be obtained in equal circumstances by following the approach presented by Jaeckle and MacGregor (2000), and if $A>R$.
- This presentation allows visualizing the different vectors $\mathbf{v}_{\mathrm{r}}^{\mathrm{T}}$ (r-th row of matrix $\mathbf{V}$), each of them corresponding to the direction of maximum variability in the latent space for the corresponding r-th CQA. This is important because they indicate that moving by one unit in the same direction as $\mathbf{v}_r$, in the latent space, will produce an increase in the expected value of the r-th CQA equal to the module of $\mathbf{v}_r$. This information can be very useful for optimization purposes, whenever such CQA is to be minimized or maximized.

Again, consider that the previous formulation assumes that the output variables Y have already been centered and scaled to unit variance. Otherwise, the expressions will vary accordingly.

Providing an Analytical Expression for the Confidence Region of Each NS

The definition of this confidence region is not trivial. Tomba et al. (Tomba et al. 2012) proposed a jackknife approach to determine the confidence regions for both the direct inversion and the NS estimation, and Bano et al. (Bano et al. 2018) developed a Bayesian-based procedure for the probabilistic definition of the DS in the latent space, but the computational cost of these approaches is high and increases quickly as the dataset size and/or the dimensionality of the latent space increase. Alternatively, the methodology presented by Facco et al. (Facco et al. 2015) accounts for the back-propagation of the uncertainty in model inversion to bracket the design space, which can be used whenever a single output is considered. None of these approaches, however, provide an analytical expression as understood in this chapter.

To achieve this goal, Palací-López et al. (2019) assume OLS-type expressions for the prediction intervals to account for the model's prediction uncertainty, which is then backpropagated to obtain the confidence region of each NS. It is important to note that this approximation is done as suggested by Faber and Kowalski (Faber and Kowalski 1996), assuming that the matrix of scores **T** obtained from fitting the PLS model is independent from **Y**, which is clearly not true (Burnham et al. 2001). Nonetheless, results obtained in this way are roughly as good as those achieved by, for example, linearization or re-sampling-based methods, or the so-called U-deviation method used in the Unscrambler, while being easier to implement (Zhang and Garcia-Munoz 2009) and allowing to obtain this analytical expression.

An example of the results achieved by considering the uncertainty in the estimation of the confidence region of the NS, as proposed by Palací-López et al. (2019), compared to the approach by Facco et al. (2015), can be seen in Figure 12.3.

An additional advantage of this approach is that it can also be applied to the constraints on the original variables transferred onto the latent space, as illustrated in Figure 12.4. Figure 12.4a highlights an apparent contradiction: even though the constraints projected onto the latent space in this example correspond to the univariate "historical limits" on the dataset (i.e., minimum and maximum values for each original variable), and therefore the projection of all observations should be expected to fall within the projection of such constraints, this is not the case. Accounting for the uncertainty in the data and model obtained, as done in Figure 12.4b, mitigates this problem. A similar effect was observed in Figure 12.3, where the shape of the confidence region for the NS changes significantly once the uncertainty in its estimation is more rigorously accounted for, even considering the approximations made.

Bracketing the Design Space versus Guaranteeing Quality

Despite the importance of bracketing the DS by considering the constraints imposed by the limits of the KS, constraints on inputs and outputs, and the confidence region

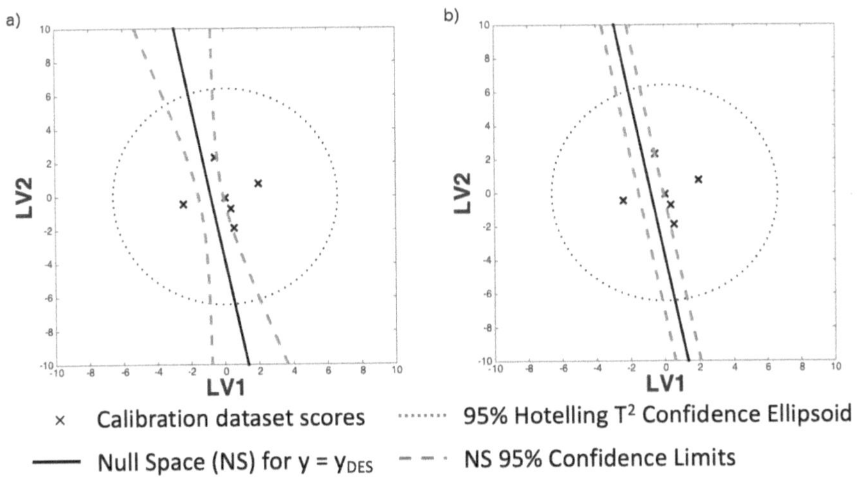

FIGURE 12.3 Delimitation of the 95% confidence region for the NS (dashed gray lines) with a) the approach by Palací-López et al. (2019) and b) the approach by Facco et al. (2015).

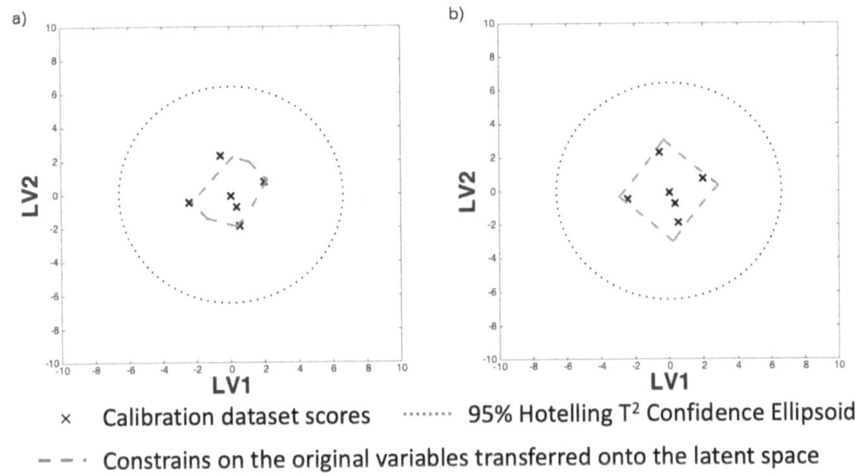

FIGURE 12.4 Representation of the projection of a dataset onto the subspace of a PLS-regression model fitted with 2 LV (black "x") together with a) univariate historical limits on the original variables, as constraints imposed on them, transferred to the same latent space (dashed gray lines) and b) 95% confidence limits for the same constraints transferred to the same latent space (dashed gray lines).

for the NS for each CQA, the relevance of doing so lies mostly in the advantage obtained by not discarding any combination of inputs that may, with a reasonable probability, lead to the desired values for the CQAs.

It is then important to mention that this advantage becomes a disadvantage when the goal is to guarantee the quality of a product and therefore to discard any combination of inputs that may, with significant probability, lead to undesired values for the CQAs. This second scenario will be addressed in more detail in Chapter 14.3, and so will not be further elaborated upon here.

12.3 OPTIMIZATION PROBLEM FORMULATION FOR INVERSE DESIGN

12.3.1 Optimization in the Original Space versus Optimization in the Latent Space

Although the merits of optimization in the latent space have already been highlighted, it is important to not dismiss potential advantages of the optimization in the original space, whenever it can be carried out.

Optimization in the latent space, for example, restricts the range of potential solutions to those that abide exactly by the correlation structure of the data used to fit the LV regression model (LVRM), as pointed by Lakshminarayanan et al. (2000). Any solution to the optimization problem formulated this way will have an SPE (as defined in Chapter 8 – "Modeling of Product Formulations Using a Latent Variable Approach") equal to zero.

Furthermore, while the reduction in the dimensionality of the space within which the optimization is carried out can significantly reduce the computational cost of carrying out the optimization, it is precisely this same reduction in dimensionality that may lead to issues in trying to optimize a process for which a large number of constraints (i.e., higher than the number of LVs) are imposed on the solution.

Optimization in the original space, on the other hand, does not suffer from these two drawbacks. Therefore, it may be reasonable, in some cases, to formulate the optimization problem in the space of the original variables, while imposing an upper limit, the SPE_{lim}, on the acceptable Euclidean distance to the latent subspace of the fitted LVRM. This way, any solution will respect to some extent, although not completely abide to, the correlation structure found by such LVRM. Note that the cautionary advice provided in Chapter 8, where it is pointed out that removing too many observations when fitting the model may lead to biased limits, is especially relevant here to properly quantify an appropriate SPE_{lim} value (usually SPE_{α} for a given confidence level 1-α).

On the other hand, as the number of original variables increases and optimization in the original space becomes practically untenable, a two-step optimization process (García-Muñoz et al. 2008) may be carried out, where the optimization problem is solved in the latent space as a first step, and then followed up by an optimization in the original space, to increase the flexibility of achievable solutions. Therefore, optimization in the original and in the latent space ought not to be seen as necessarily incompatible approaches but sometimes even as complementary ones.

Lastly,

it is also advisable to avoid extrapolating beyond the knowledge space.

For optimizing in the space of the original variables, this usually means limiting any acceptable solution to those that fall within historical limits for the input variables and, if known, respecting additional constrains related to first principles. When optimizing in the latent space, this also encompasses the definition of a maximum acceptable value for the solution's Hotelling's T^2 (HT2) as defined in Chapter 8 (again, usually HT_{α}^2 for a given confidence level 1-α).

12.3.2 Important Considerations for Latent Variables-Based Optimization Involving Multiple Variables and Objectives

Tackling the optimization problem for inverse design presents additional challenges to those related to the definition of the design space itself. Among them, two are worth mentioning because of the risks they pose if not properly addressed:

- When CQAs defined as linear combinations of outputs are involved, caution must be exercised, since optimizing a linear combination of outputs is not the same as solving a multi-objective optimization problem.
- In multi-objective optimization problems where at least one variable is to be minimized or maximized, defining appropriate, sensible minimum or maximum values as "objectives" for the corresponding CQAs is not trivial in most cases. This is partly aggravated by the scaling of the original dataset when fitting the LVRM, and its effect on the weight given to each variable in the objective function of the optimization problem formulation.

Optimizing a Linear Combination of Variables

Although, intuitively, it may seem that optimizing a linear combination of variables will provide the same results as solving a multi-objective optimization problem involving such variables and properly defining the "importance" given to each of them, it is important to consider how the optimization problem is generally formulated. Then, such misunderstandings can be easily fixed.

For simplicity's sake, given that the optimization problem may be formulated in many ways, consider in particular the objective function of the optimization problem in the latent space as formulated by Tomba et al. (2012):

$$\left[g_0 \cdot (\hat{\mathbf{y}} - \mathbf{y}_{\mathrm{DES}})^{\mathrm{T}} \cdot \Gamma \cdot (\hat{\mathbf{y}} - \mathbf{y}_{\mathrm{DES}}) + g_1 \cdot \sum_{a=1}^{A} \frac{\tau_a^2}{s_a^2} \right]$$

$$s.t.\, \hat{\mathbf{y}} = \mathbf{Q} \cdot \boldsymbol{\tau};\ \hat{\mathbf{x}} = \mathbf{P} \cdot \boldsymbol{\tau};\ \boldsymbol{\tau}^{\mathrm{T}} \cdot \Lambda^{-1} \cdot \boldsymbol{\tau} \le HT_{\alpha}^2;\ \mathbf{A}_{\tau} \cdot \boldsymbol{\tau} \le \mathbf{d}_{\tau};\ \mathbf{F}_{\tau} \cdot \boldsymbol{\tau} = \mathbf{f}_{\tau}$$

where τ is the vector of scores of the solution, and Γ is a diagonal matrix with as many rows/columns as the number of output variables, whose diagonal is a unitary vector with the relative weights given to achieving the desired value for each output; g_0 and g_1 are the weights given to each term in the objective function; $\hat{\mathbf{x}}$ and $\hat{\mathbf{y}}$ are the vectors of inputs and outputs related to τ through the PLS-regression model, fitted with A latent variables; Λ^{-1} is the inverse of the diagonal matrix of the variances of the scores (s_a^2); and $\mathbf{A}_{\tau}$, $\mathbf{F}_{\tau}$, $\mathbf{d}_{\tau}$ and $\mathbf{f}_{\tau}$ are the matrices and vectors associated to the hard constraints on the latent variables, obtained from the equivalent on the original variables as illustrated in Section 12.2.2.1. Note that the solution is also constrained so its Hotelling's T^2 will not exceed a specific value HT_α^2, given a (1-α) confidence level, but no such constraint is necessary for the SPE because the way the optimization problem is formulated in this case guarantees it to be zero.

A more general formulation of the same optimization problem, extended to consider CQAs expressed as linear combinations of output variables, is provided by Palací-López et al. (2020):

$$\left[g_0 \cdot (\mathbf{v}_0 + \mathbf{V} \cdot \boldsymbol{\tau})^{\mathrm{T}} \cdot \boldsymbol{\Gamma}_R \cdot (\mathbf{v}_0 + \mathbf{V} \cdot \boldsymbol{\tau}) + g_1 \cdot \sum_{a=1}^{A} \frac{\tau_a^2}{s_a^2} \right]$$

$$s.t. \mathbf{v}_0 = -\mathbf{d}_{\mathrm{DES}}; \ \mathbf{V} = \mathbf{A}_{\mathbf{r}} \cdot \mathbf{Q}; \ \hat{\mathbf{y}} = \mathbf{Q} \cdot \boldsymbol{\tau}; \ \hat{\mathbf{x}} = \mathbf{P} \cdot \boldsymbol{\tau}; \ \boldsymbol{\tau}^{\mathrm{T}} \cdot \Lambda^{-1} \cdot \boldsymbol{\tau} \leq HT_\alpha^2;$$

$$\mathbf{A}_\tau \cdot \boldsymbol{\tau} \leq \mathbf{d}_\tau \ ; \mathbf{F}_\tau \cdot \boldsymbol{\tau} = \mathbf{f}_\tau$$

Notice that here $\hat{\mathbf{y}} - \mathbf{y}_{\mathrm{DES}}$ has been substituted by $\mathbf{v}_0 + \mathbf{V} \cdot \boldsymbol{\tau}$, where $\mathbf{v}_0$, $\mathbf{V}$ and $\mathbf{A}_{\mathrm{r}}$ hold the same meaning as in Section 12.2.2.2, and both expressions are therefore equivalent when the CQAs are coincident with the output variables, but the second expression is more general by allowing the inclusion of CQAs that are linear combinations of the output variables.

Since $\boldsymbol{\Gamma}$ and $\boldsymbol{\Gamma}_R$ are diagonal matrices of weightings, by looking at both objective functions it is clear that the first and second formulations can only be equivalent if:

- only one linear combination of outputs has to be optimized,
- all variables in the linear combination are completely orthogonal to each other.

Therefore, holding such assumptions as a general rule would be unwise, and both formulations of the optimization problem should not be considered equivalent in most cases. An example of the solution achieved with both formulations of the optimization problem for the maximization of a CQA $d = \mathbf{a}^{\mathrm{T}} \cdot \mathbf{y}$ can be seen in Figure 12.5. In it, it is clear that optimization $d = \mathbf{a}^{\mathrm{T}} \cdot \mathbf{y}$ with the second formulation yields a solution that is much closer to the historical maximum than the multi-objective optimization of the different output variables in $\mathbf{y}$, with the elements of $\mathbf{a}^{\mathrm{T}}$ in the diagonal of the weightings matrix $\boldsymbol{\Gamma}$. This is because the outputs are not independent of each other, and therefore the first approach would not be correct, as elaborated in Palací-López et al. (2020).

Properly Defining the "Desired Values" for Variables to Minimize/Maximize

Given that quadratic programming is one of the most common approaches for solving optimization problems, using formulations of the objectives functions as the ones shown in the previous subsection, it is important to acknowledge that this optimization is based, mainly, in minimizing the squared distance between the theoretically achieved values for the different variables or CQAs considered, according

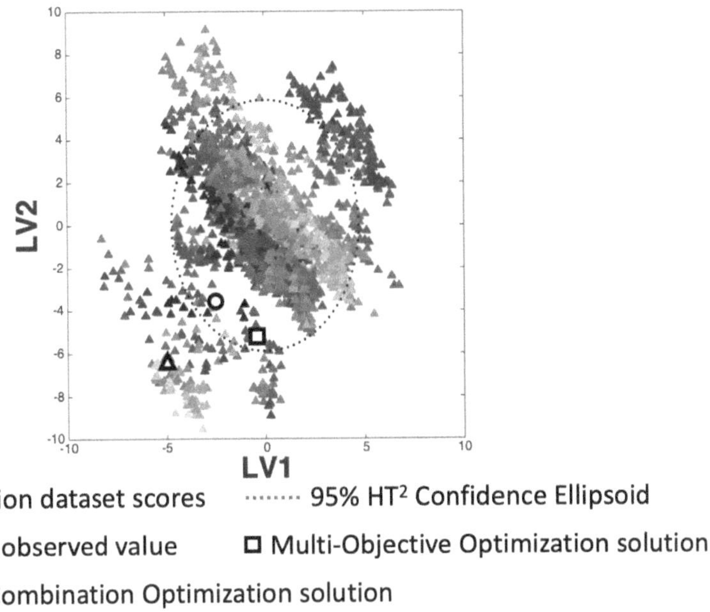

FIGURE 12.5 Projection of the calibration dataset (with grayscale-colored observations according to the value of $d = \mathbf{a}^{\mathrm{T}} \cdot \mathbf{y}$) in the subspace defined by LVs 1 and 2, highlighting the observation with the highest observed value (white-filled triangle), as well as the solution from the two optimization approaches: (i) maximizing each output separately, with the coefficients of the linear combination as the diagonal of $\mathbf{\Gamma}$ (white-filled square), and (ii) maximizing the CQA itself (white-filled circle).

to the previously obtained predictive model, and their respective "desired values." Therefore, a reasonable definition of such "desired values" is critical to obtaining equally reasonable solutions. While this is usually not a problem when actually known and realistic desired values are defined, a standardized approach is in order when a given variable or CQA is to be minimized or maximized, but no specific "desired value" for it is known a priori.

Intuitively, a reasonable "desired value" when maximizing can be thought of as one that is high enough to be not easily achieved and be worth pursuing, but not so high as to become unattainable or unfeasible. And a similar line of thought could also be reasoned in search of a sensible "desired value" when minimizing.

While apparently a good reasoning in practical terms, from a statistical point of view this can lead to some issues. Consider the following two scenarios:

- Scenario A: a reactive distillation with overall yield usually varying between 80% and 85%, and conversion between 15% and 17%.
- Scenario B: a reaction with overall yield usually varying between 40% and 50%, and conversion between 90% and 95%.

In both cases, the goal would be to maximize yield and conversion. Theoretically, both should be able to reach 100% but not any more than that. Therefore, defining the "desired value" for both yield and conversion as 100% may seem like a reasonable decision. But is it? In reality, doing such a thing would lead to the following results:

- Scenario A: because the usual variation of the conversion is less than half that of the yield (a range of 2% vs a range of 5%), and its average value much lower (16% vs 82.5%), the "number of ranges" between each average value and the desired value is much higher for the conversion than it is for the yield, namely, (100% – 16%)/(2%)=42 vs (100%-82.5%)/(5%)=3.5. Due to the scaling to unit variance performed on the original variables prior to fitting the LVRM, and assuming that Γ is the identity in the objective function of this hypothetical optimization problem, the actual importance given to achieving the "desired value" for the conversion will be roughly 12 times the one given to achieving the same 100% for the yield.
- Scenario B: predictably, a similar effect will be observed, but in this case not as pronounced as before, and with more importance being given to achieving the 100% of yield versus 100% of conversion.

To avoid this kind of situation, one can simply define the "desired values" for maximization (or minimization) as the maximum (or minimum) achievable values for the corresponding CQAs without exiting the Hotelling's T^2 confidence hyperellipsoid in the latent space, for a given confidence level, obtained after fitting the LVRM being inverted in the optimization, as suggested by Palací-López et al. (2020).

Figure 12.6 illustrates the results from trying to maximize/minimize two negatively correlated output variables when the "desired values" are defined as arbitrarily high/low ones, and when they are defined using this standardizable approach. As can be seen, when arbitrarily high/low values are used in the objective function, the solution obtained will be dominated by one of the two variables despite giving the same weight to each objective in the Γ matrix. However, when the methodology by Palací-López et al. (2020) is applied, the solution to the optimization problem lies in a middle point between maximizing/minimizing both outputs, which is expected since equal importance is given to achieving two competing objectives.

12.4 CONCLUDING CONSIDERATIONS REGARDING OPTIMIZATION AS A TOOL FOR INVERSE DESIGN

So far, this chapter has focused on model inversion through optimization for inverse design and, in particular, through quadratic programming, that is, by mathematical optimization of quadratic functions.

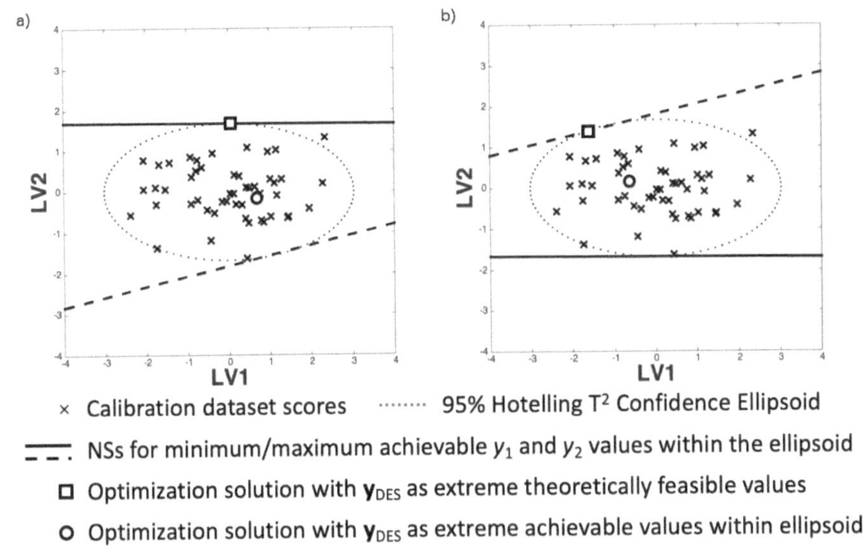

FIGURE 12.6 Projection of the calibration dataset with 50 observations into the subspace of the first two latent variables of a PLS-regression model with two output variables, and solutions of the optimization problem when a) trying to maximize both outputs, defining $\mathbf{y}_{DES}$ as a vector with the maximum feasible values (white-filled square) vs. using the approach by Palací-López et al. (2020) (white-filled circle), and b) trying to minimize both outputs, defining $\mathbf{y}_{DES}$ as a vector with the minimum feasible values (white-filled square) vs. using the approach by Palací-López et al. (2020) (white-filled circle).

To close this chapter, answers will be given for a couple of questions that may, or may not, have arisen while reading this section:

- Are there better alternatives, in the field of optimization, to quadratic programming?
- Is model inversion itself actually needed for inverse design?

Regarding the first one, the answer seems quite simple, given the existence of, for example, linear programming, in contrast with quadratic programming. There are, in fact, many potential formulations of the optimization problem that could be more efficient, or better suited, for a particular task. Two examples can be observed in this very same chapter:

- The first occurs at the beginning of Section 12.2.2, and later in Section 12.2.2, where the goals of "product discovery" and "guaranteeing quality" are presented as two different approaches that will need, inevitably, two different formulations of the constraints introduced in the optimization problem.

- The second is a bit more subtle but can be inferred from one of the implications of the analytical expression of the NS presented in Section 12.2.2: if the analytical expression of the NS contains the direction of maximum variability of a given variable, then in order to maximize or minimize it, following such direction would suffice, and so the optimization problem may be formulated as a linear one, if required at all. In fact, for the example in Section 12.3.2, it is interesting to observe the straight lines corresponding to the NSs for both (minimum and maximum for both) variables in Figure 12.6: these lines are tangential to the HT2 confidence ellipsoid, and the vectors that are orthogonal to this ellipsoid at the point where the NSs touch them follow, in fact, the same "direction of maximum variability" that the analytical expression of such NSs contain.

This does not mean that the quadratic formulation of the optimization problem should be discarded. On the contrary, many of the constraints involved even in this example are non-linear, which means that a linear formulation will be insufficient in most cases. Therefore, in many cases, a more appropriate approach will combine wisely one formulation with the other to make the best out of each formulation strength while overcoming the other's shortcomings.

With regard to the second question, one could think of model inversion as "predicting the model inputs from its outputs." So, why not simply build a predictive model, with the "outputs" as predictive (i.e., input) variables, and the "inputs" as predicted (i.e., output) variables, instead of inverting a previously fitted predictive model?

The answer is actually quite simple, at least in appearance. As an example, consider a parabolic curve of the form $y = a \cdot x^2 + b \cdot x + c$. For this curve, any given value of "x" will lead to one, and only one, value for "y." However, for any given value for "y" that meets this equation, we will find two different values of "x." This means that in order to fit a proper predictive model, there needs to be a one-to-one correspondence when going from the predictive to the predicted variables and, therefore, inverting a fitted model where this is the case and fitting a model where the predictive and predicted variables are exchanged with each other will not, generally speaking, provide equivalent results.

NOTES

1 Not to be confused with the Experimental Design Space, as defined in the Design of Experiments field.

2 Here, "analytical expressions" is used loosely to refer to a definition of specifications and equality or inequality constraints in the form of a single equation each, in a way that allows them so be used as soft or hard constrains for optimization, or as constrains that delimit a subspace of acceptable inputs/outputs.

REFERENCES

Bano, Gabriele, Pierantonio Facco, Fabrizio Bezzo, and Massimiliano Barolo. 2018. "Probabilistic Design Space Determination in Pharmaceutical Product Development: A Bayesian/ Latent Variable Approach." *AIChE Journal* 64 (7): 2438–2449. https://doi.org/10.1002/ aic.16133.

Bano, Gabriele, Pierantonio Facco, Natascia Meneghetti, Fabrizio Bezzo, and Massimiliano Barolo. 2017. "Uncertainty Back-Propagation in PLS Model Inversion for Design Space Determination in Pharmaceutical Product Development." *Computers and Chemical Engineering* 101: 110–124. https://doi.org/10.1016/j.compchemeng.2017.02.038.

Box, George E.P., J. Stuart Hunter, and William G. Hunter. 2005. *Statistics for Experimenters: Design, Discovery, and Innovation*. Second edition. John Wiley & Sons, INC.

Burnham, Alison J., John F. MacGregor, and Român Viveros. 2001. "Interpretation of Regression Coefficients under a Latent Variable Regression Model." *Journal of Chemometrics* 15 (4): 265–284. https://doi.org/10.1002/cem.680.

Faber, Klaas, and Bruce R. Kowalski. 1996. "Prediction Error in Least Squares Regression: Further Critique on the Deviation Used in The Unscrambler." *Chemometrics and Intelligent Laboratory Systems* 34 (2): 283–292. https://doi.org/10.1016/0169-7439(96)00022-6.

Facco, Pierantonio, Filippo Dal Pastro, Natascia Meneghetti, Fabrizio Bezzo, and Massimiliano Barolo. 2015. "Bracketing the Design Space within the Knowledge Space in Pharmaceutical Product Development." *Industrial and Engineering Chemistry Research* 54 (18): 5128–5138. https://doi.org/10.1021/acs.iecr.5b00863.

FDA. 2004. "Pharmaceutical CGMPs for the 21st Century – A Risk-Based Approach." *Food and Drug Administration*. http://www.fda.gov/Drugs/DevelopmentApprovalProcess/ Manufacturing/QuestionsandAnswersonCurrentGoodManufacturingPracticescGMP-forDrugs/UCM071836.

García-Muñoz, Salvador, Theodora Kourti, John F. MacGregor, Francesca Apruzzese, and Marc Champagne. 2006. "Optimization of Batch Operating Policies. Part I. Handling Multiple Solutions." *Industrial and Engineering Chemistry Research* 45 (23): 7856–766. https://doi.org/10.1021/ie060314g.

García-Muñoz, Salvador, John F. MacGregor, Debashis Neogi, Bruce E. Latshaw, and Sanjay Mehta. 2008. "Optimization of Batch Operating Policies. Part II. Incorporating Process Constraints and Industrial Applications." *Industrial and Engineering Chemistry Research* 47 (12): 4202–4208. https://doi.org/10.1021/ie071437j.

ICH. 2009, August. "Pharmaceutical Development Q8." *ICH Harmonised Tripartite Guideline* 8: 1–28.

Jaeckle, C.M., and John F. Macgregor. 1996. "Product Design through Multivariate Statistical Analysis of Process Data." *Computers & Chemical Engineering* 20 (96): S1047–S1052. https://doi.org/10.1016/0098-1354(96)00182-2.

Jaeckle, C.M., and John F. MacGregor. 2000. "Industrial Applications of Product Design through the Inversion of Latent Variable Models." *Chemometrics and Intelligent Laboratory Systems* 50 (2): 199–210. https://doi.org/10.1016/S0169-7439(99)00058-1.

Lakshminarayanan, S., H. Fujii, B. Grosman, E. Dassau, and D.R. Lewin. 2000. "New Product Design via Analysis of Historical Databases." *Computers & Chemical Engineering* 24: 671–676.

Liu, Zheng, Mark-John Bruwer, John F. MacGregor, Sweta Rathore, David E. Reed, and Marc J. Champagne. 2011. "Modeling and Optimization of a Tablet Manufacturing Line." *Journal of Pharmaceutical Innovation* 6: 170–180.

MacGregor, J.F., and M.J. Bruwer. 2008. "A Framework for the Development of Design and Control Spaces." *Journal of Pharmaceutical Innovation* 3 (1): 15–22. https://doi. org/10.1007/s12247-008-9023-5.

MacGregor, J.F., M.J. Bruwer, I. Miletic, M. Cardin, and Z. Liu. 2015. "Latent Variable Models and Big Data in the Process Industries." *IFAC-PapersOnLine* 28 (8): 520–524. https://doi.org/10.1016/j.ifacol.2015.09.020.

Palací-López, Daniel, Pierantonio Facco, Massimiliano Barolo, and Alberto Ferrer. 2019. "New Tools for the Design and Manufacturing of New Products Based on Latent Variable Model Inversion." *Chemometrics and Intelligent Laboratory Systems* 194: 103848. https://doi.org/10.1016/j.chemolab.2019.103848.

Palací-López, Daniel, Pedro Villalba, Pierantonio Facco, Massimiliano Barolo, and Alberto Ferrer. 2020. "Improved Formulation of the Latent Variable Model Inversion–Based Optimization Problem for Quality by Design Applications." *Journal of Chemometrics* 34 (6): e3230. https://doi.org/10.1002/cem.3230.

Pantelides, C.C., N. Shah, and C. Adjiman. 2009. "Design Space, Models, and Model Uncertainty, Comprehensive Quality by Design in Pharmaceutical Development and Manufacture." *American Institute of Chemical Engineering*, AIChE Annual Meeting: Nashville, TN, Paper 417f: 97.

Tomba, Emanuele, Massimiliano Barolo, and Salvador García-Muñoz. 2012. "General Framework for Latent Variable Model Inversion for the Design and Manufacturing of New Products." *Industrial & Engineering Chemistry Research* 51 (39): 12886–12900. https://doi.org/10.1021/ie301214c.

Zhang, Lin, and Salvador Garcia-Munoz. 2009. "A Comparison of Different Methods to Estimate Prediction Uncertainty Using Partial Least Squares (PLS): A Practitioner's Perspective." *Chemometrics and Intelligent Laboratory Systems* 97 (2): 152–158. https://doi.org/10.1016/j.chemolab.2009.03.007.

Section 5

Case Studies and Special Topics

Alix Schmidt and Kristin Wallace

The final section of the text, Section 5, seeks to round out the reader's understanding of the multitude of ways that data-driven decision-making can be applied to chemicals and materials research, formulation, and commercialization. Chapter 13 directs those who are eager to begin practicing the techniques put forth in previous chapters toward a series of publicly available case studies. All the studies illustrate the approach and benefits of using methods described here, and some have formulation data sets available to practice on your own or with a guided tutorial.

Chapter 14 pivots away from the materials formulation question and presents some special topics aligned with using machine learning approaches to address additional challenges you may encounter along the product commercialization path. First, what if your materials design model proposes the synthesis of a new molecule to achieve your goals – can models help? Chapter 14.1 discusses AI-guided retrosynthesis, a process that involves the use of computer-aided synthesis planning (CASP) to determine how to synthesize a chemical target from commercially available reagents. Second, following Chapter 12's presentation of model-driven process design, Chapter 14.2 presents the use of real-time statistical process monitoring to guide process operations and enable proactive interventions for process improvement. A principal components analysis (PCA) approach is illustrated for a batch monoclonal antibody process. Another task you may face when producing a new product is defining the quality specifications for incoming raw materials to enable proper formulation into quality products. Chapter 14.3 presents a partial least squares-based methodology to

DOI: 10.1201/9781003385974-17

develop meaningful multivariate specifications regions for raw material properties. Lastly, Chapter 14.4 provides a perspective on what is needed – from industry and from classrooms – to enable a digital-savvy and analytics-literate workforce in the chemicals and materials fields.

Finally, we close the book with Chapter 15, which summarizes the technical content presented and points toward the future directions of the materials informatics field to keep an eye on in the coming years.

13 Case Studies

Kristin Wallace and Alix Schmidt

13.1 INTRODUCTION

With the core technical content of this book now under your belt, a logical next step to reinforce new learnings is to review several applications or case studies. This chapter identifies and briefly describes case studies that are available in the public domain (whitepapers, vendor websites, etc.). The "case studies" listed next begin with high-level summaries and progress to more transparent and detailed cases all the way through to step-by-step software tutorials with accessible datasets. Readers are strongly encouraged to try applying the concepts presented in this book to a dataset for themselves, which can be accomplished either by following along with suggested tutorials or by replicating the results presented in detailed case studies.

13.2 AVAILABLE CASE STUDIES

1. **Digital transformation of R&D for a materials manufacturer**
   - Reference: *"Why Top Materials Company Idemitsu Partnered with Enthought to Accelerate Product Innovation using Materials Informatics."* Enthought, 2023. https://www.enthought.com/wp-content/uploads/Idemitsu_Enthought_Materials_Informatics_Case_Study.pdf
   - Details: To accelerate their search for new catalysts for carbon utilization, a materials manufacturer engaged an internal team of scientists in a digital transformation program led by Enthought. The internal team used this training to then build their own software solution to accomplish a variety of tasks from organizing historical data through to leveraging machine learning and AI to select promising candidates for lab experimentation.
2. **Reformulating food products using latent variable methods and optimization**
   - Reference: *"Food Innovation & Manufacturing with AI."* ProSensus, 2022. https://prosensus.com/food-innovation-manufacturing-with-ai/
   - Details: Includes two food datasets/applications: i) reformulating snack foods with new (more sustainable) ingredients while maintaining quality, and ii) reformulating frozen muffin batters to reach new (healthier) quality targets while maintaining taste. Data is not shared.
3. **Solvent blend discovery accelerated**
   - Reference: "*Solvent Blend Discovery Accelerated*." Citrine, 2023. https://citrine.io/wp-content/uploads/2021/04/Case-Study-Showa-Denko-Solvent-Blend.pdf

DOI: 10.1201/9781003385974-18

- Details: One hundred and fifty solvent blends were formulated from 380 available ingredients using the Citrine Platform and Sequential Learning. Fifty of those new blends met all constraints while maximizing target properties.

4. **Rapid reformulation of inks using machine learning**
   - Case study and webinar: *Machine learning for virtual experiments – ink formulation*, https://intellegens.com/machine-learning-for-virtual-experiments-ink-formulation/ (Intellegens, 2021).
   - Description: Empirical analysis can be the most resource-intensive part of the formulation R&D process: Domino Printing Sciences used machine learning-driven experimental design to reduce the number of experiments required during the COVID-19 pandemic, obtaining several ink formulations that were experimentally verified.
5. **Recycled plastics formulation using AI**
   - Webinar: "Agile Formulation of Recycle Polymers." Presented by Citrine for SPE-Inspiring Plastics Professionals (Melia and Sullivan, 2023).
   - Description: This webinar demonstrates how AI (through the Citrine platform) can increase agility for plastic compounders, reduce experimentation, and maintain quality with varying raw materials.
6. **Formulating dairy products with long shelf lives**
   - Case study: "*Developing food formulations at Yili.*" Intellegens, 2022a. https://intellegens.com/developing-food-formulations-at-yili/
   - Description: Dairy products must satisfy multiple properties over a long shelf life, achieved using balances of stabilizers, emulsifiers, and processing conditions. Using machine learning Yili identified ingredients they could drop from testing, accelerating development, as the machine learning understood the cross-interactions in the data.
7. **Development of thermoplastic materials using latent variable methods and optimization**
   - Reference: Muteki, 2006. "Mixture Product Design Using Latent Variable Methods." Doctor of Philosophy Thesis, McMaster University, 43–55. https://central.bac-lac.gc.ca/.item?id=NR29263&op=pdf&app=Library&oclc_number=232367358
   - Description: PLS modeling and optimization for formulations that include a blend of starch, lactic acid, and additives. Dataset is shared. Dataset includes 28 formulations, 12 ingredients (6 of which have not been used previously), 5 ingredient properties, 1 process conditions, and 4 quality properties.
8. **Development of additives for carbon fiber with machine learning**
   - Reference: "AI-Driven Development of Additives for Carbon Fiber Manufacturing Process." Citrine, 2021.
     https://citrine.io/wp-content/uploads/2022/03/Case-Study-Cold-start-Carbon-Fiber-Additives-Project.pdf
   - Details: A new additive recipe was formulated for carbon fiber. With a "cold start" (initially no data), the project team turned to literature and domain experts to identify 20 initial experiments to generate data and build a machine learning model in the Citrine Platform. Domain knowledge, objectives, and constraints were developed to improve model

accuracy and narrow candidate solutions. Three Sequential Learning iterations resulted in an additive recipe that met the project criteria.

9. **Development of smart biocompatible polymers using latent variable methods and optimization**
   - Reference: Tzoc Torres et al., 2014. "Designing Multi-Responsive Polymers Using Latent Variable Methods." *Polymer*, 55. https://doi.org/10.1016/j.polymer.2013.12.041
   - Details: PLS modeling and optimization of existing "failed" polymers to rapidly devise new candidates. Dataset is shared. Dataset includes 23 formulations, 9 ingredients, 3 process conditions, 3 quality properties.
10. **Accelerating catalyst development and testing with machine learning**
    - Case study and webinar: *Catalysts for clean air and life science at Johnson Matthey*, https://intellegens.com/catalysts-for-clean-air-and-life-science-at-johnson-matthey/ (Intellegens, 2023).
    - Publication: T. Whitehead, F. Chen, C. Daly, and G. Conduit. 2022. "Accelerating the Design of Automotive Catalyst Products using Machine Learning." *JMTR* 66(2), 130 (Whitehead et al., 2022).
    - Description: Developing catalysts can be a lengthy process. Several projects at Johnson Matthey showed the potential of machine learning to accelerate the process, in one case reducing the number of required tests of a product by up to 50%, or improving performance, in another case increasing yield for a key reaction for which no further improvement was thought possible.
11. **Machine learning for polymer R&D**
    - White paper: *Machine learning for polymer R&D*, Intellegens, 2022b. https://intellegens.com/machine-learning-for-polymer-rd/
    - Description: Polymer formulation involves complex, high-dimensional, non-linear interactions and so is difficult for a human to fully understand: but machine learning is able to capture these relationships and then explore polymer formulation space, predict and test potential solutions, and focus experimental work. Alchemite™, machine learning software developed by Intellegens introduced.
12. **Solubility Prediction with Deep Learning – Python Tutorial**
    - Reference: Hegde and Strachan, 2020. "Hands-on Deep Learning for Materials Science: Convolutional Networks and Variational Autoencoders." https://nanohub.org/resources/34474
    - Details: A small polymer dataset is used in this tutorial to demonstrate how Python libraries for deep learning can be used to preprocess data, train a convolutional neural network, and predict solubility.
13. **Functional Polymers – FormuSense Tutorial**
    - Reference: FormuSense by ProSensus Inc. – rapid product development software (ProSensus, 2024).
    - Details: FormuSense includes no-code tools to accelerate data assembly, predictive modeling, and optimization (inverse design). Free 30-day trials are publicly available. A functional polymers dataset, accompanied by step-by-step tutorial instructions are provided to users who activate a 30-day trial. Get started here: https://prosensus.com/new-rapid-product-development/formusense/

13.3 CONCLUSIONS

In this chapter, several case studies were referenced, with the goal of encouraging readers to review and replicate applications of the concepts presented in this book. Applications range from foods and pharmaceuticals to specialty chemicals and polymers, while solution approaches include from-scratch coding to no-code commercial software to execute dataset assembly, feature engineering, predictive modeling (linear models, random forests, gaussian processes, and neural networks) and optimization and inverse design.

With a good grasp of the core concepts presented in Chapters 1 through 13, readers are encouraged to extend their learning to the four special topics included in Chapter 14.

REFERENCES

Citrine. 2021. "AI-Driven Development of Additives for Carbon Fiber Manufacturing Process." *Citrine*. https://citrine.io/wp-content/uploads/2022/03/Case-Study-Cold-start-Carbon-Fiber-Additives-Project.pdf

Citrine. 2023. "Solvent Blend Discovery Accelerated." *Citrine*. https://citrine.io/wp-content/uploads/2021/04/Case-Study-Showa-Denko-Solvent-Blend.pdf

Enthought. 2023. "Why Top Materials Company Idemitsu Partnered with Enthought to Accelerate Product Innovation using Materials Informatics." https://www.enthought.com/wp-content/uploads/Idemitsu_Enthought_Materials_Informatics_Case_Study.pdf

Hegde, Vinay, and Alejandro Strachan. 2020. "Hands-on Deep Learning for Materials Science: Convolutional Networks and Variational Autoencoders." *nanoHUB*. https://nanohub.org/resources/34474

Intellegens. 2021. "Machine Learning for Virtual Experiments – Ink Formulation." *Intellegens*. https://intellegens.com/machine-learning-for-virtual-experiments-ink-formulation/

Intellegens. 2022a. "Developing Food Formulations at Yili." *Intellegens*. https://intellegens.com/developing-food-formulations-at-yili/

Intellegens. 2022b. "Machine Learning for Polymer R&D." *Intellegens*. https://intellegens.com/machine-learning-for-polymer-rd/

Intellegens. 2023. "Catalysts for Clean Air and Life Science at Johnson Matthey." *Intellegens*. https://intellegens.com/catalysts-for-clean-air-and-life-science-at-johnson-matthey/

Melia, Hannah, and Mark Sullivan. 2023. "Agile Formulation of Recycle Polymers." Presented by *Citrine for SPE-Inspiring Plastics Professionals*. https://www.4spe.org/i4a/pages/index.cfm?pageID=8540

Muteki, Koji. 2006. "Mixture Product Design Using Latent Variable Methods." PhD diss., McMaster University.

ProSensus Inc. 2022. "Food Innovation & Manufacturing with AI." *ProSensus Inc*.

ProSensus Inc. 2024. "FormuSense – Rapid Product Development Software." *ProSensus Inc*.

Tzoc Torres, M., E. Nichols, J.F. MacGregor, and T. Hoare. 2014. "Designing Multi-Responsive Polymers Using Latent Variable Methods." *Polymer* 55. https://doi.org/10.1016/j.polymer.2013.12.041

Whitehead, T., F. Chen, C. Daly, and G. Conduit. 2022. "Accelerating the Design of Automotive Catalyst Products Using Machine Learning." *JMTR* 66(2), 130.

14 Special Topics

14.1 AI-GUIDED RETROSYNTHESIS

Anne-Catherine Bedard and Mengjie Liu

14.1.1 INTRODUCTION TO AI-GUIDED RETROSYNTHESIS

Synthesis is the intentional construction of chemical structures. The first step of this process is synthesis planning, or the determination of how to synthesize a chemical compound from commercially available reagents (i.e., starting materials) through a series of chemically feasible reaction steps. This can be very straightforward for simple molecules but becomes exceptionally more challenging as the target grows in complexity. The concept of retrosynthesis addresses this issue.

> Retrosynthesis tackles the problem of synthesis in reverse, beginning with the desired product and choosing suitable disconnections recursively until suitable precursors are found (Figure 14.1.1).

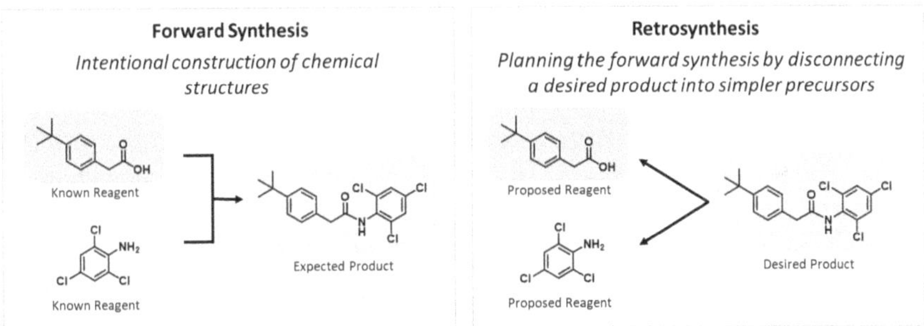

FIGURE 14.1.1 Example and definition of forward synthesis and retrosynthesis.

In fact, E. J. Corey won the Nobel Prize for his contributions to this concept in 1990. In a similar fashion to chess, retrosynthesis could potentially be conducive to computer assistance because it requires searching a large space of possible disconnections, or "moves," starting from each intermediate or state. Computational approaches date back to Corey's work on the development of LHASA (Logic and Heuristics Applied to Synthetic Analysis) in the 1970s (Corey et al. 1972), with the goal of suggesting synthetic pathways to chemist users. Such analysis was initially carried out by using expert systems utilizing hand-encoded rules. The ideal *computer-aided synthesis*

DOI: 10.1201/9781003385974-19

planning (CASP) program would take a molecular structure as input and output a sorted list of detailed reaction schemes that each connects that target to commercially available starting materials via a series of realistic chemical transformations. With the rise of computer assistance, the field of retrosynthetic software tools has grown significantly in the past few years. We are now on the path to sophisticated and automatic algorithms with the potential to provide retrosynthetic analysis with a broader application domain and better accuracy.

14.1.2 Single-Step Retrosynthesis Planning Algorithms

Fundamentally, the goal of any CASP approach is to teach a computer to apply chemical understanding and intuition in the same manner of a trained synthetic chemist approaching the problem. Retrosynthesis planning algorithms developed to tackle this problem can be broadly categorized into three types: (1A) pure template-based, (1B) template-based AI, and (2) template-free AI.

In the (1A) pure template-based approach, knowledge of chemical reactivity may be manually or automatically encoded in generalized rules, typically in the form of a reaction template, for example, Figure 14.1.2B.

Reaction templates define certain structural features along with a transformation, which allow the computer to generate the full reaction. The most common form of encoding reaction templates in modern CASP tools is the SMARTS (SMILES arbitrary target specification) language, related to the commonly used SMILES (simplified molecular-input line-entry system) specification for representing molecular structures, for example, Figure 14.1.2C. After compiling a set of reaction templates, they can be systematically applied to a target molecule to generate potential precursors (in the case of retrosynthesis). Template-based algorithms include some of the earliest retrosynthesis programs, such as Corey's LHASA.

With a template-based approach (1A and 1B), there are two primary concerns that affect the utility of its suggestions: 1) the specificity of each template and 2) the chemical space covered by the collection of templates. There is a trade-off between the specificity of a template and the likelihood that the proposed reaction will work. The more context (environment around a reacting atom) that is defined, the fewer molecules it will apply for, but the higher the likelihood that a reaction it does generate will work. For example, the template in Figure 14.1.2B includes neighbors one bond away from the reacting site, plus the nitrile group that has been identified as relevant context. Moreover, the types of chemical transformations that may be suggested are strictly limited by the types of templates that are in the database. There is no ability for a template-based approach to extrapolate beyond what has been encoded.

From a development standpoint, there is also the question of how the database of templates is created. The default method of creating the database is for chemists or others with sufficient chemical knowledge to *manually* define and curate templates.

A.

B.

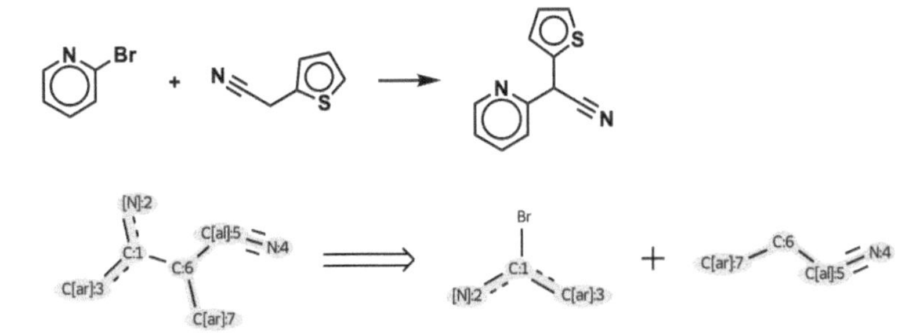

C.
```
[#7;a:2]:[c;H0;D3;+0:1](:[c:3])-[CH;D3;+0:6](-[c:7])-[C:5]#[N;D1;H0:4]>>
          Br-[c;H0;D3;+0:1](:[#7;a:2]):[c:3].[N;D1;H0:4]#[C:5]-[CH2;D2;+0:6]-[c:7]
```

FIGURE 14.1.2 A) Original reaction from literature. B) Extracted retrosynthetic reaction template with atom map numbers. Corresponding atoms between reactants and products are highlighted. C) Reaction SMARTS representation of the reaction template.

However, given that larger template sets are preferred to expand the chemical space covered by the model, this can be a very time-consuming and labor-intensive process. Among modern programs, Synthia™ (formerly Chematica) stands out as one of the major template-based algorithms that relies on manual curation.

On the other hand, automated template-extraction algorithms have also been developed to simplify this process. Generally speaking, these algorithms rely on the availability of an atom-mapped reaction dataset, meaning that the atoms have been labeled to indicate where each atom in the reactants ends up in the products. From an atom-mapped reaction, the reaction center and set of bond changes involved in the reaction can be extracted and expressed as a reaction template. Critically, the algorithm must also identify a suitable amount of context, which is required for the reaction to be successful. This can be done using heuristic rules, such as including neighbors one or two bonds away, or special function groups from a pre-defined list.

Once a sufficiently large database of templates has been compiled, the next challenge is simply the time it takes to apply all possible templates to the target molecule. This becomes an even bigger challenge for multi-step synthetic pathway searches. One solution that has been put forward is the introduction of a model which can automatically prioritize the templates. This brings us to the second approach, which is template-based AI (1B).

Template-based AI (1B) relies on a machine learning (ML) model, which is trained to identify the reaction templates that are most relevant to the target molecule.

By combining reaction templates and an ML model, we retain the interpretability of reaction templates and can significantly improve performance by only applying the most relevant templates. However, the prioritization produced by the model can be difficult to explain, and it is possible for rare transformations to be underprioritized by the model and overlooked altogether. Finally, just like the pure template-based approach (1A), this method is limited to the types of reactions that are represented in the templates, with no ability to predict completely novel reaction types.

The template-free AI (2) method aims to address the limitations of template-based methods by forgoing templates completely.

At a high level, these methods avoid any prior specification of chemical reactivity, instead relying on an ML model to learn reactivity patterns itself based on a training set of known reactions. Template-free methods can be categorized into graph-based (2A) and sequence-based (2B) approaches. Graph-based approaches operate on a molecular graph using graph-convolutional neural networks. Sequence-based approaches operate on string representations, commonly SMILES, and use methods pioneered by natural language processing (NLP) models.

14.1.3 Multi-Step Retrosynthesis Planning

While single-step retrosynthesis models are critical, the overarching goal of CASP is to propose multi-step synthesis pathways that can deconstruct a molecule back to commercially available reagents. In theory, multi-step planning can be seen as a recursive application of a single-step model; however, an exhaustive search would be expensive and may return too many undesirable pathways. Therefore,

The main challenge in developing multi-step algorithms is finding an efficient way to traverse the search space.

Since chemical reaction networks are fundamentally a kind of directed graph, such algorithms naturally derive from path-finding algorithms in graph theory.

At a high level, multi-step planning can be seen as an effort to break the target molecule into successively smaller pieces that are less synthetically complex. As such, a basic approach to multi-step planning is to score suggested single-step precursors based on some measure of their synthetic complexity and choose the least complex precursors to apply to the next iteration of single-step predictions. Many methods use heuristics based on the structural features of the molecule, such as the number of atoms, rings, or chiral centers. These can be easy to develop and calculate, and they align easily with chemical intuition. One example of a heuristic score is SAscore (Ertl and Schuffenhauer 2009). However, structural complexity may not always be equivalent to synthetic complexity – a large molecule may be easily made from available building blocks, and retro-deprotections will always increase structural complexity even if advantageous for the pathway.

To address this discrepancy, alternative ML models have been developed. One such model is SCscore (Coley et al. 2018), which was trained on reaction data under the assumption that reactants should be considered less synthetically complex than products on average. Another model is RAscore (Thakkar et al. 2021), which was trained on whether an existing multi-step planner could successfully generate a synthesis pathway to the target.

With the implementation of one of these scoring methods, it is straightforward to envision a multi-step planning algorithm that iteratively runs a one-step prediction for a molecule, selects the optimal precursor suggestion based on the scorer, and then repeats the process with the precursors until commercially available starting materials are reached. Such an algorithm can be considered a "best-first" search. Such an algorithm could involve only the selection and expansion steps shown in Figure 14.1.3, where scores for newly found precursors are calculated at the end of the expansion step. In algorithms with more complex selection algorithms, an update step is used to propagate information obtained from the expansion that serves as inputs for the selection step.

One of the leading multi-step algorithms is Monte-Carlo Tree Search (MCTS), which has notably been used in AI algorithms for board games, such as AlphaGo (Fu 2016). Chemical synthesis planning can similarly be framed as a "game" where decisions are made as to which retrosynthetic step is taken in each round to eventually "win" by reaching buyable precursors. Segler et al. (2018) applied the MCTS algorithm to chemical synthesis with the incorporation of three separate machine learning models: an expansion policy for choosing retrosynthetic steps, a rollout policy for estimating the value of a precursor, and an in-scope filter to exclude reactions suggested by the expansion policy which are unlikely to work in the lab.

The retrosynthetic tree builder algorithm in the ASKCOS software (Coley et al. 2021) uses a neural-guided best-first search, with a scoring function that incorporates the probability given to a template from the template-based expansion policy network. This is combined with an in-scope filter inspired by Segler et al.'s work to eliminate retrosynthetic steps that are unlikely to work.

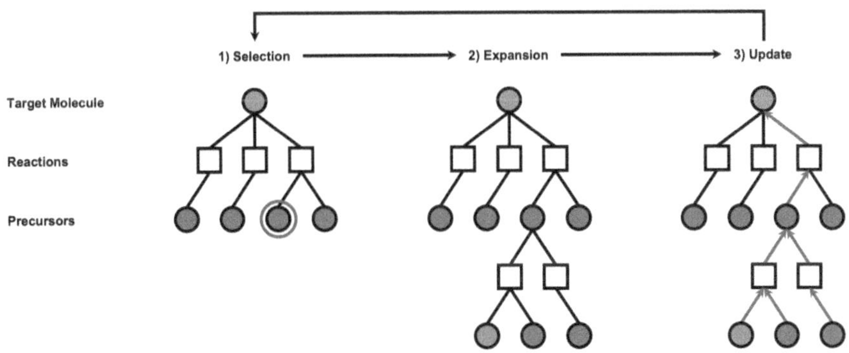

FIGURE 14.1.3 Generalized procedure of a multi-step planning algorithm that iteratively grows a reaction network by applying a one-step retrosynthesis model. In each iteration, the algorithm will 1) select a precursor molecule, 2) expand the precursor using the one-step model, and 3) propagate newly obtained information up the tree to inform the next selection step.

Another best-first search algorithm that has been developed is Retro-* (Chen et al. 2020), which stands out by employing a trained value network to predict the synthetic cost of a given precursor. This method reduces the computational expense associated with the rollout phase of MCTS by approximating it with a single neural network prediction.

The formulation of the multi-step planning problem also lends itself well to reinforcement learning. Simply put, the synthetic cost of a precursor can be estimated based on the routes that are found by the multi-step algorithm, which in turn improves by having a more accurate model for synthetic cost. Wang et al. (2020) demonstrated one such approach using MCTS with a value network trained using reinforcement learning.

The key benefit of reinforcement learning approaches is to address the lack of data by generating new routes on its own.

14.1.4 Forward Synthesis Planning

With a holistic approach to multi-step retrosynthesis now accessible, the next challenge for CASP is to engage in reaction conditions recommendation as well as forward reaction predictions.

It is important to understand that a synthetic route based on retrosynthetic templates does not guarantee a successful forward synthesis.

Templates are inherently unaware of functional groups present in the rest of the molecule, so it is very common for a proposed retrosynthetic disconnection to be impossible in the forward direction. Forward analysis is such a crucial part of retrosynthesis evaluation that even the very first retrosynthesis program, LHASA, could identify functional group incompatibilities that might lead to a lack of specificity or selectivity.

The MIT team developed a predictive model that assesses candidates in terms of the molecules that contribute atoms to the final product, that is, the reaction core. Although major improvements have been made in this field, lots remain to be addressed: 1) consideration of the reaction conditions, such as reactant concentration, catalyst identity, catalyst concentration, reagent concentrations, solvent(s), temperature, pressure, or reaction time; 2) lack of reaction context arising from data scarcity; 3) unideal reagent representation leading to poor use of information (Coley et al. 2017). Figure 14.1.4 depicts an example of both a successful (14.1.4A–B) and unsuccessful (14.1.4C–D) forward prediction highlighting the current challenges. Figure 14.1.4B shows a ring-forming reaction between an aryl alkyne and an oxime that leads to the desired isoxazole with a probability of 100.0%. Figure 14.1.4D

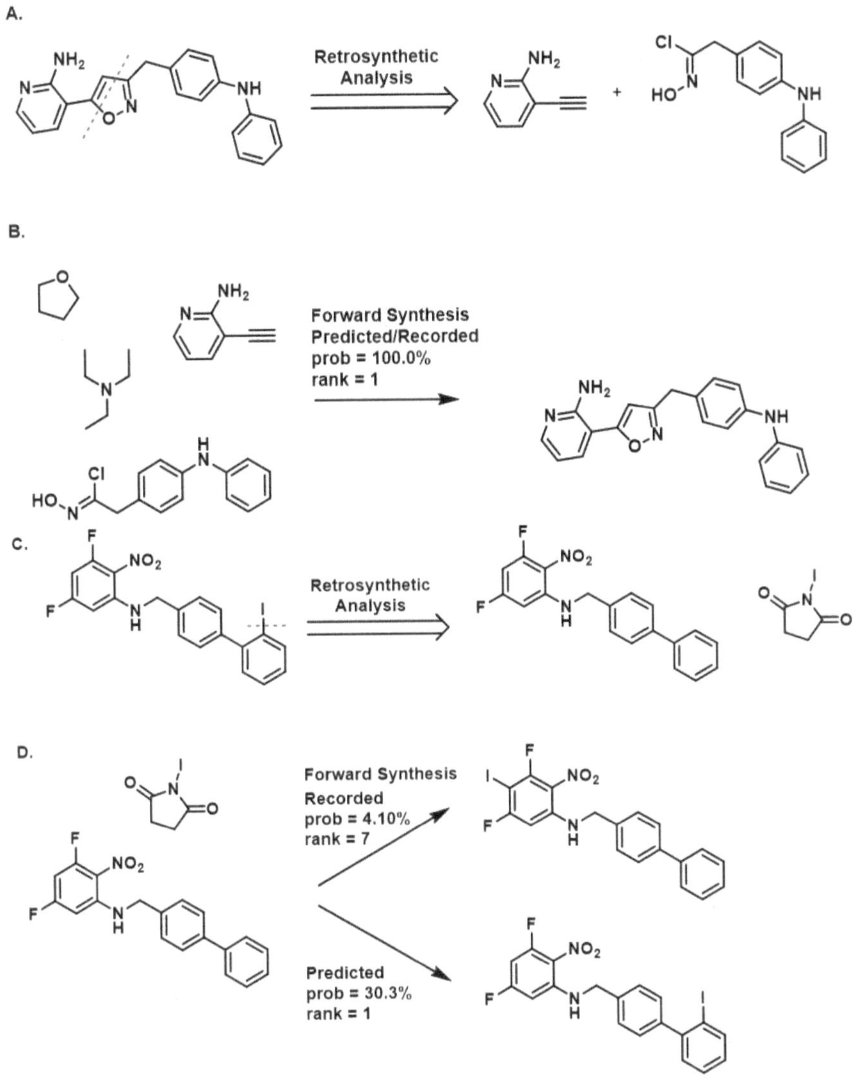

FIGURE 14.1.4 Retrosynthetic Analysis proposed by ASKOS (A and C) and corresponding forward synthesis results (recorded experimentally vs. predicted; B and D). Figure adapted from Coley et al. (2017).

depicts an iodination reaction by N-iodosuccinimide (NIS), where due to the presence of three aromatic rings there are multiple plausible reactive sites. The recorded outcome with an assigned rank of 7 (4.10% probability) was actually the recorded outcome of the reaction. In this case, the trained model does not properly capture the impact of directing groups. This is likely due to an insufficient input featurization or an insufficient number of reaction examples from which to learn this nuance.

14.1.5 Case Study: ASKCOS

In one demonstration of CASP software, Levin et al. (2022) used ASKCOS to perform hybrid chemo-enzymatic synthesis planning. As mentioned previously, ASKCOS uses a template-based approach for the one-step retrosynthesis model. In this work, the authors combined two independently trained one-step models: one using templates extracted from Reaxys data (non-enzymatic reactions) and one using templates extracted from BKMS data (enzymatic reactions).

Figure 14.1.5 shows one example in which ASKCOS was able to recover a published experimental pathway for the synthesis of fluoropyridinyl involving both enzymatic and synthetic steps. The particular enzymatic transformation used in the experimental route was not part of the training dataset for the one-step model, but the model was able to generalize the transformation.

This case demonstrates a few advantages of template-based retrosynthesis. The scope of the tool was easily extended by training a new model on a completely different reaction domain. The model was able to extrapolate from the training set to similar but different molecules and reactions. In addition, the predicted transformations can be easily traced back to the supporting literature. In the case of the enzymatic step in the recovered model, the authors found that the original literature source for the transformation used a similar enzyme to the one used in the published route for fluoropyridinyl.

However, the work also highlighted some limitations of the multi-step algorithm. The set of target molecules for which the model successfully found routes depended heavily on the choice of one-step models. When both synthetic and enzymatic models

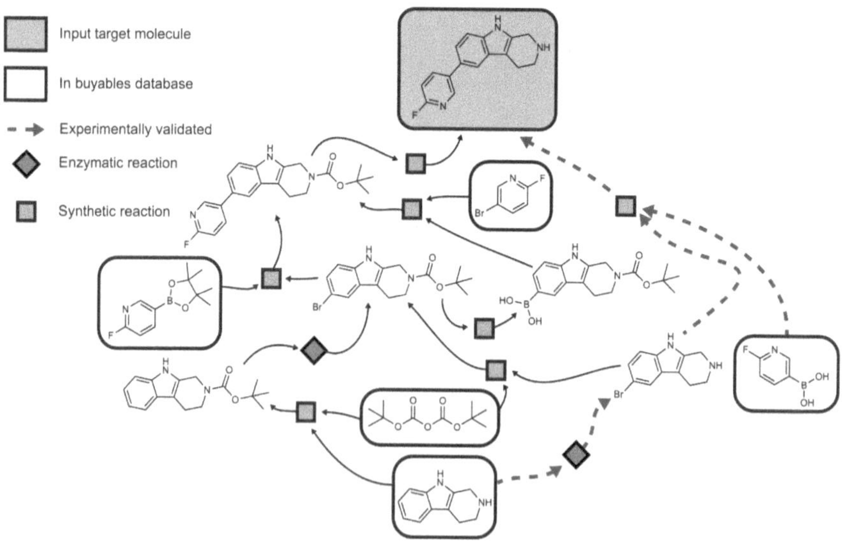

FIGURE 14.1.5 Selected predicted pathways for hybrid synthesis of fluoropyridinyl. The recovered experimental pathway is indicated with dashes. Reproduced with permission from Levin et al. (2022).

were used together, the multi-step algorithm found some new pathways that couldn't be found using one of the models alone but also failed to find pathways for some targets that either of the models solved alone. This demonstrates the complexity of multi-step planning and the inability to exhaustively search the entire chemical space.

14.1.6 Case Study: Chematica

In 2016, Grzybowski reported an expert encoded tool called Chematica that was able to provide de novo syntheses for non-trivial molecular targets (Szymkuć et al. 2016). This tedious encoding strategy allows for representation of molecular context, that is, recognition of incompatible functional groups. Chematica was purchased by MilliporeSigma and is now licensed under the name Synthia. In a demonstration of the advancement of CASP software, Grzybowski and his team showed the synthesis of medicinally relevant targets that were planned by computer and successfully executed by scientists in the laboratory. The team reports eight examples in this publication which all synthesized were solely designed by the CASP software in less than 20 minutes and synthesized in approximately three to four months (Klucznik et al. 2018).

One example reported 5β/6β-hydroxylurasidone (Figure 14.1.6), the main active metabolite of lurasidone, a US Food and Drug Administration-approved atypical antipsychotic drug for the treatment of schizophrenia and bipolar disorder was investigated. The objective for Chematica in this exercise was to find a synthesis that would avoid the only known (and patented route). All in all, the route proved easily scalable to multigram quantities and gave an overall yield of ~55%; that is, two and a half times higher than reported in the patented route. Of note, the protection of the alcohol group in the first step was identified by the software to avoid undesirable substitution later on in the synthesis.

Alternatively, several groups have been trying to use known databases such as REAXYS or SciFinder to scrape information and train AI to define rules (Coley et al. 2019).

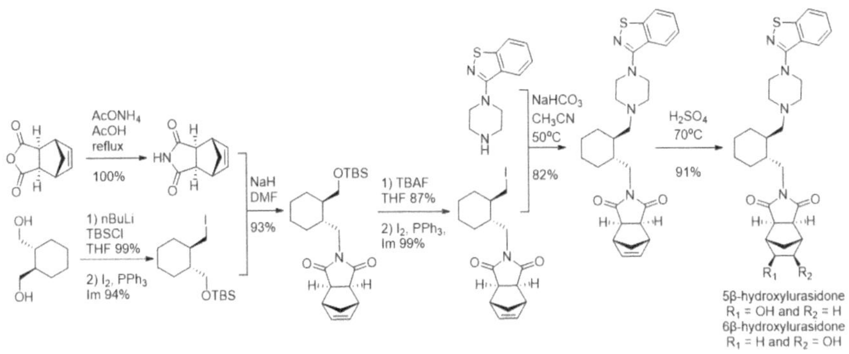

FIGURE 14.1.6 Laboratory synthesis of 5β/6β-hydroxylurasidone as proposed by Chematica. Figure adapted from Klucznik et al. (2018).

14.1.7 Outlook for CASP

Current state-of-the-art CASP programs already demonstrate remarkable success in finding reasonable pathways for synthesizing organic molecules. Among the various approaches, there is no clear winner, and the quality of predictions is ultimately dependent on the quality and quantity of the input data. Template-based approaches are still competitive with template-free approaches, though there is more potential for the latter to continue improving with rapid advances in machine learning. ChatGPT, based on the same transformer architecture that underlies many template-free retrosynthesis models, has demonstrated the incredible possibilities given by large model sizes and training datasets.

However, current CASP programs are still not quite at a level where they can fully replace an experienced chemist. Problematic reactions that are obvious to a chemist are often much more difficult for the computer to identify. Though the models can sometimes suggest rare or unique reactions that a chemist may not have considered, such cases can be buried beneath hundreds of obvious or implausible suggestions. As such,

> Current CASP programs can be viewed primarily as a tool to assist chemists in generating ideas for difficult syntheses or in high-throughput application as a quick screen for synthesizability.

For CASP programs to become more useful, there are improvements needed to single-step and multi-step retrosynthesis algorithms as well as forward synthesis planning tools, all of which are active areas of research. For single-step algorithms, there is still room for improving model awareness of features such as stereochemistry, long-range effects of functional groups, and 3D steric effects. Such features are challenging for existing models which use 1D or 2D representations and may consider reaction centers of limited size. For multi-step algorithms, comparing pathways and identifying the best pathway is a major challenge, as is quantifying route strategy. Existing algorithms can easily generate hundreds or thousands of routes, and it is challenging to select the most viable ones to present to a chemist or to use in an automated synthesis setup. For forward synthesis planning, reaction condition prediction needs to be more accurate and quantitative to facilitate automated synthesis, and improvements to language models suggest that experimental procedures could possibly be generated as well. Selectivity considerations and prediction of side products and impurities is also an open area of research, constrained by the limited amount of data.

REFERENCES

Chen, B., C. Li, H. Dai, and L. Song. 2020. "Retro*: Learning Retrosynthetic Planning with Neural Guided A* Search." *Proceedings of the 37th International Conference on Machine Learning, in Proceedings of Machine Learning Research* 119, 1608–1616. https://proceedings.mlr.press/v119/chen20k.html

Coley, Connor W., Regina Barzilay, Tommi S. Jaakkola, William H. Green, and Klavs F. Jensen. 2017. "Prediction of Organic Reaction Outcomes Using Machine Learning." *ACS Central Science* 3 (5), 434–443. https://doi.org/10.1021/acscentsci.7b00064

Coley, Connor W., L. Rogers, W.H. Green, and K.F. Jensen. 2018. "SCScore: Synthetic Complexity Learned from a Reaction Corpus". *Journal of Chemical Information and Modeling* 58 (2), 252–261. https://doi.org/10.1021/acs.jcim.7b00622

Coley, Connor W., et al. 2019. "A Robotic Platform for Flow Synthesis of Organic Compounds Informed by AI Planning." *Science* 365 (6453), eaax1566.

Coley, Connor W., et al. 2021. "ASKCOS v0.4.1." *GitHub*. https://github.com/ASKCOS/ASKCOS

Corey, E.J., W.T. Wipke, R.D. Cramer, and W.J. Howe. 1972. "Computer-Assisted Synthetic Analysis. Facile Man-Machine Communication of Chemical Structure by Interactive Computer Graphics." *Journal of the American Chemical Society* 94 (2), 421–430.

Ertl, P., and A. Schuffenhauer. 2009. "Estimation of Synthetic Accessibility Score of Drug-Like Molecules Based on Molecular Complexity and Fragment Contributions." *Journal of Cheminformatics* 1 (1). https://doi.org/10.1186/1758-2946-1-8

Fu, M.C. 2016. "AlphaGo and Monte Carlo Tree Search: The Simulation Optimization Perspective." *Winter Simulation Conference (WSC)*, 659–670.

Klucznik, T., et al. 2018. "Efficient Syntheses of Diverse, Medicinally Relevant Targets Planned by Computer and Executed in the Laboratory." *Chem* 4 (3), 522–532. ISSN 2451-9294, https://doi.org/10.1016/j.chempr.2018.02.002

Levin, I., M. Liu, C.A. Voigt, and C.W. Coley. 2022. "Merging Enzymatic and Synthetic Chemistry with Computational Synthesis Planning." *Nature Communications* 13 (1). https://doi.org/10.1038/s41467-022-35422-y

Segler, M., M. Preuss, and M. Waller. 2018. "Planning Chemical Syntheses with Deep Neural Networks and Symbolic AI." *Nature* 555, 604–610. https://doi.org/10.1038/nature25978

Szymkuć, S., E.P. Gajewska, T. Klucznik, K. Molga, P. Dittwald, M. Startek, M. Bajczyk, and B.A. Grzybowski. 2016. "Computer-Assisted Synthetic Planning: The End of the Beginning." *Angewandte Chemie International Edition* 55 (20), 5904–5937. https://doi.org/10.1002/anie.201506101

Thakkar, A., V. Chadimová, E.J. Bjerrum, O. Engkvist, and J.L. Reymond. 2021. "Retrosynthetic Accessibility Score (RAscore)-Rapid Machine Learned Synthesizability Classification from AI Driven Retrosynthetic Planning." *Chemical Science* 12 (9): 3339–3349. https://doi.org/10.1039/d0sc05401a

Wang, X., Y. Qian, H. Gao, C.W. Coley, Y. Mo, R. Barzilay, and K.F. Jensen. 2020. "Towards Efficient Discovery of Green Synthetic Pathways with Monte Carlo Tree Search and Reinforcement Learning." *Chemical Science* 11 (40): 10959–10972. https://doi.org/10.1039/d0sc04184j

14.2 REAL-TIME STATISTICAL PROCESS MONITORING

Anastasia Nikolakopoulou, Ou Yang and Gabriele Bano

14.2.1 Introduction

Traditionally, drugs have been manufactured via chemical synthesis. Over the past 30 years, biological drugs that are derived from biological organisms have been gaining more ground. Biological drugs made up 45% of all 20,109 drugs across the pipeline, meaning all stages of development, from the preclinical stage up to launch to the market, as of January 2022 (Intelligence 2022). While drugs developed via chemical synthesis still consist of more than 50% of the pipeline, the biological drugs share has been consistently increasing; in 2015 biologic drugs amounted to just below 30% of all drugs in development. Among biological drugs, monoclonal antibodies (mAbs) hold the leading position in the pipeline with 2,681 drugs under development in 2022, a 7.9% increase compared to 2021 (Intelligence 2022). Additionally, mAbs amount to 10% of the active compounds listed in the top 25 therapeutic categories in 2022 (Intelligence 2022). Monoclonal antibodies are a class of proteins able to target foreign particulate matter and eliminate it from the body (Reichert 2001). Their high specificity and reduced immunogenicity are the primary reasons behind the high interest in these proteins as therapeutic agents for a wide range of diseases from cancers, and autoimmune, to inflammatory diseases (Yang et al. 2000; Buss et al. 2012; Hong et al. 2018).

Manufacturing of therapeutic mAbs consists of a series of unit operations divided in upstream and downstream processing (Hong et al. 2018). In upstream processing, a series of steps facilitates cell culture and scale-up. Typically, cell culture expansion in flasks of gradually increasing sizes is followed by cell culture expansion in bioreactors of increasing sizes. The largest bioreactor is the production bioreactor where the therapeutic protein is expressed and then harvested. Cells, cell debris, and other impurities are removed via centrifugation. In downstream processing, the harvested therapeutic protein is purified in a series of steps, including various chromatography, filtration, and diafiltration techniques, as well as a viral inactivation step if the process involves mammalian cells (Hong et al. 2018).

MAbs are commonly manufactured in batch or fed-batch bioreactors, although continuous operations have also been increasingly attracting attention (Fisher et al. 2016; Yee et al. 2018; Luo et al. 2023). The mode of operation is chosen by the manufacturers based on economic reasons (Xu et al. 2017; Yang et al. 2019), protein stability, and equipment availability (Yee et al. 2018). In a batch operation, the cell culture and the necessary nutrients (media) for the cell growth are charged to the bioreactor at the beginning of the operation and the therapeutic protein is harvested at the end of the operation, after several days. In contrast, in a continuous cell culture method, such as in a perfusion bioreactor, the culture medium is continuously provided to the bioreactor while toxic metabolites and the product are regularly removed. Fed-batch bioreactors, which are commonly encountered in an industrial setting, are charged with the cell culture and media on the first day of the operation, while nutrient supplements are provided to the system in scheduled intervals to promote cell growth and protein production, and the drug is harvest at the end of the operation (Yang et al. 2000).

Various factors impact the operation of a bioreactor system. Appropriate oxygenation, agitation, and tight temperature and pH control are paramount to the cells' growth and metabolism (Ozturk 1996). N_2, O_2, and CO_2 (for pH control) are provided to the bioreactor at the top (overlay) or at the bottom through a ring sparger and/or a micro sparger, ensuring that proper mass transfer of the gases is taking place through the liquid phase. Anti-foam solution is added to prevent excessive foaming due to gas bubbles formed inside the bioreactor at the gas/liquid interface. Base solution addition is used for pH control additionally to CO_2 addition, depending on which pH region the bioreactor is operating at. The culture media design is of utmost importance to maintain a high cell count within the bioreactor. The addition strategy of the media has a major impact on the nutrient profile inside the bioreactor and its careful design aims to prevent depletion of certain nutrients, for example, glucose or glutamine, and accumulation of inhibitory metabolites such as lactate and ammonia (Lu et al. 2013). Protein concentration and viable cell density are most commonly used as proxies for productivity (Luo et al. 2023).

Due to this high complexity of bioreactor operations, enhancing productivity and improving product quality need to be facilitated by the deployment of process optimization (Lu et al. 2013), monitoring (Li et al. 2018; Tulsyan et al. 2021), and control methodologies (Craven et al. 2014; Yee et al. 2018; Luo et al. 2023). Mathematical models are used to predict the dynamic behavior of the cells and are critical to the success of such methodologies. In these mathematical models, operating variables that can be adjusted during the operation, for example, temperature, dissolved oxygen (DO), agitation speed, pH, and volume of medium feed added to the fed-batch bioreactor, are treated as system *inputs* (Luo et al. 2023). Variables such as the protein concentration (titer), viable cell density, as well as the concentration of metabolites such as lactate, ammonia, and nutrients including glucose reflect the state of the system at different times and will be referred to as system *states*. The subset of system states, typically measured daily in biomanufacturing operations, will be referred to as system *outputs*.

The effect of the bioreactor system inputs to the system states and outputs can be quantified via different types of mathematical representations, depending on the degree of understanding of the system. Mechanistic models can capture mechanisms that take place at a fundamental level during the mAb production, such as the cellular metabolism (Sha et al. 2018) or the post-translation modification of glycosylation which impacts the mAb efficacy (Jimenez del Val et al. 2011). These models are accompanied by a high degree of complexity and if parameter estimation is carried out carefully their predictive capabilities can be extrapolative. However, their development is resource intensive. On the other hand,

Data-driven models, such as neural networks, or gaussian processes, can formalize statistical correlations observed between the process inputs and outputs based on data collected during the biomanufacturing operations (Chtourou et al. 1993; Tulsyan et al. 2021).

In this chapter, we demonstrate how data-driven models based on principal component analysis can be developed and utilized for real-time process monitoring of bioreactor operations. The process of interest is a fed-batch cell culture running over 12-day batches.

14.2.2 Case Study

In this case study, we demonstrate how production scale bioreactor data for monoclonal antibody manufacturing from 30 batches were used to construct statistical

TABLE 14.2.1
Operating Variables Used in Each Model

Type of Model	Type of Variable	List of Variables
Trajectory model	Dynamic, online	pH, air flow through ring sparger, O_2 flow through microsparger, vessel pressure, microsparger pressure, vessel weight, anti-foam weight, base weight, cell growth time
PCA model	Static, online	DO, overlay/microsparger air flow, N_2 flow through ring sparger, overlay/ring sparger CO_2 flow, O_2 flow through ring sparger, agitator speed, jacket temperature, reactor temperature, vent filter 1&2 temperature
Trajectory model	Dynamic, offline	VCC, cell viability, pH, CO_2 partial pressure, O_2 partial pressure, glucose concentration, lactate concentration

models for on-line process monitoring (MacGregor 1994) in order to timely identify excursions in the operations and prevent batch failures. The data were grouped into two categories based on the frequency of collection. Measurements for variables such as DO, overlay air flow, and O_2 flow through the microsparger, referred to as *online variables*, were collected every 360 seconds. Variables such as viable cell concentration (VCC), cell viability, and glucose concentration, referred to as *offline variables*, were collected once a day.

The progression of offline and online variables in the batch was monitored via two different types of statistical models, namely trajectory and principal component analysis (PCA) models (Westad et al. 2015). Trajectory models are used to monitor the time evolution of dynamic variables, for example, pH, total airflow through the ring sparger, VCC, and cell viability. PCA models are used to monitor the progression of static variables whose values are expected to remain constant or within a specified range during the batch, for example, DO, reactor temperature, and agitator speed. The variables involved in the development of each type of model are listed in Table 14.2.1.

Methodology

In an industrial setting, it is common that the available data quality and quantity are not sufficient to generate a model suitable for a specific application. In a statistical process monitoring application context, when a manufacturing process is still new and consequently the number of historical batches is low, generating statistical models to assess nominal versus abnormal operating conditions based on historical data is challenging. In-silico data generation, using stratified resampling and Gaussian state-space model training based on historical nominal data, has been proposed as a remedy to this problem and is also adopted as a solution in this case study (Tulsyan et al. 2019).

Thirty in-silico trajectories of the dynamic, online variables were generated using three high-frequency real batches (Gasparini et al. 2024). For the static, online variables, each of the three high-frequency trajectories was used to separately generate ten in-silico trajectories, in order to maintain the same level of variability in the in-silico trajectories compared to the original real batch data. The in-silico trajectory generator used a resampling interval of 7,200 seconds (2 hours). Additionally, a nominal data set and a data set with faults were generated as testing cases for both offline and online variables. The faults were introduced as random excursions acting on the nominal values.

In-silico data generation for the offline variables was challenging due to the low frequency of available data in the real batches (daily measurements over a 12-day culture). The low-frequency offline variables were linearly interpolated to generate high-frequency sequences (120 samples per batch), which were consequently used to generate in-silico batches with 24 samples per batch, similar to the online data. Thirty in-silico trajectories of the dynamic, offline variables were generated in total from three real batches. The faulty data set was constructed by introducing random perturbations in the bioreactor glucose concentration and accounting for its effect to the other offline process variables using a mechanistic model developed in-house. The mechanistic model was calibrated using one real batch.

The in-silico data were generated in MATLAB 2023b and used to develop trajectory and PCA models for the dynamic and static variables, respectively, in Aspen Unscrambler. Trajectory models for the dynamic online and offline variables were constructed separately.

Results and Discussion

The results for all models are visualized in Figures 14.2.1–14.2.27 with the help of scores plots, while the process monitoring performance is assessed through Hotelling's T^2 and F-residuals diagnostics charts.

The scores plots graph the observations in the principal component space. Increasing distance between points indicates an increased difference in process conditions.

In the diagnostics charts, points above the indicated limit in Hotelling's T^2 statistics plots suggest process conditions outside the normal batch variation as defined by the model. Points above the indicated limit in the F-residuals plots suggest that the process correlation structure has been disrupted, therefore, flagging the observation of a type of event not accounted for in the training set.

Online PCA Model

Thirty in-silico batches were used to train the PCA model for the static, online process variables (Figure 14.2.1).

Each cluster in Figure 14.2.1 indicates slightly different operating conditions among the historical data.

Cross-validation was used to validate the model as shown in Figure 14.2.2, where the solid line represents the explained variance as a function of the number of principal components included in the model for the calibration data set and the dashed line

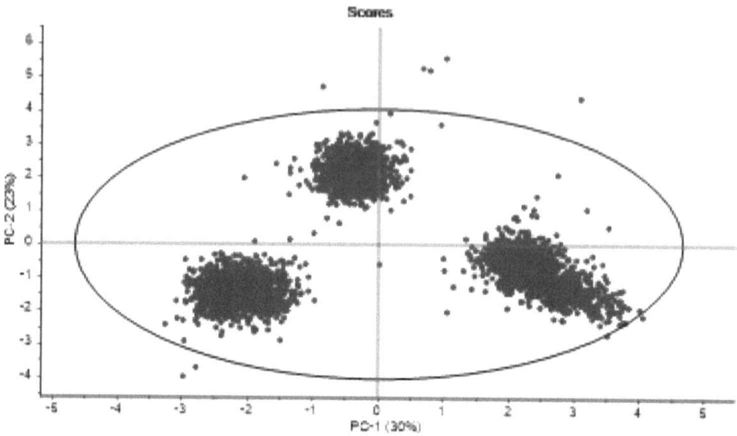

FIGURE 14.2.1 Scores plot of the first two principal components for the static, online process variables PCA model. The dots represent the nominal process instances in the training data set and the black line represents the 95% confidence interval.

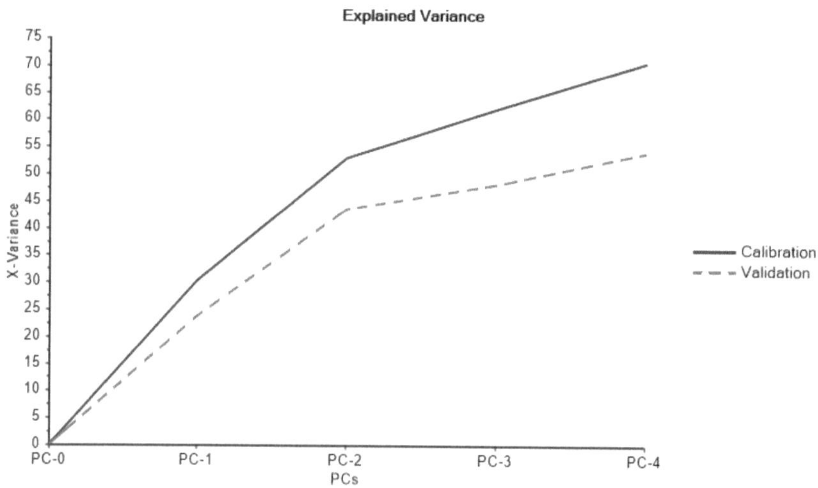

FIGURE 14.2.2 Explained variance as a function of the principal components included in the PCA model for the static, online process variables for the calibration (solid line) and validation (dashed line) data sets.

corresponds to the validation data set. The validation set was selected using random shuffling in Aspen Unscrambler. Four principal components, capturing 75% of the variance, were included in the final model.

The nominal data set scores plots (black dots) are superimposed with the testing data sets scores plots (gray dots) in Figure 14.2.3 and Figure 14.2.4. While the first testing batch seems to operate within the nominal operating conditions (Figure 14.2.3), the second testing case displays signs of excursion with multiple

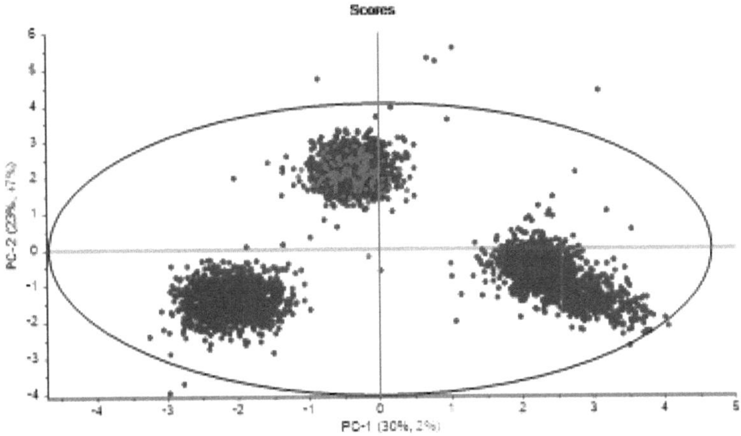

FIGURE 14.2.3 Scores plot of the two first principal components for the static, online process variables PCA model. The black dots represent the nominal process instances in the training data set and the gray dots the first testing data set. The ellipse represents the 95% confidence interval. The dots overlap because the first testing data set is collected under nominal operating conditions.

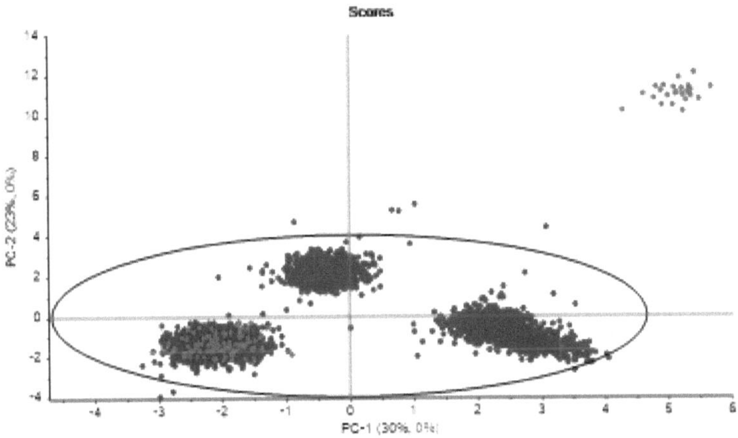

FIGURE 14.2.4 Scores plot of the two first principal components for the static, online process variables PCA model. The black dots represent the nominal process instances in the training data set and the gray dots the second testing data set. The ellipse represents the 95% confidence interval. Some of the gray dots overlap with the black dots pointing to a nominal operation while a subset of the gray dots are outside the nominal operating region (top right).

operating points lying outside the 95% confidence interval region indicated with the black line (Figure 14.2.4).

The contribution of each batch sample to the first and second principal components' Hotelling's T^2 statistic for the second testing data set is shown in Figures 14.2.5 and 14.2.6, respectively. The dashed line in the figures marks the Hotelling's T^2 critical limit, which is calculated based on the *F*-test (p value < 0.05) and can be used to identify outliers. Figures 14.2.5 and 14.2.6 indicate that samples labeled as number 70 to number 95 were responsible for the excursion of the operation outside the nominal operating range as determined by the model.

We can further investigate which static, online process variables were responsible for the excursion of each sample.

For example, the high Hotelling's T^2 value for sample number 86 of the second testing data set can be attributed to the variable "Temp Control" (see Figure 14.2.7), which is associated with the bioreactor temperature. This analysis extended to all the samples manifesting an excursion can give insights into fault identification.

Operators can use the method outlined in this section to identify operating issues related to static, online process variables when consecutive outliers are detected during the biomanufacturing operation.

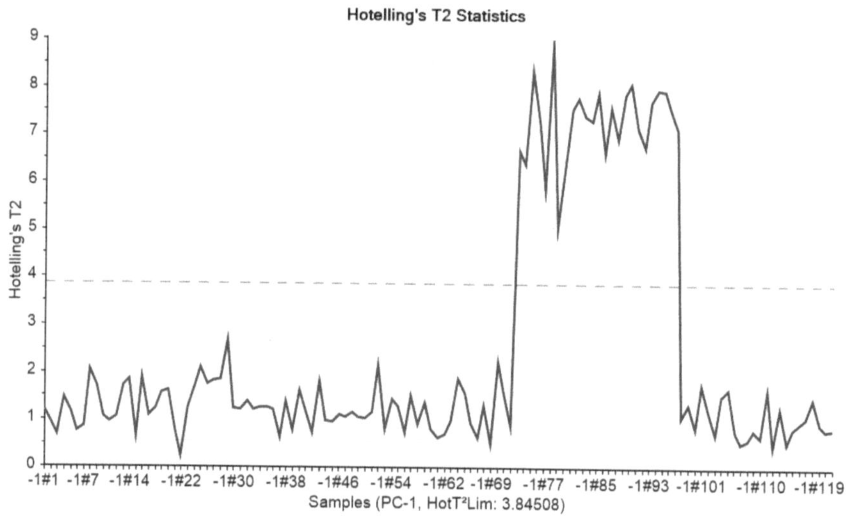

FIGURE 14.2.5 Hotelling's T^2 statistic for each batch sample's contribution to the first principal component for the second testing data set (solid line). The dashed line represents the Hotelling's T^2 critical limit calculated based on the *F*-test (p value = 0.05). Samples labeled as number 70 to number 95 lie above the dashed line indicating an excursion of the operation outside the nominal operating range.

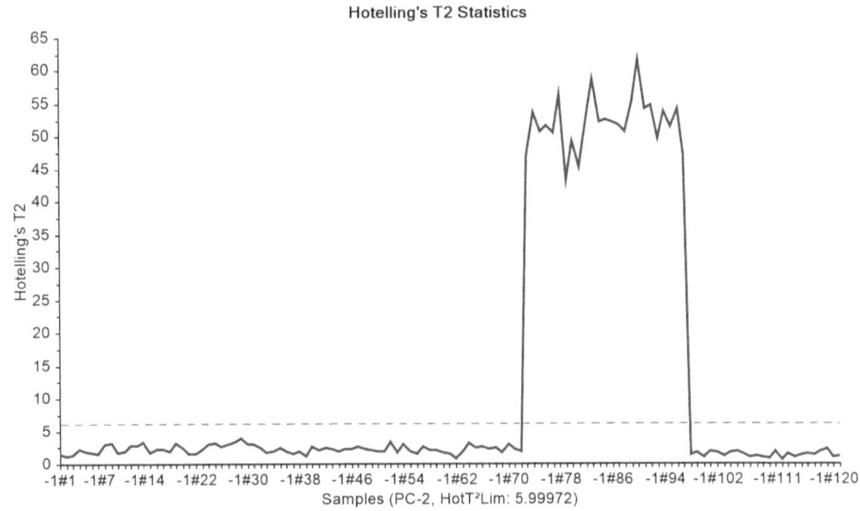

FIGURE 14.2.6 Hotelling's T^2 statistics for each batch sample contribution to the second principal component for the second testing data set (solid line). The dashed line represents the Hotelling's T^2 critical limit calculated based on the *F*-test (p value = 0.05). Samples labeled as number 70 to number 95 lie above the dashed line indicating an excursion of the operation outside the nominal operating range.

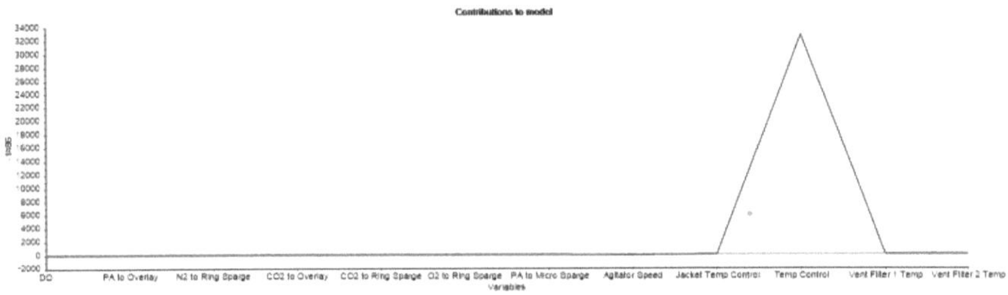

FIGURE 14.2.7 Contribution of each static, online process variable to Hotelling's T^2 statistic for sample number 86 from the second testing data set. Temperature control appears to have a bigger contribution compared to the other static, online process variables.

Online Trajectory Model

Thirty in-silico batches were used to train the trajectory model for the dynamic, online process variables (Figure 14.2.8).

Cross-validation was used to validate the model as shown in Figure 14.2.9, where the solid line represents the explained variance in the data as a function of the number of principal components included in the model for the calibration data set and the dashed line corresponds to the validation data set. Two principal components were included in the final model, accounting for a total of about 55% of the variance in the training data set.

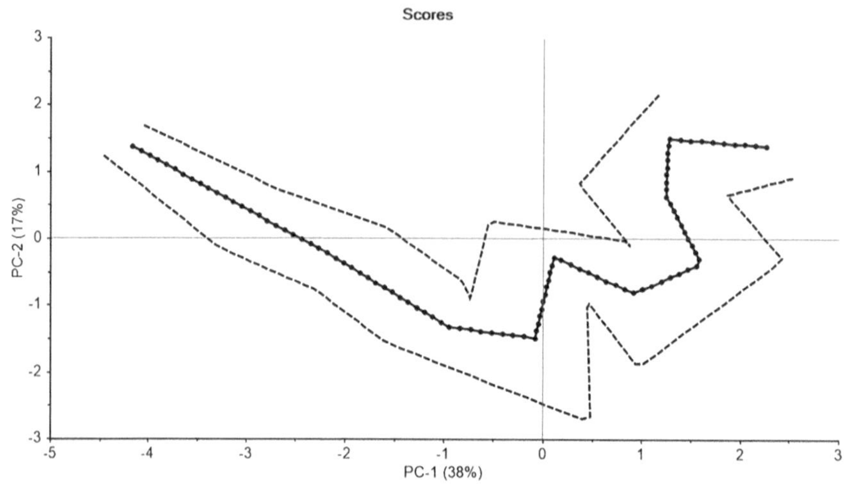

FIGURE 14.2.8 Scores plot of the two first principal components for the dynamic, online process variables trajectory model. The dotted line represents the nominal trajectory in the training data set and the dashed lines represent the 95.5% confidence intervals.

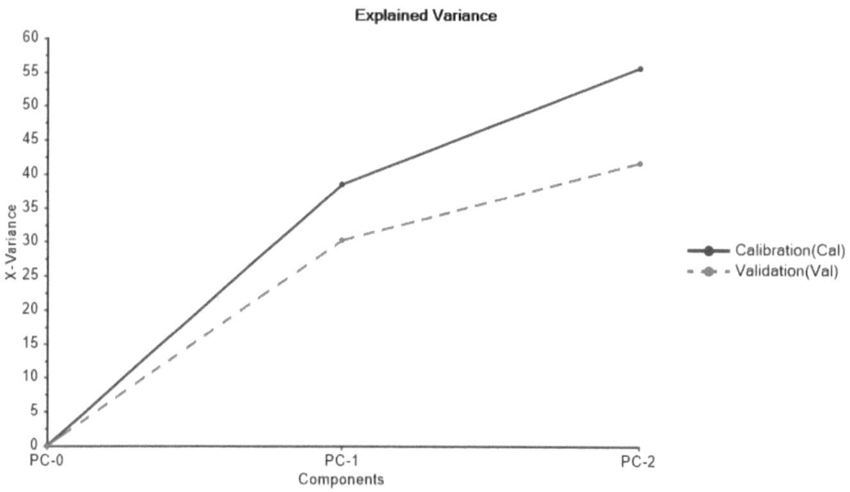

FIGURE 14.2.9 Explained variance as a function of the number of principal components included in the trajectory model for the dynamic, online process variables for the calibration (solid line) and validation (dashed line) data sets.

The nominal data set scores plot (nominal trajectory) is superimposed with the first and second testing data sets scores plots (process trajectory) in Figures 14.2.10 and 14.2.13, respectively. The dashed lines represent the 95.5% confidence interval (2 standard deviations). While in the first testing batch (Figure 14.2.10) very few

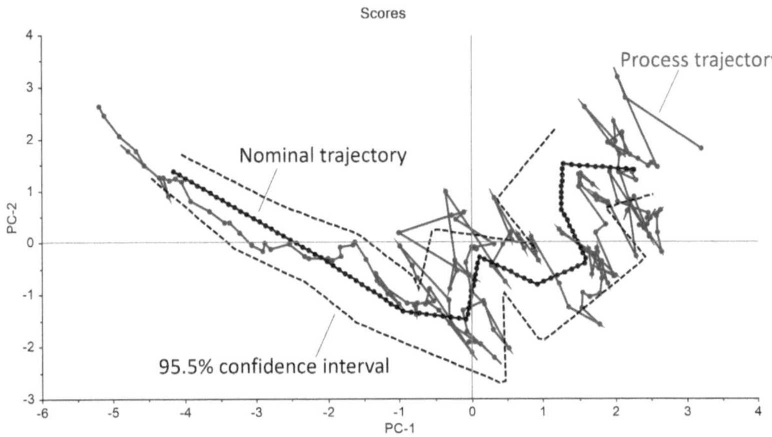

FIGURE 14.2.10 Scores plot of the two first principal components for the dynamic, online process variables trajectory model. The black dotted line represents the nominal trajectory in the training data set and the dashed lines represent the 95.5% confidence intervals. The process trajectory corresponds to the first testing data set which lies for the most part within the 95.5% confidence intervals.

FIGURE 14.2.11 Hotelling's T^2 statistic for each batch sample's contribution to the second principal component for the first testing data set (dotted line). The solid line represents the Hotelling's T^2 critical limit calculated based on the *F*-test (p value = 0.05). The Hotelling's T^2 statistic for almost every sample lies below the Hotelling's T^2 critical limit.

points cross outside the confidence intervals region, in the second testing batch (Figure 14.2.13) many points are located significantly outside of the confidence interval region, which implies an excursion from the nominal trajectory.

The Hotelling's T^2 statistic for each batch sample's contribution to the first two principal components retained in the model were calculated for the two testing data (Figures 14.2.11 and 14.2.14). The individual samples that cross the Hotelling's T^2 critical limit are responsible for deviations related to the nonconformity of these samples to the model. The second testing data set excursion from the nominal trajectory which was observed in Figure 14.2.13, is also corroborated by Figure 14.2.14, where samples between the 34th and 58th trajectory instances are observed to substantially cross outside the bounds of the Hotelling's T^2 critical limit. The same samples are pointed out as responsible for an excursion by the *F*-residual plot corresponding to the second training data set (Figure 14.2.15). The *F*-residual plot corresponding to the first training data set (Figure 14.2.12) agrees with prior observations that the first training data set does not manifest any notable abnormalities compared to the nominal batch.

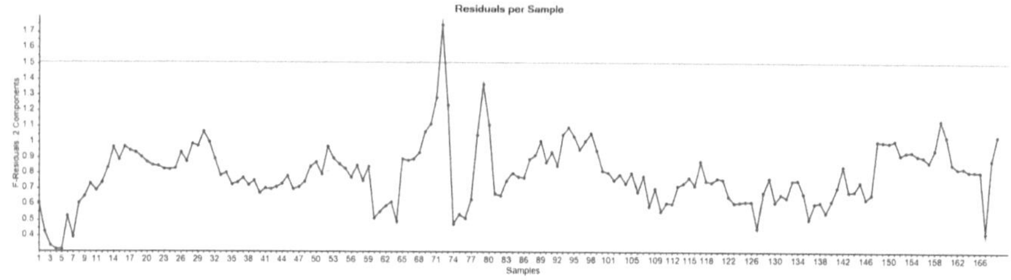

FIGURE 14.2.12 *F*-residuals (dotted line) for each batch sample's contribution to the second principal component of the first testing data set. The solid line represents the critical limit calculated based on the *F*-test (p value = 0.05). The *F*-residuals for almost every sample lie below the critical limit calculated based on the *F*-test.

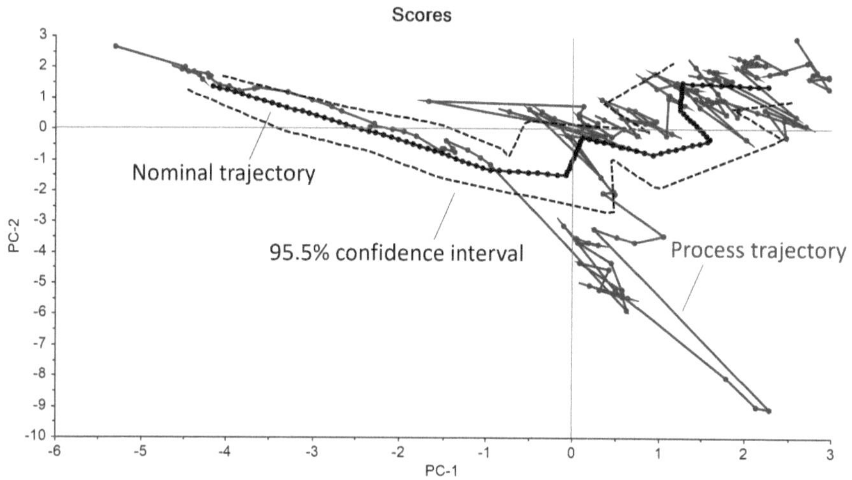

FIGURE 14.2.13 Scores plot of the two first principal components for the dynamic, online process variables trajectory model. The black dotted line represents the nominal trajectory in the training data set, and the dashed lines represent the 95.5% confidence intervals. The process trajectory corresponds to the second testing data set. Segments of the dotted line lie outside the 95.5% confidence intervals.

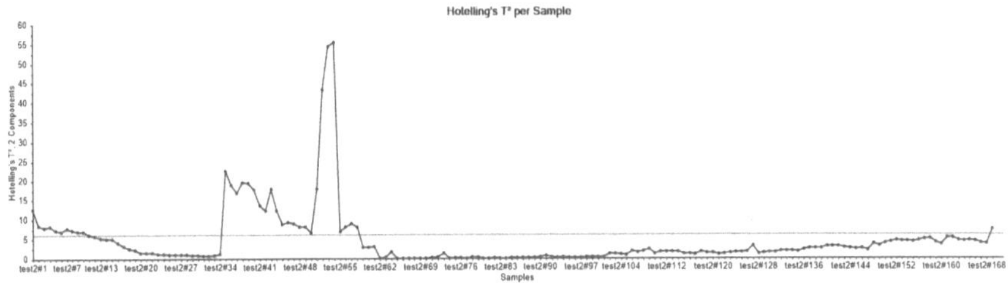

FIGURE 14.2.14 Hotelling's T^2 statistic for each batch sample's contribution to the second principal component for the second testing data set (dotted line). The solid line represents the Hotelling's T^2 critical limit calculated based on the F-test (p value =0.05). The Hotelling's T^2 statistic for some samples lies above the Hotelling's T^2 critical limit.

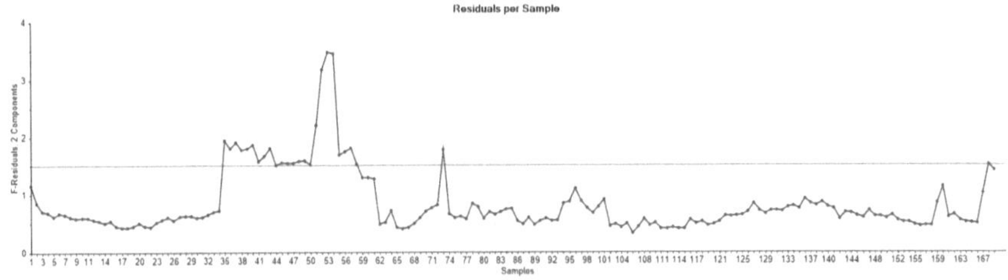

FIGURE 14.2.15 F-residuals (dotted line) for each batch sample's contribution to the second principal component of the second testing data set. The solid line represents the critical limit calculated based on the F-test (p value = 0.05). The F-residuals for some samples lie above the critical limit calculated based on the F-test.

We can further investigate which dynamic, online process variables were responsible for the excursion of each sample in the trajectory of the second testing data set.

For example, the high Hotelling's T^2 value for sample number 54 of the second testing data set can be mostly attributed to the process variables associated with the air flow through the ring sparger and the O_2 flow through the microsparger (see Figure 14.2.16). Similarly, the process variables contributing to the F-residuals for sample number 54 of the second testing data are associated mostly with O_2 flow through the microsparger and to some extent with the pH, the anti-foam addition as well as the base addition (see Figure 14.2.17). A higher F-residual due to these process variables implies that the process is manifesting an excursion in an operating region not observed in the training data and can be used complementary to the Hotelling's T^2, which quantifies the extent to which the process is close to the nominal trajectory.

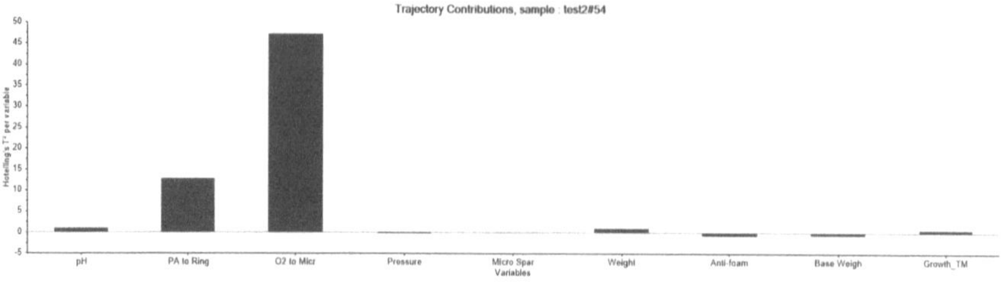

FIGURE 14.2.16 Contribution of each dynamic, online process variable to the Hotelling's T^2 statistic of sample number 54 from the second testing data set.

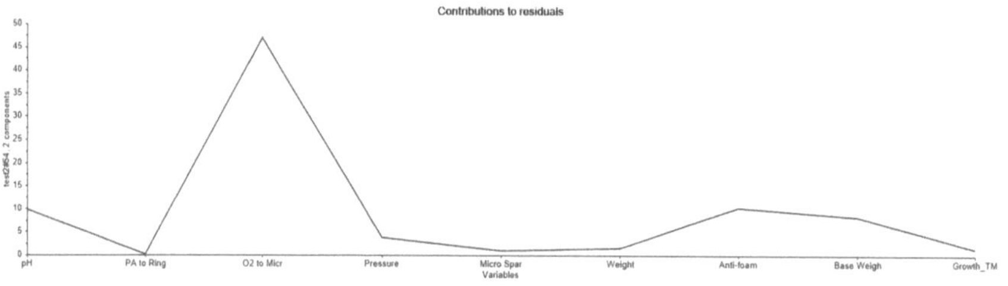

FIGURE 14.2.17 Contribution of each dynamic, online process variable to the *F*-residuals statistic of sample number 54 from the second testing data set.

Operators can use the method outlined in this section to identify operating issues related to dynamic, online process variables when consecutive outliers are detected during the biomanufacturing operation.

Offline Trajectory Model

Thirty in-silico batches were used to train the trajectory model for the dynamic, offline process variables (Figure 14.2.18), as outlined in Section 14.2.2.1.

Cross-validation was used to validate the model as shown in Figure 14.2.19, where the solid line represents the explained variance in the data as a function of the number of principal components included in the model for the calibration data set and the dashed line corresponds to the validation data set. Two principal components were included in the final model, accounting for a total of about 75% of the variance in the training data set.

The nominal data set scores plot (nominal trajectory) is superimposed with the first and second testing data sets scores plots (process trajectory) in Figures 14.2.20 and 14.2.23, respectively. The dashed lines represent the 95.5% confidence intervals (2 standard deviations). In the first testing batch, very few points cross outside the confidence intervals region (Figure 14.2.20); however, in the second testing batch, a greater number of batch samples are located significantly outside of the confidence interval region, which implies an excursion from the nominal trajectory (Figure 14.2.23).

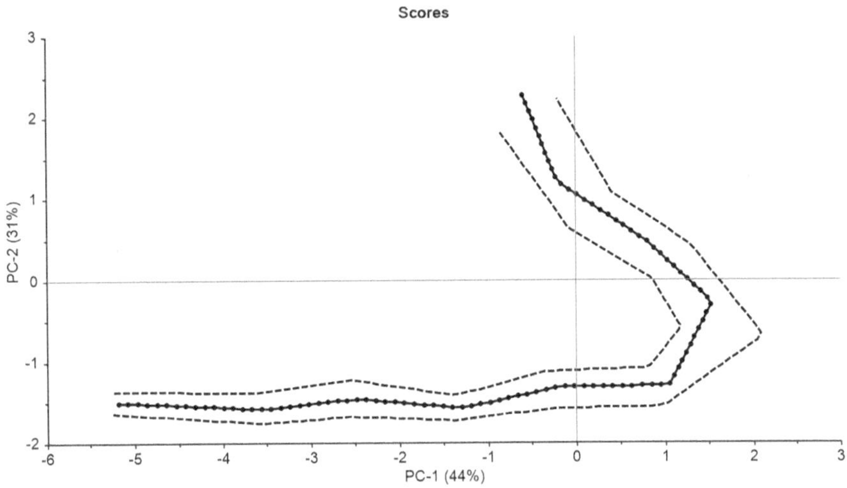

FIGURE 14.2.18 Scores plot of the two first principal components of the dynamic, offline process variables trajectory model. The dotted line represents the nominal trajectory in the training data set and the dashed lines represent the 95.5% confidence interval.

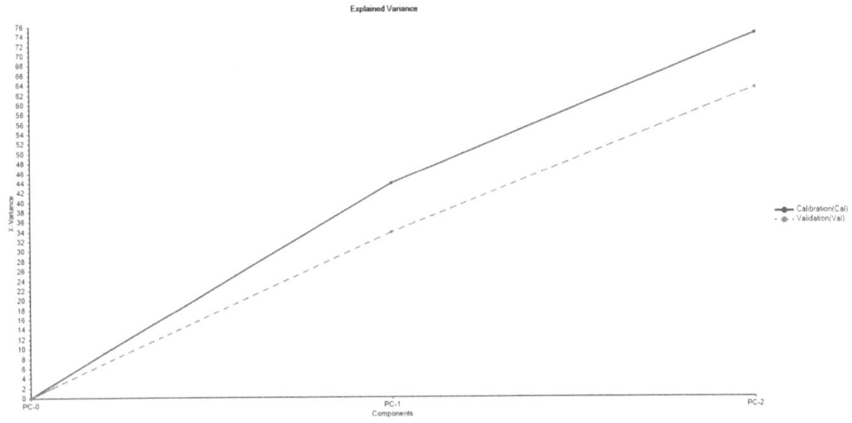

FIGURE 14.2.19 Explained variance as a function of the number of principal components included in the trajectory model for the dynamic, offline process variables for the calibration (solid line) and validation (dashed line) data sets.

The Hotelling's T^2 statistic for each batch sample's contribution to the first and second principal components retained in the model were calculated for the two testing data sets respectively (Figures 14.2.21 and 14.2.24). The individual samples that cross the Hotelling's T^2 critical limit are responsible for significant deviations of these samples from the center of the model. The second testing data set manifested an excursion from the nominal trajectory in Figure 14.2.23. This finding is also supported by Figure 14.2.24, where the bounds of the Hotelling's T^2 critical

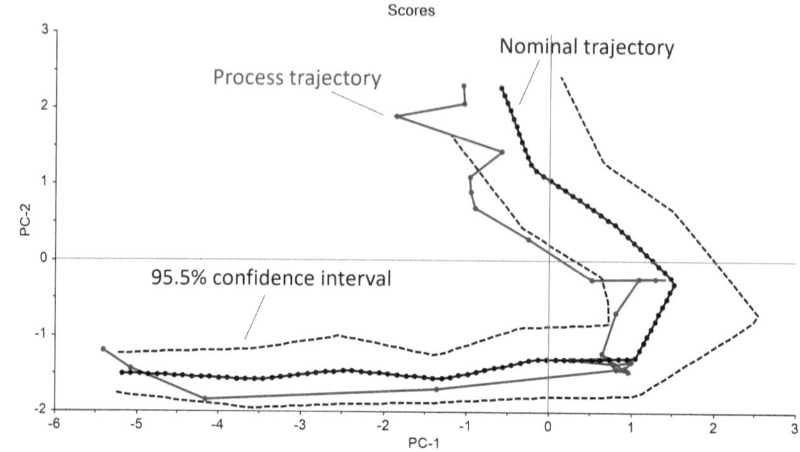

FIGURE 14.2.20 Scores plot of the two first principal components of the dynamic, offline process variables trajectory model. The black dotted line represents the nominal trajectory in the training data set and the dashed lines represent the 95.5% confidence intervals. The process trajectory corresponds to the first testing data set which lies for the most part within the 95.5% confidence intervals.

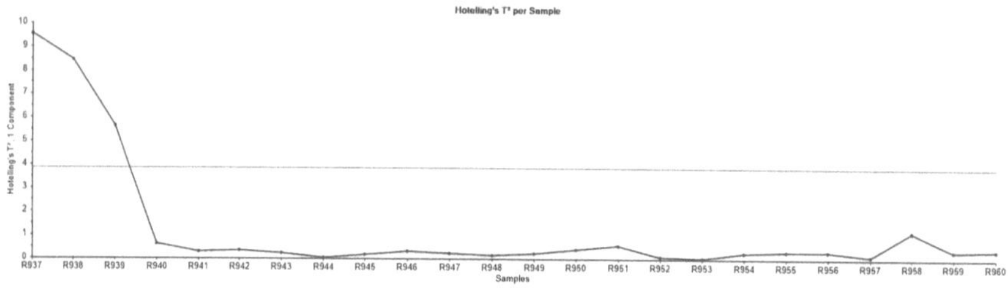

FIGURE 14.2.21 Hotelling's T^2 statistic for each batch sample's contribution to the first principal component of the first testing data set (dotted line). The solid line represents the Hotelling's T^2 critical limit calculated based on the F-test (p value = 0.05). The Hotelling's T^2 statistic for almost every sample lies below the Hotelling's T^2 critical limit.

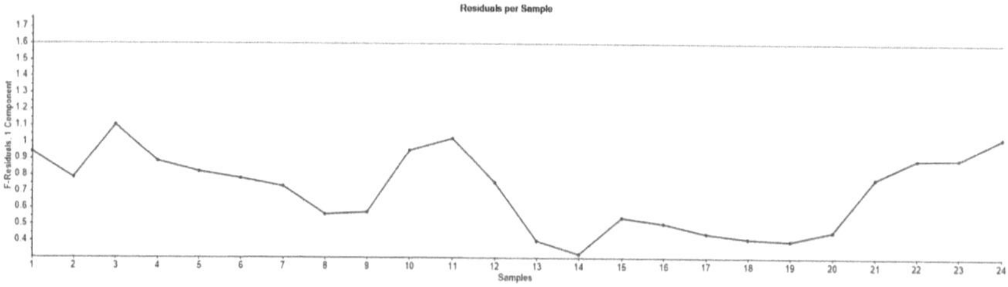

FIGURE 14.2.22 F-residuals (dotted line) for each batch sample's contribution to the first principal component of the first testing data set. The solid line represents the critical limit calculated based on the F-test (p value = 0.05). The F-residuals for all samples lie below the critical limit calculated based on the F-test.

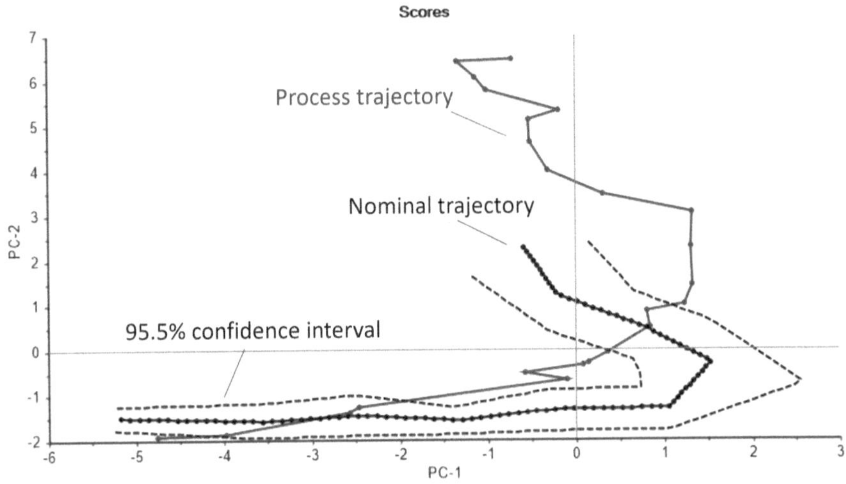

FIGURE 14.2.23 Scores plot of the two first principal components of the dynamic, offline process variables trajectory model. The black dotted line represents the nominal trajectory in the training data set and the dashed lines represent the 95.5% confidence intervals. The process trajectory corresponds to the second testing data set. Segments of the dotted line lie outside the 95.5% confidence intervals.

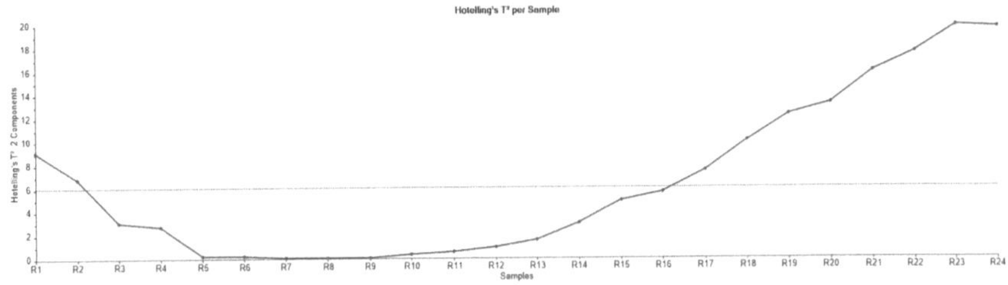

FIGURE 14.2.24 Hotelling's T^2 statistic for each batch sample's contribution to the second principal component of the second testing data set (dotted line). The solid line represents the Hotelling's T^2 critical limit calculated based on the F-test (p value = 0.05). The Hotelling's T^2 statistic for some samples lies above the Hotelling's T^2 critical limit.

limit are crossed starting from the 17th sample of the trajectory and onward. The F-residual plot corresponding to the second training data set (Figure 14.2.25) starts showing signs of potential deviation earlier compared to the Hotelling's T^2 statistic (Figure 14.2.24), thus illustrating the importance of using both statistics metrics to gain insights into process excursions. The Hotelling's T^2 and F-residual plots from the first testing data set (Figures 14.2.21 and 14.2.22, respectively) verify that the process is operating normally as observed in the corresponding trajectory plot (Figure 14.2.20).

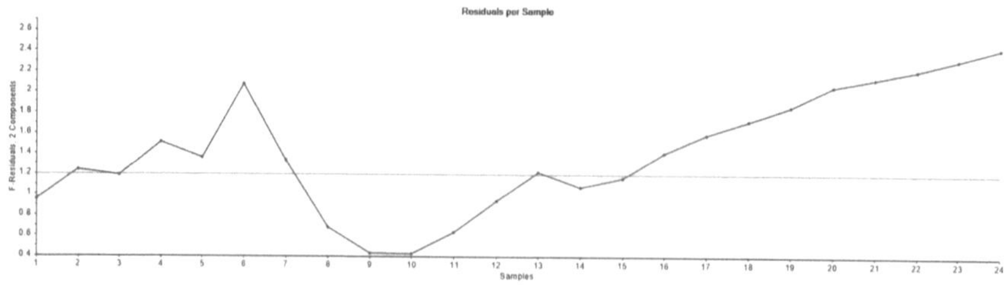

FIGURE 14.2.25 *F*-residuals (dotted line) for each batch sample's contribution to the second principal component of the second testing data set. The solid line represents the critical limit calculated based on the *F*-test (p value = 0.05). The *F*-residuals for some samples lie above the critical limit calculated based on the *F*-test.

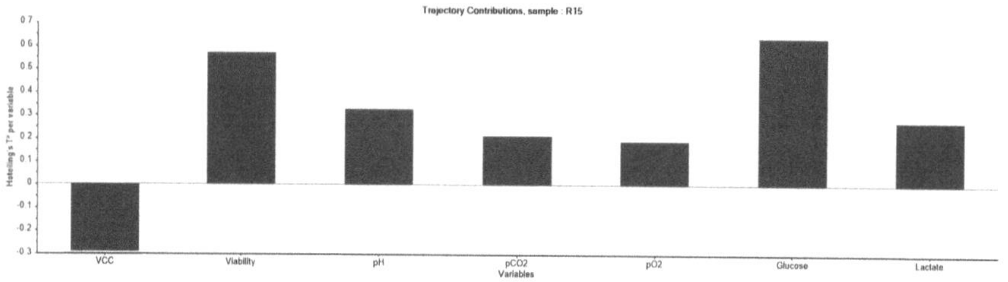

FIGURE 14.2.26 Contribution of each dynamic, offline process variable to the Hotelling's T^2 statistic of sample number 15 from the second testing data set.

> The dynamic, offline process variables primarily responsible for the deviation of each sample in a faulty trajectory can be identified by examining Hotelling's T^2 and *F*-residuals contributions.

For example, the high Hotelling's T^2 value for sample number 15 of the second testing data set can be mostly attributed to the glucose concentration and the cell viability (Figure 14.2.26). VCC, pH, and lactate concentration contribute substantially to the Hotelling's T^2 as well. This is expected due to the high correlation among these offline variables, which is captured by the model. The *F*-residuals for sample number 15 of the second testing data are associated mostly with the glucose concentration and VCC and to a lesser extent with the lactate concentration (Figure 14.2.27). Therefore, by using the *F*-residuals we gain the insight that the glucose concentration is mostly responsible for the trajectory's deviation from the historical batches, which is a useful insight for fault identification.

Operators can use the method outlined in this section to identify operating issues related to dynamic, offline process variables when consecutive outliers are detected during the biomanufacturing operation.

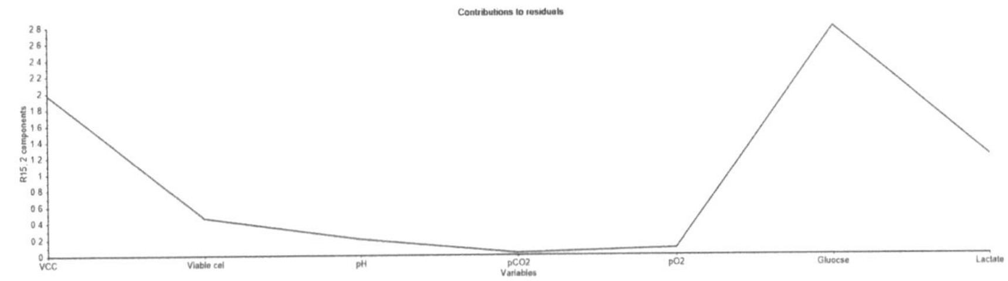

FIGURE 14.2.27 Contribution of each dynamic, offline process variable to the F-residuals statistic of sample number 15 from the second testing data set.

14.2.3 Conclusion

In this chapter, multivariate statistical models combined with metrics such as the F-residuals and Hotelling's T^2 were used to successfully detect excursions from nominal operating conditions during a production bioreactor run. Three different multivariate statistical models were built in Aspen Unscrambler to monitor the evolution of dynamic (online and offline) and static, online process variables. Trajectory models, which define the nominal operating region and corresponding confidence intervals, were constructed for the dynamic variables. PCA models, which contain the nominal operating region and its corresponding confidence intervals, were constructed for the static variables using nominal training data sets. Two test batches were examined for each model category. In all cases, when the batch ended up in a region outside of the nominal, the event was detected using various statistics tests. When multiple consecutive deviations from the nominal historical data were detected, they were flagged as an excursion. This approach is very powerful in a practical industrial setting because persisting deviations can be flagged to operators, ensuring timely intervention, and avoiding faults, hence leading to more efficient biomanufacturing operations.

14.2.4 Outlook

In this chapter, we demonstrated how real-time multivariate statistical process monitoring is important for the analysis of system performance and can facilitate raising fault alarms during process excursions by applying this methodology to cell culture data.

A natural extension of process monitoring is computer-aided decision-making to determine the system input values which will maintain the bioreactor on an optimal operating trajectory even under the existence of disturbances.

Advanced control algorithms, for example, model predictive control (MPC) have yet to be deployed in a biomanufacturing commercial production setting, but ongoing efforts indicate that closed-loop control will find its application in the pharmaceutical industry in the near future (Craven et al. 2014; Yee et al. 2018; Luo et al. 2023).

REFERENCES

Buss, N. A. P. S., S. J. Henderson, M. McFarlane, J. M. Shenton, and L. de Haan. 2012. "Monoclonal antibody therapeutics: history and future." *Current Opinion in Pharmacology* 12(5): 615–622.

Chtourou, M., K. Najim, G. Roux, and B. Dahhou. 1993. "Control of a bioreactor using a neural network." *Bioprocess Engineering* 8(5): 251–254.

Craven, S., J. Whelan, and B. Glennon. 2014. "Glucose concentration control of a fed-batch mammalian cell bioprocess using a nonlinear model predictive controller." *Journal of Process Control* 24(4): 344–357.

Fisher, A. C., S. L. Lee, D. P. Harris, L. Buhse, S. Kozlowski, L. Yu, M. Kopcha, and J. Woodcock. 2016. "Advancing pharmaceutical quality: an overview of science and research in the U.S. FDA's Office of Pharmaceutical Quality." *International Journal of Pharmaceutics* 515(1): 390–402.

Gasparini, L., A. Benedetti, G. Marchese, C. Gallagher, P. Facco, and M. Barolo. 2024. "On the use of machine learning to generate in-silico data for batch process monitoring under small-data scenarios." *Computers & Chemical Engineering* 180: 108469.

Hong, M. S., K. A. Severson, M. Jiang, A. E. Lu, J. C. Love, and R. D. Braatz. 2018. "Challenges and opportunities in biopharmaceutical manufacturing control." *Computers & Chemical Engineering* 110: 106–114.

Intelligence, C. P. 2022. Pharma R&D Annual Review 2022 Navigating the Landscape.

Jimenez del Val, I., J. M. Nagy, and C. Kontoravdi. 2011. "A dynamic mathematical model for monoclonal antibody N-linked glycosylation and nucleotide sugar donor transport within a maturing Golgi apparatus." *Biotechnology Progress* 27(6): 1730–1743.

Li, M.-Y., B. Ebel, C. Paris, F. Chauchard, E. Guedon, and A. Marc. 2018. "Real-time monitoring of antibody glycosylation site occupancy by in situ Raman spectroscopy during bioreactor CHO cell cultures." *Biotechnology Progress* 34(2): 486–493.

Lu, F., P. C. Toh, I. Burnett, F. Li, T. Hudson, A. Amanullah, and J. Li (2013). "Automated dynamic fed-batch process and media optimization for high productivity cell culture process development." *Biotechnology and Bioengineering* 110(1): 191–205.

Luo, Y., V. Kurian, L. Song, E. A. Wells, A. S. Robinson, and B. A. Ogunnaike. 2023. "Model-based control of titer and glycosylation in fed-batch mAb production: modeling and control system development." *AIChE Journal* 69(4): e18075.

MacGregor, J. F. 1994. "Statistical process control of multivariate processes." *IFAC Proceedings Volumes* 27(2): 427–437.

Ozturk, S. S. 1996. "Engineering challenges in high density cell culture systems." *Cytotechnology* 22(1): 3–16.

Reichert, J. M. 2001. "Monoclonal antibodies in the clinic." *Nature Biotechnology* 19(9): 819–822.

Sha, S., Z. Huang, Z. Wang, and S. Yoon. 2018. "Mechanistic modeling and applications for CHO cell culture development and production." *Current Opinion in Chemical Engineering* 22: 54–61.

Tulsyan, A., C. Garvin, and C. Undey. 2019. "Industrial batch process monitoring with limited data." *Journal of Process Control* 77: 114–133.

Tulsyan, A., H. Khodabandehlou, T. Wang, G. Schorner, M. Coufal, and C. Undey. 2021. "Spectroscopic models for real-time monitoring of cell culture processes using spatiotemporal just-in-time Gaussian processes." *AIChE Journal* 67(5): e17210.

Westad, F., L. Gidskehaug, B. Swarbrick, and G. R. Flåten. 2015. "Assumption free modeling and monitoring of batch processes." *Chemometrics and Intelligent Laboratory Systems* 149: 66–72.

Xu, S., J. Gavin, R. Jiang, and H. Chen. 2017. "Bioreactor productivity and media cost comparison for different intensified cell culture processes." *Biotechnology Progress* 33(4): 867–878.

Yang, J.-D., Y. Angelillo, M. Chaudhry, C. Goldenberg, and D. M. Goldenberg. 2000. "Achievement of high cell density and high antibody productivity by a controlled-fed perfusion bioreactor process." *Biotechnology and Bioengineering* 69(1): 74–82.
Yang, O., S. Prabhu, and M. Ierapetritou. 2019. "Comparison between Batch and Continuous Monoclonal Antibody Production and Economic Analysis." *Industrial & Engineering Chemistry Research* 58(15): 5851–5863.
Yee, J. C., M. S. Rehmann, G. Yao, S. W. Sowa, K. L. Aron, J. Tian, M. C. Borys, and Z. J. Li. 2018. "Advances in process control strategies for mammalian fed-batch cultures." *Current Opinion in Chemical Engineering* 22: 34–41.

14.3 RAW MATERIAL SPECIFICATIONS

Joan Borràs-Ferrís, Daniel Palací-López, Carl Duchesne and Alberto Ferrer

14.3.1 Introduction

Problems derived from the use of univariate raw material specifications

Raw materials properties (Z) are usually considered as critical input parameters because their variability has an impact on Critical Quality Attributes (CQAs) of the final product. Thus, as commented by Duchesne and MacGregor (Duchesne and MacGregor 2004), the development of specification regions for raw materials is crucial to ensure the desired quality of the product.

> Despite their importance, specifications are usually defined in an arbitrary way based mostly on subjective past experience, instead of using a quantitative objective description of their impact on CQAs.

Furthermore, in many cases, univariate specifications on each property are designated, with the implicit assumption that these properties are independent from one another. As a consequence, significant amounts of raw materials whose properties are correlated may be misclassified, as appropriate or otherwise, when univariate specifications are considered, as it is shown in Figure 14.3.1.

Let us consider a raw material with two correlated properties, Z1 and Z2 (see Figure 14.3.1), used in the manufacturing of a particular product with final product quality Y. The elliptical region "A" is the true multivariate region in Z1 and Z2 such that any batch of raw material used with Z1 and Z2 properties falling within it will provide good product quality (i.e., within the Y quality specification limits). On the contrary, raw material batches with properties outside this elliptical region correspond to unacceptable raw material batches, as they lead to poor product quality (i.e., outside the Y quality specification limits). The square region "B" corresponds to the univariate specification region when accepting the same variance on each individual property as the multivariate region. In this case, accepting raw material batches with properties outside region "A" and inside region "B" leads to manufacturing products with final product quality Y outside its specification limits. To avoid this, companies

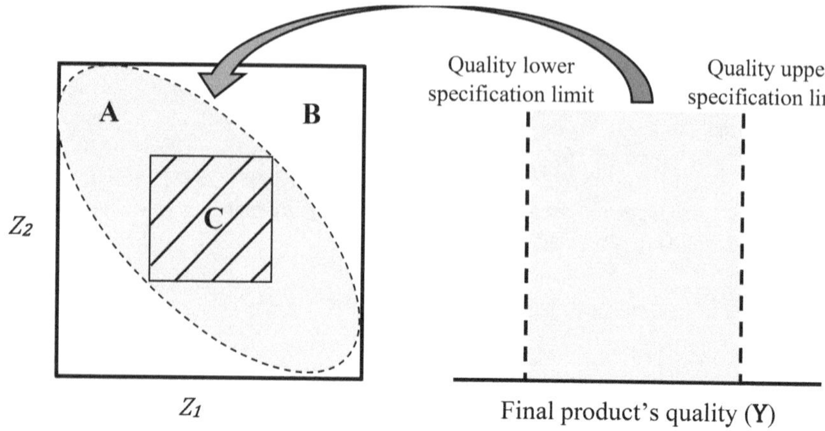

A. True multivariate region.

B. Univariate specification region containing the true multivariate region.

C. Univariate specification region containing only acceptable raw material.

FIGURE 14.3.1 Problem of using univariate specifications on correlated raw material properties (Z1 and Z2).

are forced to shrink the univariate specifications from region "B" to the region "C," at the cost of rejecting acceptable raw material batches (i.e., those outside region "C" but inside region "A"). Another consequence of setting these more restrictive univariate specifications is an increase in costs in the acquisition of raw material batches with tighter variations in their properties. Multivariate specifications provide, therefore, much insight into what constitutes acceptable raw material batches when their properties are correlated (as usually happens).

The Design Space in the Raw Material Paradigm

The advantages of being able to define a precisely meaningful multivariate raw material specification region are enormous. They allow increasing the number of potential suppliers, by allowing a wider range of raw material properties, without compromising the CQAs of the final product.

> Defining meaningful specifications means finding a multivariate window of raw material properties given a desired set of CQAs.

In this regard, such a window refers to the subspace lest likely to fall outside of the true design space as commented in Chapter 12, the so-called High-Confidence Design Space (HCDS) of raw materials. Thus, the HCDS of raw materials (hereinafter only HCDS) is defined as the multidimensional combination and interaction of raw material properties that have been demonstrated to provide assurance of quality.

The HCDS would allow evaluating the capability of the raw material batches of producing product with CQAs within specification limits, before producing a single unit of the product, and based on that information, making a decision about accepting or rejecting the batch of raw materials offered by a given supplier. This is totally different to existing approaches that evaluate (and also accept or reject) raw material batches based on their raw material properties but not on the desired final product properties.

Development of a Proper Control Strategy under the Quality-by-Design Approach

Let us consider a two-dimensional latent space, and a Lower Specification Limit (LSL) is considered for one CQA. Thus, although working in the null space associated with the LSL leads to a predicted value of CQA between specifications, it might yield out of specifications values due to prediction uncertainties. In this sense, Borràs-Ferrís et al. (2022) proposed to define the HCDS as the region in the latent space where any batch of raw material with particular properties, once processed, yield a final product with CQA prediction interval inside specifications. Thus, when calculating the confidence limit for the multiple solutions along the null space of the LSL, a non-linear boundary is obtained: Lower Specification Confidence Limit (LSCL).

Figure 14.3.2 shows the HCDS as the intersection region delimited by the lower specification confidence limit, and the Hotelling T^2 confidence ellipsoid.

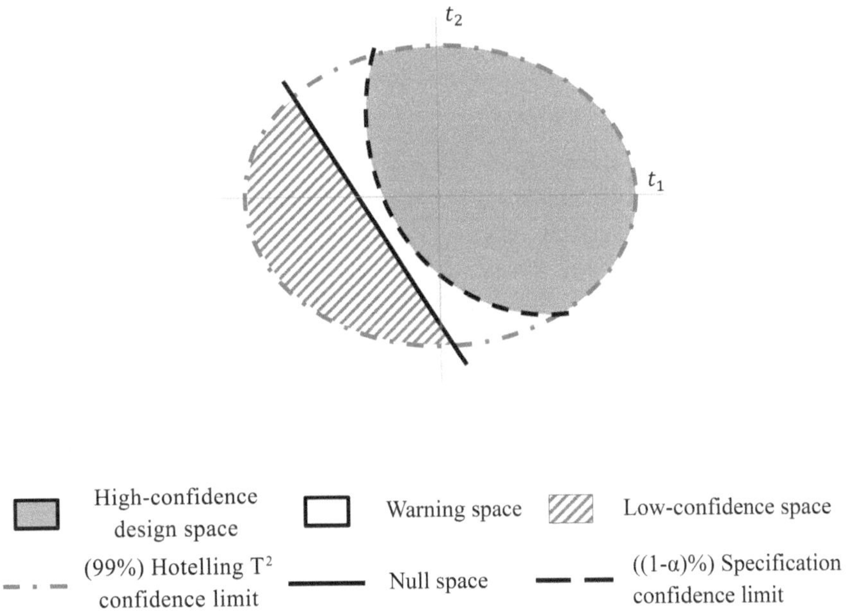

FIGURE 14.3.2 Defining the high-confidence design space, warning space, and low-confidence space in the two-dimensional latent space.

From a frequentist probabilistic interpretation, batches belonging to the HCDS in Figure 14.3.2 are expected to produce a final product with CQAs within specification limits with a confidence level equal or higher than 1-α. The HCDS is a potential opportunity to establish real-time release testing, which is defined as the ability to evaluate and ensure the acceptable quality of final product based on input variables (e.g., raw material properties) without using end-product testing (ICH Harmonised Tripartite 2009).

Additionally, the intersection between the region bounded by the null space corresponding to the LSL, and the Hotelling's T^2 confidence region, but outside the HCDS, defines the so-called warning space (Figure 14.3.2). Note that, although this space does not belong to the HCDS as defined, it does not necessarily imply the rejection of batches. In fact, batches lying within the warning space lead to predicted values between specifications, but they result in prediction intervals for the CQA partially outside of specifications, given the predefined confidence level 1-α. Namely, there is no assurance of quality due to the prediction uncertainty, and, hence, real-time release testing is not feasible. Instead, end-product testing may be employed, which usually involves undertaking specific lab-testing procedures on samples of the final product. This could be interesting when rejecting all batches in the warning space is not affordable. Finally, the low-confidence space leads to predicted values outside specifications. Although batches lying within this subspace may lead to response values between specifications, most of the time such values are expected to be outside.

Therefore, following the proposed approach, the knowledge space is split into three regions: high-confidence design space, warning space, and low-confidence space, providing a strategy where real-time release or end-product testing can be used as needed.

14.3.2 Industrial Case Study

Description of the Dataset

Historical data collected from a maize cereal extraction process is used to illustrate the proposed methodology presented by Borràs-Ferrís et al. (2022). The data available in this case are a compilation of eight raw material properties (Z) of maize: promatest value, protein, acid value, specific weight, burnt grain, broken grain, starch and extractable lipids, and one response variable y (extraction yield of starch slurry). These variables are easily registered in order to assess the feasibility of a raw material batch. In total, 969 historical batches/observations were measured: Z (969 × 8) and y (969 × 1). Besides, a lower specification limit of –2/3 is considered for the response variable.

Performance of the Multivariate Raw Material Specifications

Leave-one-out cross-validation was used for selecting the number of PLS components. Thus, two LVs were chosen to fit a PLS model (R^2_{Zcum} =38.72%, R^2_{Ycum} =29.31% and Q^2_{Ycum} =28.23%) using the 969 calibration observations. The R^2_{Zcum} and R^2_{Ycum} values (goodness of fit) give, respectively, the percentage of the total sum of squares of y and z, that are explained by the fitted PLS model, while the Q^2_{Ycum} (goodness of prediction) gives the percentage of the total sum of squares of the response that can be predicted with the PLS model by cross-validation. None of the historical observations exhibit any unusual behavior caused by special cause process variations based on SPE and T^2 charts (charts not shown).

Figure 14.3.3 illustrates the 99% Hotelling T^2 confidence limit, the NS associated with the LSL, and its 90% confidence limit when considering the prediction uncertainty (i.e., the Low Specification Confidence Limit, LSCL). The intersection of all confidence regions, defined by their limits, yields the HCDS (i.e., the proposed multivariate raw material specifications in the latent space) within which there is assurance of obtaining superior or equal yields to -2/3 with at least 90% confidence level.

To evaluate the performance of the definition of the multivariate raw material specification region, a diagnostic test is carried out by leave-one-out cross-validation. In particular, type I risk, type II risk, and the negative predictive value are calculated for the HCDS. Type I risk is defined as the proportion of truly acceptable batches of raw materials that is rejected by the customer under a given specification region; type II risk consists of the proportion of truly unacceptable batches of raw materials that is

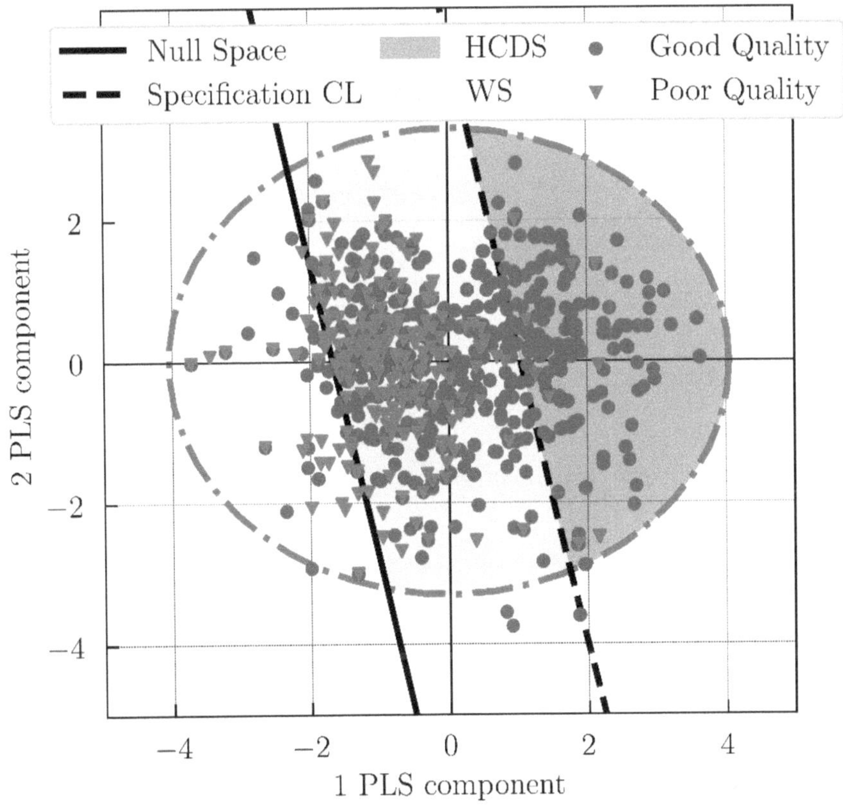

FIGURE 14.3.3 Case Study: Graphical definition of the High-Confidence Design Space (HCDS) and Warning Space (WS) by showing calibration data.

accepted by the customer under a given specification region; the negative predictive value is the proportion of batches that actually result in a good product out of all those within the HCDS, and, hence, this metric is directly connected to the definition of the multivariate raw material specifications itself.

The HCDS is defined with at least a 90% confidence level of obtaining superior or equal yields to -2/3. Thus, one would expect that, of the batches lying within the HCDS, 90% or more would be acceptable batches (the negative predictive value for the HCDS is 96.34%). On the other hand, the HCDS leads to 66.95% type I risk and 3.57% type II risk. This means that if only batches lying within the HCDS are accepted, 3.57% of unacceptable batches of raw materials will be accepted at the expense of rejecting 66.95% of acceptable batches. These results are the consequence of the low PLS goodness of prediction (Q^2_{Ycum} =28.23%) in this case study, due to the fact that historical data presents a low signal to noise ratio. Alternatively, one could accept batches lying within the warning space knowingly that the negative predictive value in such space would be 67.70% and, hence, an end-product test should likely be required. Another option would be to balance the type I and type II risks by modifying the confidence level of the HCDS. Figure 14.3.4 shows the HCDS for different

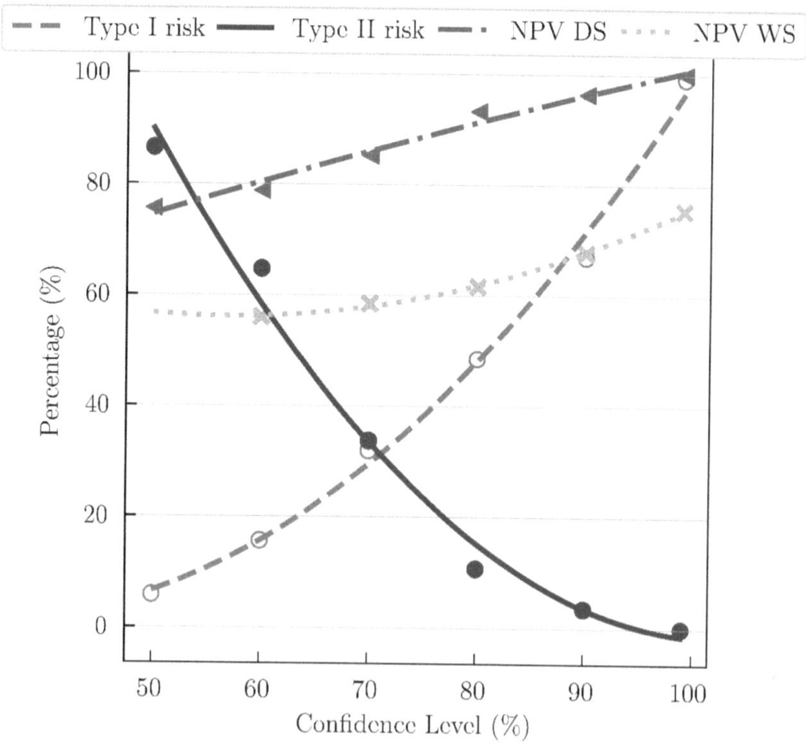

FIGURE 14.3.4 Case Study: type I risk, type II risk, and the Negative Predictive Value for the High-Confidence Design Space (NPV HCDS), and the Negative Predictive Value for the Warning Space (NPV WS) vs. confidence level.

confidence levels. The corresponding type I risk, type II risk and the negative predictive value for the HCDS, and the negative predictive value for the warning space are shown in Figure 14.3.4. Note that, the 50% confidence level case corresponds to the HCDS without considering the uncertainty.

Figure 14.3.4 shows that as confidence level increases, a tighter HCDS is spanned, thereby, the type II risk is reduced at the expense of increasing the type I risk, as is shown in Figure 14.3.4. Therefore, the confidence level of the HCDS must be chosen according to the users by balancing the consequences of having type I and type II errors in their processes, and the total amount of such errors. Besides, for all cases, the negative predictive value is equal or higher than its corresponding confidence level as expected.

Influence of Goodness of Prediction in the Multivariate Raw Material Specifications

In order to investigate how PLS goodness of prediction Q^2_{Ycum} affects the performance of the DS, a simulation study is carried out. In these simulations we assume that the true model relating Z and y is, indeed, the one calculated on the calibration set. Hence, individual values of y are obtained using the latent model where independent random noise, from a normal distribution with zero mean and standard deviation σ, is added. By modifying the value of such standard deviation, one can create simulated datasets yielding PLS models with different goodness of prediction. Figure 14.3.5 shows the HCDS with 90% confidence level of obtaining superior or equal yields to -2/3 for different datasets simulated by using a standard deviation of 0.025, 0.1, 0.5, and 1, yielding Q^2_{Ycum} of 92.38%, 74.86%, 39.87%, and 20.44%, respectively. Nine hundred and sixty-nine batches have been simulated for each dataset.

Figure 14.3.5 shows that the lower the noise standard deviation, the higher the goodness of prediction and, consequently, the clearer the discrimination between acceptable and unacceptable raw materials. Besides, regardless of the goodness of prediction, the proposed method defines the multivariate specification region, given the same confidence level (90%). As can be seen, lower values for the goodness of prediction result in a narrower multivariate specification region where more acceptable material is rejected to guarantee such confidence level. This will affect the type I and type II risks as shown in Figure 14.3.6.

Figure 14.3.6 shows that with moderate/high values of Q^2_{Ycum} (σ <=0,1), it is feasible to obtain HCDS with high confidence level and low type I and II risks, and high negative predictive value. For example, given desired yields equal or superior to -2/3, the HCDS with 90% confidence level and $\sigma = 0.025$ ($Q^2_{Ycum} = 92.38\%$) leads to 10.63% type I risk, 0.89% type II risk, and 99.87% negative predictive value. However, with low values of Q^2_{Ycum} (σ > = 0.5) it is more critical to consider the prediction uncertainty for guarantying quality (i.e., high negative predictive value in the HCDS) at the expense of increasing the type I risk.

Note that the apparently bad performance for low values of Q^2_{Ycum} is solely due to the nature of the data and not the methodology, as noise refers to random variation with no pattern, and therefore usually unavoidable and unpredictable.

In case of desiring to increase the signal-to-noise ratio of the data sets, some process excitation is needed. Multivariate design of experiments can be used such that it provides the greatest amount of additional information with respect to the

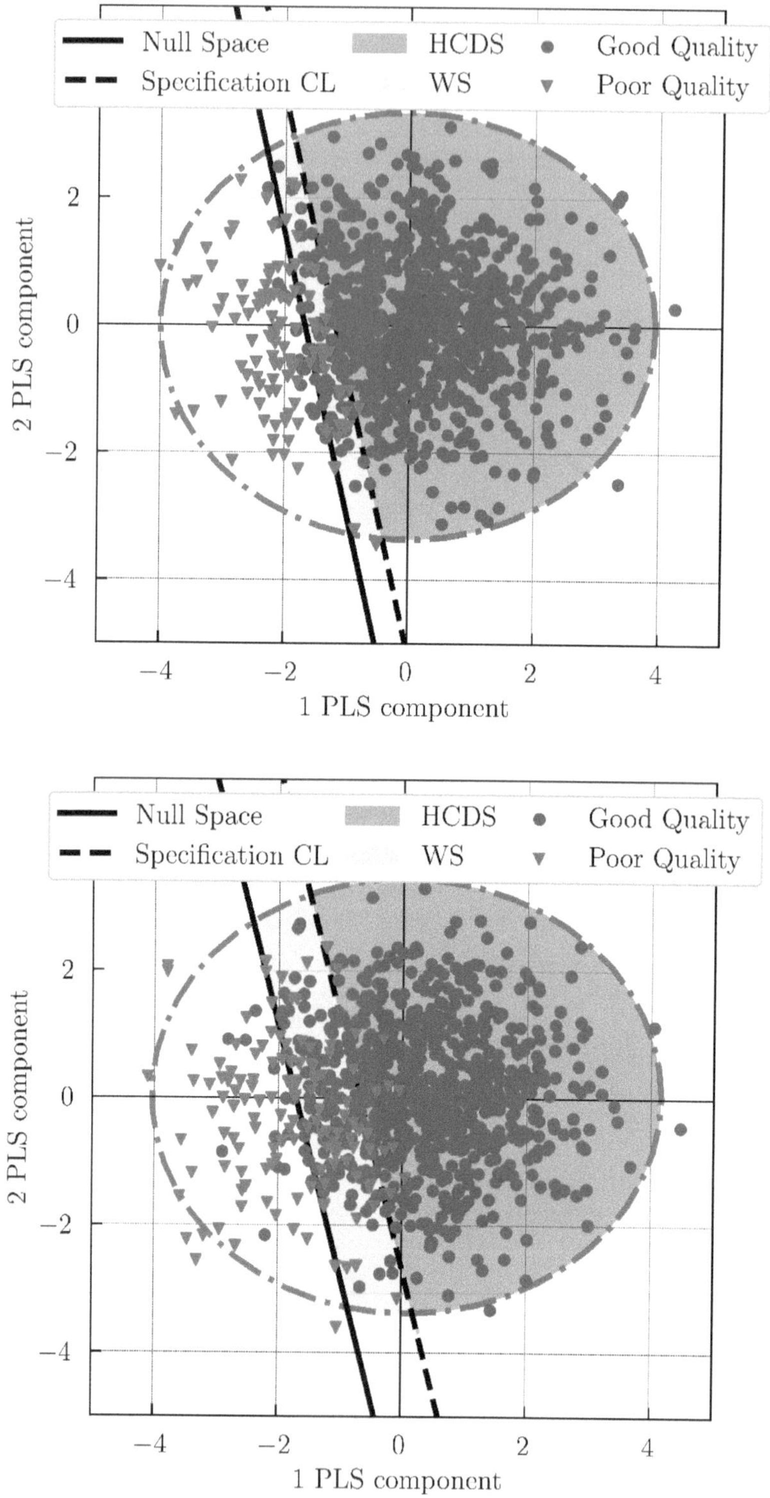

FIGURE 14.3.5 Case Study: High-Confidence Design Space (HCDS) and Warning Space (WS) for simulated data with different noise variability: a) $\sigma = 0.025$; b) $\sigma = 0.1$; c) $\sigma = 0.5$; and d) $\sigma = 1$.

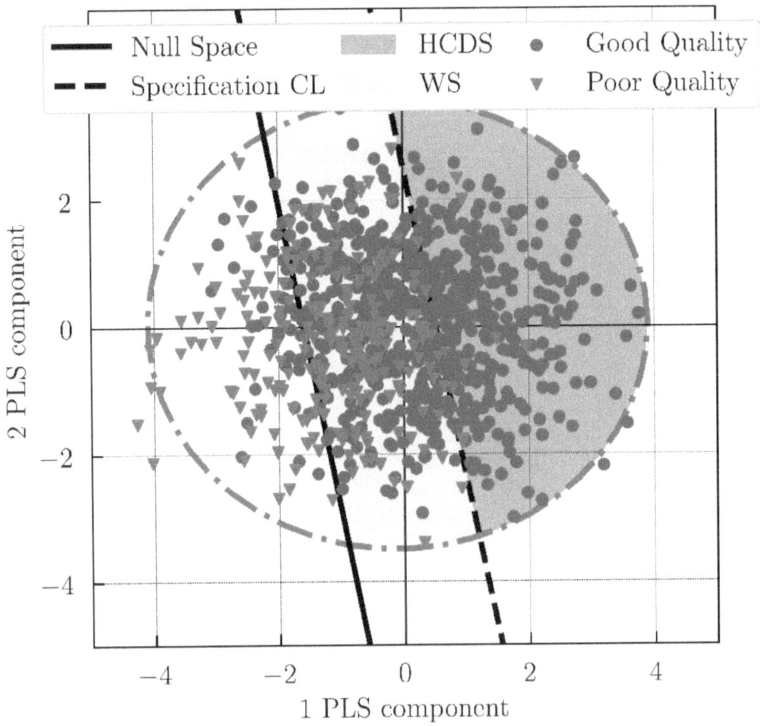

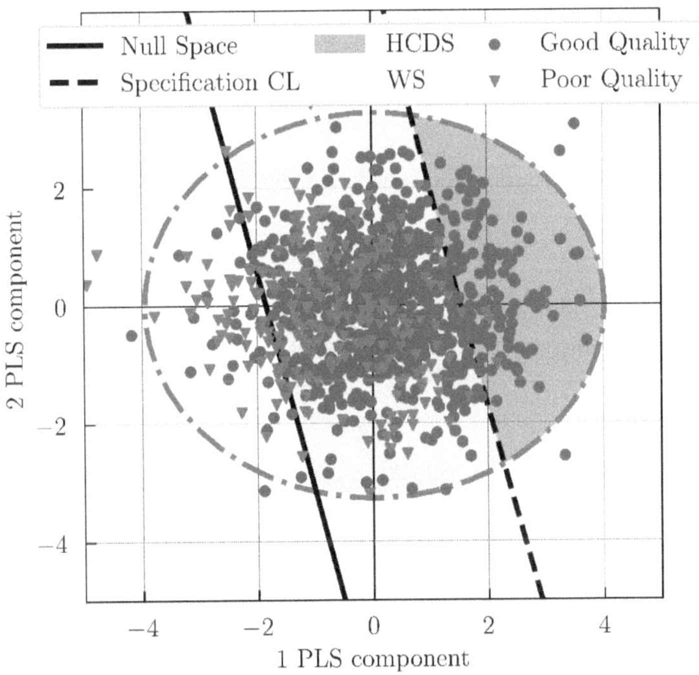

FIGURE 14.3.5 (Continued)

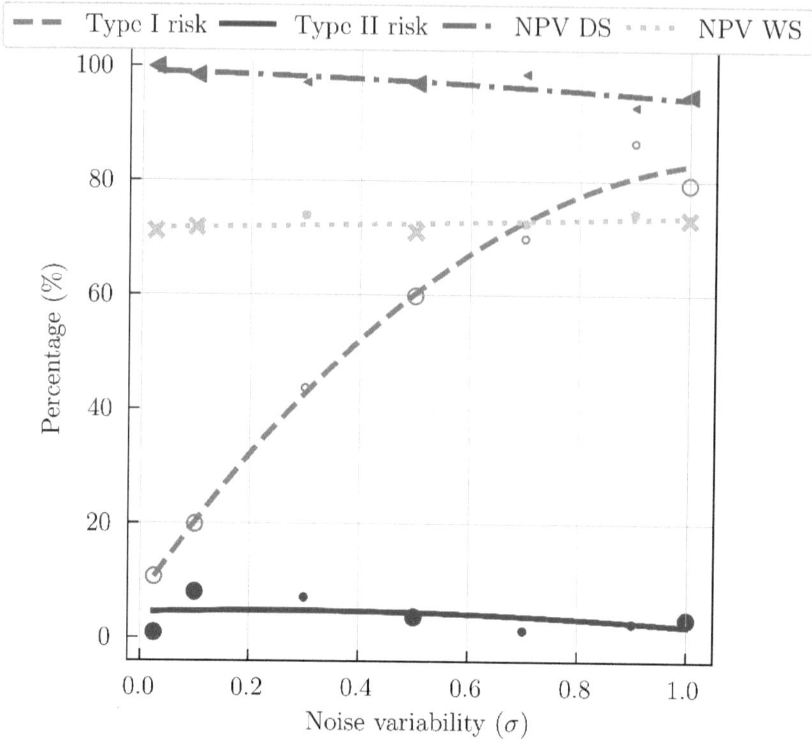

FIGURE 14.3.6 Case Study: type I risk, type II risk, and the Negative Predictive Value for the High-Confidence Design Space (NPV HCDS), and the Negative Predictive Value for the Warning Space (NPV WS) with 90% confidence level vs. noise variability: **0.025**, **0.1**, 0.3, **0.5**, 0.7, 0.9, and **1**. Bold values refer to variability used in Figure 14.3.5 and are shown bigger.

information available in the existing dataset (Wold et al. 2004). Considering these new observations from experimentation in addition to the historical/happenstance data will improve the estimation of the HCDS (i.e., wide multivariate specification region with high confidence level and low type I and type II risks will be obtained).

REFERENCES

Borràs-Ferrís, Joan, Daniel Palací-López, Carl Duchesne, and Alberto Ferrer. 2022. "Defining Multivariate Raw Material Specifications in Industry 4.0." *Chemometrics and Intelligent Laboratory Systems* 225. doi: 10.1016/j.chemolab.2022.104563.

Duchesne, Carl, and John F. MacGregor. 2004. "Establishing Multivariate Specification Regions for Incoming Materials." *Journal of Quality Technology* 36 (1): 78–94. Taylor & Francis. doi: 10.1080/00224065.2004.11980253.

ICH Harmonised Tripartite. 2009. "Guidance for Industry Q8(R2) Pharmaceutical Development." *International Conference on Harmonisation of Technical Requirements for Registration of Pharmaceuticals for Human Use*, Rockville, USA. https://database.ich.org/sites/default/files/Q8_R2_Guideline.pdf

Wold, Svante, Mats Josefson, Johan Gottfries, and Anna Linusson. 2004. "The Utility of Multivariate Design in PLS Modeling." *Journal of Chemometrics* 18 (34): 156–165. Wiley-Blackwell. doi: 10.1002/cem.861.

14.4 CHALLENGES AND BARRIERS TO DATA-DRIVEN FORMULATION DESIGN AND CHARACTERIZATION

Arthi Jayaraman, Shizhao Lu and Alix Schmidt

14.4.1 Need for More Investment in Interdisciplinary Education

As chemical industries start to embrace the power of AI in their R&D divisions, it is critical for universities to invest in educational programs that train the workforce appropriately to handle various interdisciplinary/intercultural aspects of R&D. Currently, most universities offer graduate degrees (Masters, Doctorates) in specific disciplines where the graduate students build upon their technical background knowledge that they gained during their undergraduate education and expand their skill set during graduate education while staying mostly within the boundaries of their own discipline. For AI-driven formulations R&D to succeed, it is vital for researchers to be comfortable with regular interactions with others from completely different technical backgrounds and associated cultures. For example, it is obvious that a team of computer scientists, data scientists, chemical/materials engineers, chemists, and electrical/electronics engineers is required to build the digital-centric lab of the future. If researchers from each of these diverse disciplines remained in their own silos during their graduate training, they will likely never learn about unique cultures in other disciplines versus their own (e.g., nomenclature, importance of proceedings versus publications, varying practices of sharing open-source codes) and technical communication style and jargon in other disciplines.

Teams in R&D with high diversity in individuals' technical expertise will have better success if the members of the team, despite having expertise in specific disciplines, know how to communicate across disciplines or practices prevalent in other disciplines.

For example, for a computer scientist/data scientist (e.g., mathematicians, computer engineers) to customize methods that suit the domain scientists' (e.g., chemists, chemical engineers) problems, we must minimize cultural and language barriers at the very least. This can be achieved through interdisciplinary education and project-based coursework that force interdisciplinary collaboration. It is the authors' experience that project-based courses (e.g., Computing and Data Science Training for Materials Innovation, Discovery, and AnalyticS (NRT-MIDAS 2024)) that are a requirement for an interdisciplinary graduate certificate/degree and bring together students from different departments, serve as valuable platforms to simulate real-life team dynamics in R&D toward technical problem solving. If academic institutions invest personnel (e.g., faculty members) and resources (e.g., classroom space, laboratories) in interdisciplinary degree or certificate programs that complement existing pure disciplinary strength, they will serve the students well and better prepare students for real-life careers, especially in AI-driven industries.

14.4.2 Availability of Data and Open-Source Codes

Except for a handful of cases described in this text, most ML methods rely on having large training data sets (i.e., 1,000 or greater training data points). However, most researchers working with formulation design and characterization and without automated hardware tend to generate 100 or less materials with associated performance characterization for a publication, such as microscopy images or no more than 10–15 scattering profiles. Furthermore, in most cases these data are not available openly or with appropriate metadata for other researchers to use for the purpose of machine learning model training. These data silo challenges were thoroughly discussed in Chapter 4 and affect both industrial and academic data generation. Thus,

> The biggest challenge for the development and use of many of the machine learning methods described earlier is the access to *openly available* datasets with well-curated data of formulated materials.

Readers working with formulations-related data who may want to see examples of good publications that share well-curated data sets of characterization are directed to one such noteworthy publication with openly available characterization data of block copolymer morphologies (Doerk et al. 2023). The reason the authors select this example manuscript is because it served as training data for the authors of Chapter 6 to create a new model without any communications with Doerk et al. The characterization data from Doerk et al. served as training data for the new variational autoencoder (PairVAE) approach developed by Lu and Jayaraman (2023) for the generation of bottleneck characterization data of materials from complementary high throughput characterization data. Openly available high-throughput small angle X-ray scattering (SAXS) data and SEM data of block copolymer assembled morphologies observed by Doerk et al. were used to train the encoders and decoders of the PairVAE, which after successful training generates a SAXS profile from an SEM image as input or an SEM image from SAXS profile as input. Without the open and easy access to both small angle scattering, and microscopy paired data from Doerk et al., such method development by Lu and Jayaraman would not have been possible.

In a 2023 editorial "For chemists, the AI revolution has yet to happen" in the journal *Nature*, the authors state:

> *"Any AI system is only as good as the data it is trained on."*

When machine learning models like neural networks are trained, the available data for training and testing must be large data sets that are balanced and dependable (i.e., superior quality data). For formulation development to be guided by ML and AI tools, it is critical to ensure the availability of data – from experiments, theory, and simulations, current and past, and from both successful (i.e., yielding desired results) and unsuccessful (i.e., working experiment with less than desired outcomes, or negative outcome) experiments. The editorial emphasizes this need for large open available data with this: "*A generalist generative-AI system such as ChatGPT, developed by OpenAI in San Francisco, California, is simply data-hungry. To apply such a generative-AI system to chemistry, large number of data points would be needed.*" Machine learning has succeeded in the field of protein engineering and bioinformatics with open-source tools like Alphafold (Jumper et al. 2021) because of the open collection of protein structures in protein databanks. The ordered secondary and tertiary structures that proteins form or the crystalline structures that inorganic materials form (e.g., BCC or FCC lattice structures) have an advantage in that they are unique and can be reproducibly formed in different laboratories without significant user/processing effects; this is however a major challenge in soft materials and formulations where structures are amorphous (not perfectly crystalline or with repetitive patterns) and naturally have a distribution. Storing finite number of datasets that faithfully represent the broad distribution of structures within a formulation can be a barrier as compared to the ease of storing coordinates of one secondary structure of a protein with a specific sequence of amino acids.

One way to increase data available to researchers could be the data mining techniques discussed in Chapter 5 to enable users to automate the laborious extraction of relevant data from past experiments (e.g., from old laboratory notebooks), past research papers (e.g., converting figures to tables, extraction of terms in method sections for processing details), and collecting information in various existing databases. Another solution is for publications to have a requirement to submit data and provide it in a standard format, such as crystallography journals, which require submission of data to the Cambridge Crystallographic Data Centre (https://www.ccdc.cam.ac.uk/). Similar efforts are being undertaken for thermophysical property data (Chirico et al. 2013).

However, formulations and soft materials data are much more varied than crystallographic data. Effort is underway to create a standard data model – the data, labels, and metadata associated with each material – for materials data across industry and academia (Dima et al. 2016). Using common labels and reporting common metadata such as test method parameters will accelerate efforts to combine data sources from across the field. There are also data structures such as graph database storage, which enable a more flexible structure for adding additional properties and metadata at a later time – "future-proofing" the database for future development. Those interested in helping to define or propagate data collection best practices in materials science are referred toward the Materials Research Data Alliance (MaRDA), "a community-led network focused on connecting and integrating U.S. materials research data infrastructure to realize the promise of open, accessible, and interoperable materials data. Each of these elements are aligned with the goals of the Materials Genome Initiative (MGI)" (MaRDA 2024). All researchers are encouraged to consider publishing materials formulation data in accordance with FAIR principles – findable, accessible, interoperable, and reusable (Wilkinson et al. 2016).

REFERENCES

Chirico, Robert D., et al. 2013. "Improvement of Quality in Publication of Experimental Thermophysical Property Data: Challenges, Assessment Tools, Global Implementation, and Online Support." *Chemical and Engineering Data* 58 (10): 2699–2716. https://doi.org/10.1021/je400569s.

Dima, A., et al. 2016. "Informatics Infrastructure for the Materials Genome Initiative." *JOM* 68: 2053–2064. https://doi.org/10.1007/s11837-016-2000-4.

Doerk, Gregory S., Aaron Stein, Suwon Bae, Marcus M Noack, Masafumi Fukuto, and Kevin G. Yager. 2023. "Autonomous Discovery of Emergent Morphologies in Directed Self-assembly of Block Copolymer Blends." *Science Advances* 9 (2): eadd3687.

Jumper, John, Richard Evans, Alexander Pritzel, Tim Green, Michael Figurnov, Olaf Ronneberger, Kathryn Tunyasuvunakool, Russ Bates, Augustin Žídek, and Anna Potapenko. 2021. "Highly accurate protein structure prediction with AlphaFold." *Nature* 596 (7873): 583–589.

Lu, Shizhao, and Arthi Jayaraman. 2023. "Pair-Variational Autoencoders (PairVAE) for Linking and Cross-Reconstruction of Characterization Data from Complementary Structural Characterization Techniques." *arXiv preprint arXiv:2305.16467.*

MaRDA. "About MaRDA." https://www.marda-alliance.org/about/ (accessed January 24, 2024).

NRT-MIDAS. "NRT: Computing and Data Science Training for Materials Innovation, Discovery, AnalyticS (MIDAS)." https://sites.udel.edu/midas-nrt/ (accessed January 20, 2024).

Wilkinson, Mark D., et al. 2016. "The FAIR Guiding Principles for Scientific Data Management and Stewardship." *Scientific Data* 3: 160018. doi: 10.1038/sdata.2016.18.

15 Conclusion

Alix Schmidt and Kristin Wallace

It is our hope that the lessons from this book have convinced you to see the digital-centric lab vision presented in Chapter 1 not as a fantasy but instead as a goalpost of accelerated innovation within reach.

> Researchers and formulation scientists in an enormous variety of industries – consumer electronics, personal care products, automotive and mobility, industrial and construction products, packaging, pharmaceuticals and healthcare, and even inorganic materials and alloys – have a research revolution at their fingertips, thanks to the advanced insights, rapid innovation, and value-driven decisions that digital transformation enables.

Unlocking this revolution first and foremost requires addressing people and non-technical challenges.

To conclude this book, we will first review the data science approaches discussed herein, then we will spend a moment to prepare you for the next step of your digitalization journey by noting some potential challenges as well as highlighting the new research directions to keep an eye on over the coming years.

15.1 RECAPITULATION OF KEY CONCEPTS FOR DATA-DRIVEN FORMULATION

After you have won the support of your organization to embark on a digital transformation journey, the technical work begins. To guide a brief summary of the data-driven formulation concepts presented in this book, we will revisit the project flow diagram from Chapter 2, which should be even more meaningful after having digested the material within each chapter.

Starting in Chapter 3, and further emphasized throughout the book, it was discussed that the first step in Figure 15.1 – fully understanding the problem to be solved with data – is a critical prerequisite to assessing the data quality and relevance, setting up the machine learning problem and selecting algorithms, and constructing the optimization framework for inverse design. The key aspects of this exercise are finding projects that have a clear path of action from data-driven decision to new value, which have supportive and open-minded subject-matter experts, and have sufficient amount and quality of data aligned to the problem to make successful models. Several important exploratory data analysis (EDA) techniques were presented

DOI: 10.1201/9781003385974-20

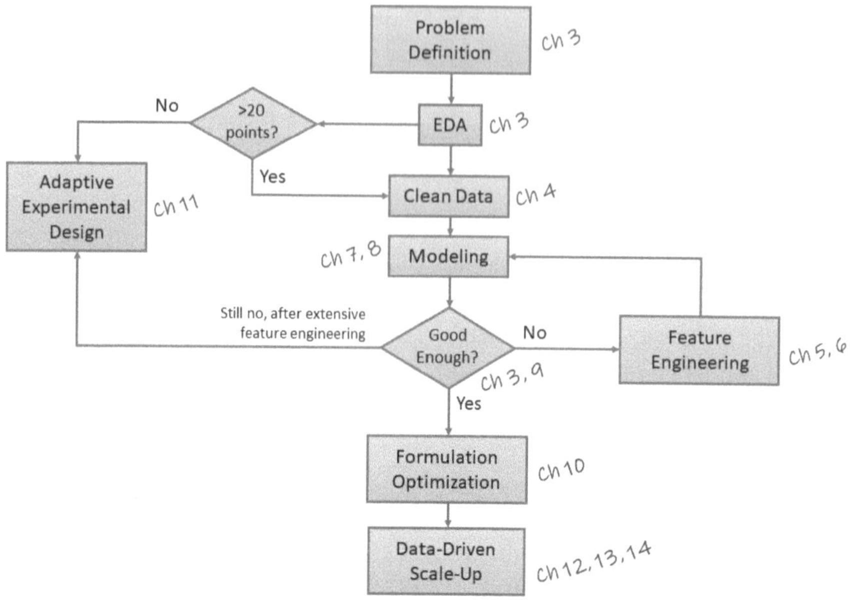

FIGURE 15.1 Revisiting Figure 2.3 – a formulation inverse design project flowchart including problem definition, EDA (exploratory data analysis), cleaning data, adaptive experimental design, predictive modeling, feature engineering, model validation, formulation optimization, and data-drive scale-up.

to assess the data and its relevance to the decisions to be made, including identifying your variables and their distributions, inspecting the correlation structure of the variables, visualizing the space using dimensionality reduction, and testing some preliminary models.

This initial evaluation of the data can help to indicate if the next steps should be to move forward with predictive modeling or if new data is needed to provide sufficient overlap between the information within the training data and the information needed to accomplish the project goal. If data quality, quantity, and initial model feasibility all look good, the team may proceed to formal data cleaning; if additional data is needed, an adaptive experimental design approach provides an efficient mechanism to identify the most informative and helpful experiments to add to the data set. In traditional design, the entire experimental plan is locked down at inception, whereas adaptive experimental design offers the flexibility to make iterative adjustments based on emerging data. This adaptability empowers teams to balance the risk and reward of exploring new design spaces versus exploiting current knowledge of the system, all while avoiding spending too much time pursuing unhelpful directions, thereby expediting the optimization process.

Materials formulation data tends to have several common issues. Chapter 4 presented the steps involved in preparing a formulation data set: identifying all relevant types of data, locating and accessing historical data points, consolidating and structuring all data, and identifying and resolving data quality problems. Most challenges are related to consolidating data from data silos – most commonly formulations for the same type of product made in different labs using different nomenclature, test methods, and data structures. It is also important to mitigate the effects of small data sets, missing data, heavily correlated variables, or insufficiently varying variables, either by fixing the data or by accounting for the issues in modeling.

Formulation ratio data – the available ingredients and composition of each formulation – presents some unique numerical challenges to modeling due to its sparse nature. In cases where it is not possible to overcome some of the data quality challenges present in the formulation data, models can be enhanced by including domain-relevant descriptors of the materials, as discussed in Chapters 5 and 6. Subject-matter experts will heavily guide the identification of these highly informative variables, and the team will need to decide which ones are useful and inexpensive enough to pursue. Methods for acquiring descriptors include directly measuring physical properties of the system, using physics-based mechanistic models or simulations to generate data, engineering features from processing data, accessing public cheminformatics data or mining data from literature, using alternative forms of representing formulation ingredients, or using machine learning to generate numerical features from advanced characterization such as microscopy.

Statistical and machine learning algorithms to capture the valuable patterns within the historical data were presented in Chapters 7 and 8. While it is straightforward to peruse the general machine learning literature and make a list of algorithms to blindly test, or even to use an automated procedure to do the same (i.e., "AutoML"), the authors here present hard-earned know-how about the failure modes and special considerations needed for formulation modeling using these off-the-shelf algorithms. These include how to identify what is causing poor model accuracy (such as measurement noise, incorrect data granularity, global vs. local accuracy), ways to handle non-linearity in high dimensional spaces, avoiding overfitting of unbalanced ingredient data, and managing small data using advanced model architectures. To close out the modeling section, in Chapter 9 we reviewed ways to build trust in your model through *in silico* and experimental validations as well as verification versus subject-matter expertise via model interpretability methods.

The optimization section focused on using a predictive model of formulation performance to most effectively make business decisions. In Chapter 10, the work of the project team to translate the project goals to mathematical expressions for the decision variables (i.e., the formulation design), objective function, and constraints to potential solutions is presented. From there, a variety of search strategies help to navigate the large pool of potential formulations to identify preferred solutions. Chapter 11 offers an optimization-driven solution, called adaptive experimental design, to efficiently explore a design space. To conclude this section, Chapter 12 used model inversion to define the combination – out of all possible combinations of process conditions, ingredient ratios, and their properties – that minimizes the risk of obtaining unacceptable product characteristics or process outputs.

Lastly, to provide the reader with a holistic and well-rounded understanding of using data for new materials design and production, we provided a handful of case studies in Chapter 13 and special topics in Chapter 14. The special topics included emerging methods for data-driven retrosynthesis, managing product quality through model-based statistical process control, and using model information to help specify the raw material quality specifications needed for reliable production of the new product. We hope that these brief discussions can inspire further consideration of digital transformation beyond the formulation lab to production and even supply chain opportunities.

15.2 PREVIEW OF ADDITIONAL DIGITALIZATION CHALLENGES

In reading about the digital-centric lab of Chapter 1, you may have noticed that predictive modeling and formulation design was only a small part of a much bigger puzzle. Indeed, we do not use the word *transformation* lightly; there are an immense number of changes to the usual research and development workflows to achieve such seamless, data-driven research. The following sections will discuss two facets of digital transformation you should be aware of.

15.2.1 Digitalization

The vast majority of the changes in digital transformation are based not on data science but on modernization of data architecture, thorough data engineering to automate data transfer and reduce errors, extensive software engineering for human and machine interfaces, and all of the skills and culture change required to fully capitalize on this technology. While all of those topics were outside the scope of this text, the authors want to emphasize the equal importance of these aspects relative to the more sensational machine learning elements. An organization that invests in predictive formulation without digitalization will see an initial acceleration in product development, but it will not exponentially grow over time – adding new formulations back into the model will be just as time-consuming as the initial modeling effort. The model will not *learn*. Additionally, it will be difficult to leverage the approach to new product spaces, meaning scaling the technology will be slow and tedious instead of compounding acceleration.

Thus, it is our recommendation to plan for the digitalization of lab workflows and project management processes to occur alongside predictive formulation efforts.

Challenges include resourcing software developers and managing the communication gap between lab workflow experts (e.g., chemists, chemical engineers) and the computer scientists who will automate them. There are many organization-specific decisions to be made, such as standardization of test methods and their metadata, data accessibility and governance policies, and behavioral expectations around sharing one's data. These decisions can be political as much as technical, and they will require significant time and effort from everyone: data scientists, domain experts and lab users, and sponsoring management included. Reaching out to external partners with expertise in R&D digital transformation can facilitate the process.

15.2.2 Machine Learning Operations (MLOps)

There is one digitalization element worth specifically highlighting due to its utility for predictive formulation programs: machine learning operations (MLOps).

> Machine learning operations is the set of workflows, best practices, technology, and governance required to sustainably implement machine learning.

It encompasses several sub-disciplines under the "XOps" banner: data operations (DataOps), development operations (DevOps), and model operations (ModelOps). Data operations pertain to the automated processing and governance of data, which is critical for repeatable machine learning; keeping track of where training data was sourced from, how it was transformed, and how to process new data for ingestion allows for transparent and reproducible creation of a particular model. Data science, like all science, is founded on reproducibility.

DevOps is a relatively mature concept pertaining to modern software development practices. The complexity and scale of modern software necessitated a move away from implementing all changes in major version upgrades and toward a more continuous improvement of software applications. This continuous integration and continuous deployment (CI/CD) approach requires operational discipline, tools for versioning and collaborating on code such as git, and work process management technologies developed by platforms such as GitHub. It also includes components pertaining to the infrastructure for deploying models, such as containerization, creation, and management of application programming interfaces (APIs), and the provisioning of cloud computing resources using technologies such as Kubernetes.

Finally, ModelOps is a newer field pertaining to the management of machine learning models using the principles of DevOps and DataOps, with the goals of making the development and deployment of machine learning models reproducible, auditable, and manageable at scale. Components of ModelOps include tracking the performance of model experiments with varying training data sets and hyperparameters, version control of models using registries, and the automated monitoring of model performance over time as data drifts. For a comprehensive introduction to MLOps, see Kreuzberger et al. (2023).

15.3 EMERGING TRENDS IN PREDICTIVE FORMULATION

Next, we share a perspective on the research trends in formulation inverse design that we expect to be most impactful in the coming years.

First, autonomous labs for fully automated materials discovery. While this may sound far-fetched, it is already a reality (Stach et al. 2021; Kusne et al. 2020;

Lookman et al. 2019; Stein and Gregoire 2019; Montoya et al. 2022; Szymanski et al. 2023). The technology involves the combination of high throughput robotics and the adaptive experimental design (AED) methods described in Chapter 11. As discussed, AED techniques such as Bayesian adaptive learning help to dynamically select the next best experiment to explore a design space based on each new piece of information. When paired with a way to run each experiment, such as a molecular simulation model, the system can autonomously explore very large spaces without human intervention. However, in cases where there is no computational substitute for lab experimentation, autonomy will come from robotics.

As you can imagine, in addition to the data science challenges involved in creating the surrogate model and acquisition function within the AED program, there are a host of additional challenges in data management and integration, automated materials characterization, physical control of the material synthesis, enabling safe operation, and defining the goals and bounds of the system. This type of autonomous discovery represents the pinnacle of digital transformation. However, note the word "discovery."

Exploration of unknown spaces is the strength of this approach and as such it will in no way replace careful research by humans, whose strengths are in creative problem solving, determining underlying mechanistic behavior for causal understanding, joining knowledge from disparate research fields, and translating the needs of a customer or society at large into scientific problems.

The next major research trend to highlight is hybrid models: methods for combining the strengths of data-driven models with those of fundamental mechanistic models. As discussed in this text, data-driven modeling methods can have significant limitations, especially pertaining to the scope and quality of the training data. However, they are excellent at capturing complex observed behavior, that is, the real world, without limiting assumptions. Physics-based fundamental models can be limited by their assumptions or necessary simplifications, but they are much better at extrapolation beyond current observations due to their mechanistic foundations. Hybrid modeling seeks to intelligently combine both types to make powerful models for very specific applications, and it has been readily adopted by many other fields of engineering (Kurz et al. 2022). Of course, development of hybrid modeling in materials science and particularly formulation science will be highly application specific (Bock et al. 2021; Yang et al. 2016).

We expect that the growing body of mesoscale models enabled by modern high-performance computing will provide growing opportunities for hybrid formulation modeling.

The last emerging trend is materials representation and foundation models. The feature engineering approaches described in this text are along the lines of classic machine learning: features are created by the data scientist, and the learning algorithm identifies which ones are important and the form of their correlation to the output variable. However, in the deep learning paradigm, learning algorithm identifies both the form of the model *and* the features simultaneously from the raw data. This approach solved a critical challenge for problems such as image analysis and natural language processing, where the modeling inputs were unstructured (i.e., non-tabular) data and thus difficult or impossible to use with standard machine learning algorithms. In general, deep learning architectures accomplish this task by creating a numerical embedding – essentially a fixed-size feature vector – to represent the original unstructured input data.

The "best" embedding representation of the input data is determined by supervised learning against some objective. This approach has now been extended to all types of data under the term representation learning.

As discussed in Chapters 5 and 7, one of the challenges of modeling formulations is the difficulty in representing the constituent materials in domain-relevant but machine-readable way. It is now common practice to use representations for molecular discovery and materials property prediction problems (Gómez-Bombarelli et al. 2018; Agrawal and Choudhary 2019; Wu et al. 2018; Ward et al. 2016; Fung et al. 2021; Goodall and Lee 2020). In addition to small molecule structure graphs, representation learning is also being explored to provide more effective featurization of polymer materials, which have historically been very challenging to incorporate into machine learning data sets due to their size distribution and microstructure (Aldeghi and Coley 2022; Park et al. 2022; Antoniuk et al. 2022). Obtaining effective material representations, which can initially require significant amounts of data, may later improve the modeling of smaller experimental data sets via transfer learning, as discussed in Chapter 7 and demonstrated by Kong et al. (2021). Extending representation learning of a data set to an ever-increasing number of applications creates increasingly general embeddings for further transfer learning. At some point, the size of the training data and the breadth that the embeddings represent create what is called a foundation model; in natural language processing, models been trained on massive quantities of text data, so-called Large Language Models (LLMs), such that they can now readily generate language to solve a variety of tasks (e.g., summarization, translation, and stylization).

While a foundation model trained on materials is still a long way off due to the limited quantity of data, representing materials such that LLMs can predict properties is already being investigated (Xu et al. 2023; Qiu et al. 2024). This especially makes sense for polymers, since LLMs use mechanisms such as transformers and self-attention to excel at sequence prediction: given this set of words, what is the most probable next word? One can readily see how prediction of a monomer sequence is a natural extension. Indeed, using embeddings and deep learning to propose new solutions (e.g., text, polymer, molecular graph) – termed Generative AI (artificial intelligence) – is the next step. Fuhr and Sumpter provide a review (2022).

We have already experienced how these impressive developments have affected society at large through generative text, images, and even music. Data-driven materials science, formulation design, and new product development will not be far behind.

15.4 CLOSING REMARKS

Your digital transformation journey awaits. Perhaps it has already begun. Most likely, so has your competitor's. As you navigate your course, do not rest on your laurels.

> Digital transformation is a journey, not a destination. Scientific advancements will perpetually accelerate the pace and power of data-driven product formulation.

Bon voyage!

REFERENCES

Agrawal, Ankit, and Alok Choudhary. 2019. "Deep materials informatics: applications of deep learning in materials science." *MRS Communications* 9 (3): 779–792. https://doi.org/10.1557/mrc.2019.73

Aldeghi, Matteo, and Connor W. Coley. 2022. "A graph representation of molecular ensembles for polymer property prediction." *Chemical Science* 13: 10486–10498. https://doi.org/10.1039/D2SC02839E

Antoniuk, Evan R., Peggy Li, Bhavya Kailkhura, and Anna M. Hiszpanski. 2022. "Representing polymers as periodic graphs with learned descriptors for accurate polymer property predictions." *Journal of Chemical Information and Modeling* 62 (22): 5435–5445. https://doi.org/10.1021/acs.jcim.2c00875

Bock, Frederic E., Sören Keller, Norbert Huber, and Benjamin Klusemann. 2021. "Hybrid modelling by machine learning corrections of analytical model predictions towards high-fidelity simulation solutions." *Materials* 14 (8): 1883. https://doi.org/10.3390/ma14081883

Fuhr, Addis S., and Bobby G. Sumpter. 2022. "Deep generative models for materials discovery and machine learning-accelerated innovation." *Frontiers in Materials* 9. https://doi.org/10.3389/fmats.2022.865270

Fung, V., et al. 2021. "Benchmarking graph neural networks for materials chemistry." *NPJ Computational Materials* 7: 84. https://doi.org/10.1038/s41524-021-00554-0

Gómez-Bombarelli, Rafael, Jennifer N. Wei, David Duvenaud, José Miguel Hernández-Lobato, Benjamín Sánchez-Lengeling, Dennis Sheberla, Jorge Aguilera-Iparraguirre, Timothy D. Hirzel, Ryan P. Adams, and Alán Aspuru-Guzik. 2018. "Automatic chemical design using a data-driven continuous representation of molecules." *ACS Central Science* 4 (2): 268–276. https://doi.org/10.1021/acscentsci.7b00572

Goodall, R.E.A., and A.A. Lee. 2020. "Predicting materials properties without crystal structure: deep representation learning from stoichiometry." *Nature Communications* 11: 6280. https://doi.org/10.1038/s41467-020-19964-7

Kong, Shufeng, Dan Guevarra, Carla P. Gomes, and John M. Gregoire. 2021. "Materials representation and transfer learning for multi-property prediction." *Applied Physics Reviews* 8: 021409. https://doi.org/10.1063/5.0047066

Kreuzberger, D., N. Kühl, and S. Hirschl. 2023. "Machine learning operations (MLOps): overview, definition, and architecture." *IEEE Access* 11: 31866–31879. https://doi.org/10.1109/ACCESS.2023.3262138

Kurz, S., et al. 2022. "Hybrid modeling: towards the next level of scientific computing in engineering." *Journal of Mathematics in Industry* 12: 8. https://doi.org/10.1186/s13362-022-00123-0

Kusne, A.G., et al. 2020. "On-the-fly closed-loop materials discovery via Bayesian active learning." *Nature Communications* 11: 5966. https://doi.org/10.1038/s41467-020-19597-w

Lookman, T., et al. 2019. "Active learning in materials science with emphasis on adaptive sampling using uncertainties for targeted design." *NPJ Computational Materials* 5: 21. https://doi.org/10.1038/s41524-019-0153-8

Montoya, Joseph A., et al. 2022. "Toward autonomous materials research: recent progress and future challenges." *Applied Physics Reviews* 9: 011405. https://doi.org/10.1063/5.0076324

Park, Jaehong, Youngseon Shim, Franklin Lee, Aravind Rammohan, Sushmit Goyal, Munbo Shim, Changwook Jeong, and Dae Sin Kim. 2022. "Prediction and interpretation of polymer properties using the graph convolutional network." *ACS Polymers Au* 2 (4): 213–222. https://doi.org/10.1021/acspolymersau.1c00050

Qiu, Haoke, Lunyang Liu, Xuepeng Qiu, Xuemin Dai, Xiangling Ji, and Zhao-Yan Sun. 2024. "PolyNC: a natural and chemical language model for the prediction of unified polymer properties." *Chemical Science* 15: 534–544. https://doi.org/10.1039/D3SC05079C

Stach, Eric, et al. 2021. "Autonomous experimentation systems for materials development: a community perspective." *Matter* 4 (9): 2702–2726. https://doi.org/10.1016/j.matt.2021.06.036

Stein, Helge S., and John M. Gregoire. 2019. "Progress and prospects for accelerating materials science with automated and autonomous workflows." *Chemical Science* 10: 9640–9649. https://doi.org/10.1039/C9SC03766G

Szymanski, N.J., et al. 2023. "An autonomous laboratory for the accelerated synthesis of novel materials." *Nature* 624: 86–91. https://doi.org/10.1038/s41586-023-06734-w

Ward, L., A. Agrawal, A. Choudhary, and C. Wolverton. 2016. "A general-purpose machine learning framework for predicting properties of inorganic materials." *NPJ Computational Materials* 2: 16028. https://doi.org/10.1038/npjcompumats.2016.28

Wu, Zhenqin, Bharath Ramsundar, Evan N. Feinberg, Joseph Gomes, Caleb Geniesse, Aneesh S. Pappu, Karl Leswing, and Vijay Pande. 2018. "MoleculeNet: a benchmark for molecular machine learning." *Chemical Science* 9: 513–530. https://doi.org/10.1039/C7SC02664A

Xu, C., Y. Wang, and A. Barati Farimani. 2023. "TransPolymer: a transformer-based language model for polymer property predictions." *NPJ Computational Materials* 9: 64. https://doi.org/10.1038/s41524-023-01016-5

Yang, Dong, Zhanqiang Liu, Xiaoping Ren, and Peng Zhuang. 2016. "Hybrid modeling with finite element and statistical methods for residual stress prediction in peripheral milling of titanium alloy Ti-6Al-4V." *International Journal of Mechanical Sciences* 108–109: 29–38. https://doi.org/10.1016/j.ijmecsci.2016.01.027

Index

FSC
www.fsc.org
MIX
Papier | Fördert gute Waldnutzung
FSC® C083411

Kolibri 360 GmbH
Ferdinand-Jühlke-Straße 7
99095 Erfurt, Deutschland
produktsicherheit@kolibri360.de